# Assembling and Repairing Personal Computers

Second Edition

## Dan L. Beeson
Pima Community College

Prentice Hall
*Upper Saddle River, New Jersey*     *Columbus, Ohio*

**Library of Congress Cataloging-in-Publication Data**

Beeson, Dan L.
  Assembling and repairing personal computers / Dan L. Beeson.—
2nd ed.
   p. cm.
  Includes bibliographical references and index
  ISBN 0-13-081949-2
  1. IBM-compatible computers—Maintenance and repair. I. Title.
TK7887.B44  2000
621.39'16—dc21            98-50976

Publisher: Charles E. Stewart, Jr.
Associate Editor: Kate Linsner
Production Editor: Alexandrina Benedicto Wolf
Production Supervision: Custom Editorial Productions, Inc.
Cover Design Coordinator: Karrie Converse-Jones
Cover Photo: Image Bank
Cover Designer: Linda Fares
Production Manager: Deidra M. Schwartz
Marketing Manager: Ben Leonard

This book was set in Times Roman by Custom Editorial Productions, Inc., and was printed
and bound by Banta/Harrisonburg. The cover was printed by Phoenix Color Corp.

© 2000, 1997 by Prentice-Hall, Inc.
Pearson Education
Upper Saddle River, New Jersey 07458

Printed in the United States of America

10 9 8 7 6 5 4 3 2 1

ISBN 0-13-081949-2

Prentice-Hall International (UK) Limited, *London*
Prentice-Hall of Australia Pty. Limited, *Sydney*
Prentice-Hall Canada Inc., *Toronto*
Prentice-Hall Hispanoamericana, S. A., *Mexico*
Prentice-Hall of India Private Limited, *New Delhi*
Prentice-Hall of Japan, Inc., *Tokyo*
Prentice-Hall (Singapore) Pte. Ltd., *Singapore*
Editora Prentice-Hall do Brasil, Ltda., *Rio de Janeiro*

*To my three darling daughters,*
*Tracy Ann,*
*Rose Alisa,*
*and Rachel Lena,*
*the joys of my life,*
*and my wonderful wife, Liz*

# Preface

With the many computer repair books now available, why one more? The major goal of this book is to provide practical, immediate help to the user (the nontechnician) and the beginning student, as well as the experienced personal computer repair technician in assembling new systems or repairing broken systems by identifying failed subassemblies and replacing them in the field. Users do not have to read the technician sections—and they may skip over the more technical information. The technician should use the basic troubleshooting sections to repair some of the simpler problems of the failed unit. The more detailed technical sections are written for those who attempt difficult repairs.

The decision to repair, compared to replacing a whole unit, is a personal one. With the many chips being used in today's computer systems, many with more than 100 pins, documentation of just the pinouts is becoming extremely difficult to obtain, much less stock, for a number of different boards or devices. Also, some chips have leads fifty thousandths of an inch apart, making it extremely difficult to even clearly see the leads, much less neatly desolder, replace, and resolder these chips. These are replaced only by trained electronic technicians with specialized soldering equipment. Therefore, it is often more pragmatic to just quickly replace the failing board or device.

The scope of this book is restricted to Industry Standard Architecture (ISA) IBM compatibles or "clones." Much information is given for foundational knowledge. The documentation allows you to quickly determine which block, device, or board is failing and the best way to get the system back into operation. Then you can decide whether to play or pay! With some motherboards costing $70, power supplies $30, and keyboards $17, you can't afford to spend much time searching for a failure on a board only to trace it down to a chip you can't really replace anyway. The appendixes list depot repair shops where a technician can send a board or device—that is, if the repair price plus shipping is less than the cost of a new board or device. Sometimes you may just want to pay for a unit (e.g., a new keyboard) and fix the system quickly. Other times you may want to troubleshoot to a deeper level to actually repair the device or board, or to experiment and consolidate knowledge.

A word about safety. Be aware that static electricity can destroy a PC. Grease from fingers can disrupt a connection. Flexing a circuit board can break a stripe (electrical copper path)! Dust can create new current paths. Don't destroy the PC you set out to repair. Be aware that the power supply and the display monitor have dangerous high voltages that can deliver current that can kill. Do not open these units unless you have had training in electronics.

Finally, included with this text are four programs to help you diagnose a personal computer system. *SysChk 2.42* is a shareware program that gives IRQs, DMAs, and I/O addresses. *BurnIn 4.5* is a shareware program that tests most features of a computer system. *Wintune 2.0*, from *Windows* magazine, tunes up your Windows setup. You must have a 3.5" disk drive to install these shareware programs.

To install SYSCHK and BURNIN to your hard drive, insert the disk supplied with this book in your 3.5" floppy drive. You must be in the root directory at the C:\> prompt. Type the following command lines, and press the Enter key at the end of each. Replace the capital letter A with B if your 3.5" drive is not the A:\ drive.

```
c:\dos\xcopy A:*.*/s/v
```

You should already be in Windows to install WINTUNE to your hard drive. From the Program Manager menu bar, just click on File, select Run, and then type the following:

```
a:setup
```

The very best of all diagnostic programs I've seen is *The Troubleshooter* by Forefront Direct at (800) 653-4433, www.ffg.com; a time-limited version is included. No installation needed. Just boot the disk.

**Acknowledgments**

I would like to express my appreciation to the reviewers of this edition: Ronald J. Davis, Glen G. Graham, and Fauzan Qazi.

Dan L. Beeson

# Brief Contents

# Contents

# 1

# First Things First

**OBJECTIVES**

After completing this chapter you should be able to:

- Draw a block diagram of a generic personal computer
- Utilize block diagrams for troubleshooting a computer system
- List tools and supplies to set up an assembly and repair shop
- Convert bits of bus widths to kilobytes or megabytes of possible memory
- Measure voltage and resistance with a digital meter
- List rough prices of subassemblies of 9 computer systems
- Recite order of assembly with intermediate testing
- Diagnose problem to specific subassembly

## 1.1 BLOCK DIAGRAMS— KEYS TO A FAILED SYSTEM

A personal computer system has many components or blocks that work together. Block diagrams showing how major functional units are interrelated are very helpful in troubleshooting a system that has failed. These block diagrams can be drawn on many levels, from a very few major blocks to numerous, more detailed blocks of minor components. Figure 1.1 shows a basic block diagram of a personal computer system. While it has little detail, it can be of great use in diagnosing a failed system.

Without proper operation of the power supply, no other circuit has a chance to operate properly. It receives its power from the wall socket. The fan is powered by the supply, so if it is running, the power supply is probably all right, even if the system seems dead. If the fan is not working, the power supply is either bad or being loaded down by the motherboard, an I/O card, or a drive. The keyboard is directly connected to the motherboard and does not go through a driving circuit board or card. Therefore, if the keyboard is not working, there is no controller card to blame, and the keyboard itself is the likely suspect.

While some systems do have disk drive controllers, parallel printer ports, and video driver circuits on the motherboard, most don't. These circuits that are added on to the motherboard are called **peripherals.** It is wise to keep the circuit cards for these peripherals separate from the motherboard, not only to give the ability to upgrade by changing circuit cards, but also to prevent a minor peripheral failure from requiring an entire new motherboard.

The motherboard power light-emitting diode (LED) is usually mounted on the front panel of the case. If it is on, it shows the motherboard is receiving power from the power supply. If it is not on, and the fan is running, the motherboard is likely to have a problem.

**FIGURE 1.1  System block diagram**

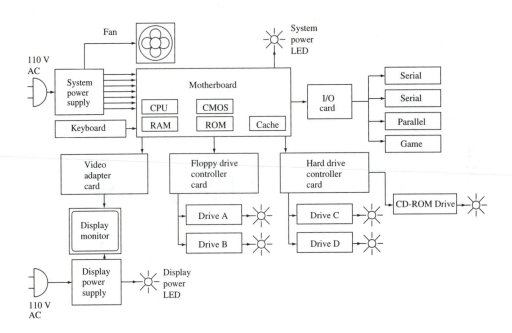

Like the system power supply, the display power supply receives its power from the wall socket. There is a power supply inside the display that lights the display's LED. If this LED is on, it means the power supply for the display is working, even if the display is blank. That throws the suspicion on the video adapter card. If this LED is off, either there is no power coming from the wall socket, or the display is bad. Therefore the computer may be running fine, even though the display is blank.

The LEDs on the floppy and hard drives come on only when the drive is accessed. They are not on all the time. If they are on all the time, there is a problem with the drive or drive controller card. If there are two drives connected to one controller card, and one drive works and the other doesn't, the nonworking drive is probably bad, since the controller is operating the other drive. If you are trying to read the directory of a drive with the DOS DIR command, and the LED does not light up, you know you have either a drive or controller problem.

The serial, parallel, and game ports may be separate cards, but most systems have a card for input and output, called the I/O card, that controls all of these. Since these are inexpensive, grouping these does not share the problems of placing them on the motherboard. If any of these fail, it is usually the I/O card. This analysis may have shown the value of even simple block diagrams.

## 1.2  SYSTEM IDENTIFICATION

One of the first things to do when building or repairing a system is to thoroughly verify exactly what you are working with. Identifying the system may seem difficult, but as you progress through this course, come back to this section now and then to track your progress. All of the following will be described in detail later, so if you don't understand it now, just make a mental note of it for later. Get in the habit of noting the following every time you open a system.

Is the CPU an 8088, 8086, 80286, 80386, 80486, 80586, or 80686 processor? Is it an SX, DX, DX2, or DX4? What kind of expansion slots does it have? Note motherboard switches and jumpers. What size of cache? What kind of memory does it use (DIP, SIP, SIMM, or DIM), and how much? What drives are installed? What kind of display? What adapter cards and peripherals are installed? If the system boots, go into the CMOS setup (if it has one). Copy it down or do a PrintScrn to get a hard copy. It is extremely wise to write down these findings. They may save you hours later on. What version of DOS is installed (use the DOS VER command)? If you have version 6 or above, and if the system works at all, try to run a program included in DOS called MSD. The command is MSD/F FILENAME, in which the filename is any name of eight characters or fewer that you choose. This prints a detailed technical report about what is in your system. You may later print this file out with the command PRINT FILENAME. This allows you to keep documentation for the future.

If the system has Windows 95 or Windows 98, click on Start, Settings, Control Panel, System, Device Manager, and Print. In the report type box, choose all devices and system summary, and then click OK. This sequence will print a thorough report on the system. Have plenty of paper in the printer. In a typical system this report can be twelve pages long. For a shorter, two-page version, instead of choosing All devices and system summary in the sequence above, choose *System Summary* from the report type box.

## 1.3 TOOLS AND SUPPLIES REQUIRED

To troubleshoot and repair a computer system that has been functioning requires more knowledge than equipment. Repairing at a deeper level requires more equipment along with that knowledge. The highest or most basic level of repair is identifying a failing device or card and replacing it. This level includes correcting interrupt conflicts and software problems. This is called **basic card swapping.** One level deeper is using some diagnostic tools to attempt to repair the device or card. This is called **simple signal checking.** To do any repairs beyond these, you need to be a trained electronic technician. A list of standard tools is given in Table 1.1.

What supplies you stock depends on what level you want to function. The supplies listed in Table 1.2 are minimum for building and upgrading systems. You should check local

**TABLE 1.1  Tools**

| Basic Card Swapping Tools |
|---|
| #2 Phillips screwdriver |
| 1/4 inch nut driver |
| Needlenose pliers |
| Flashlight |
| Antistatic wristband |
| Chip remover (or small flat-bladed screwdriver) |
| Air spray and small paint brush (to remove dust) |
| Contact cleaner and rag (or pencil eraser) to clean card edge connectors |
| Disk head-cleaning kit |

| Simple Signal Checking Tools (voltages, resistances, waveforms) |
|---|
| Digital logic probe |
| Multimeter (voltage and resistance scales) |
| Oscilloscope (optional) |
| Solder kit (repairing broken traces on PC boards)<br>          (replacing components in power supplies)<br>          (replacing components in displays) |
| Freeze spray (to be used sparingly and wisely) |

| Software Tools (programs to help diagnose and repair the system) |
|---|
| DOS system disk with CHKDSK.EXE, MEM.EXE, FDISK.EXE,<br>     SCANDISK.EXE, MSD.EXE, DEFRAG.EXE, and<br>     FORMAT.EXE on 360K and 720K disks (bootable) |
| Hard disk low level format program |
| Floppy disk alignment disk |
| Diagnostic program |

| The Most Important Tools of All |
|---|
| Documentation |
| Knowledge |

**TABLE 1.2    Supplies**

| Type | Description |
|------|-------------|
| Screws | Metric machine panhead Phillips  3 mm x 6 mm<br>Metric machine panhead Phillips  4.7 mm x 10 mm<br>Machine panhead Phillips          6/32" x 1/4" |
| Memory | SIMM 1 MB |
| Cables, Floppy | Edge card to two edge cards<br>Header pin to two edge cards<br>Header pin to two header pins |
| Cables, Hard Drive | 20-pin data cable<br>Header pin to one edge card<br>34-pin control cable<br>Header pin to two edge cards<br>IDE header pin to two header pins |
| Cable, Power Supply | 3.5" drive adapter<br>4 pin standard to 4 pin mini |
| Rails | To slide mount drives in AT type cases |
| Frames | 5.25" frame to mount 3.5" floppy drives<br>5.25" frame to mount 3.5" hard drives<br>(includes front bezel and LED) |
| Bezels | Blank front plastic panels to fill empty slots |

sources, then price mail-order ones. It is extremely helpful to have a good, reasonably priced local source. For mail-order sources, see Appendix B.

Beyond these tools and supplies, anything else is a luxury. Stock one each of functioning used adapter cards of every description. This can simplify your job tremendously. No matter how good a technician you are, by the time you hook up a scope and check signals, the card swapping is done (this is on a working system that failed). On the other hand, if someone has "played" with interrupts, jumpers, and switches (as in putting together a new system), no amount of card swapping will bring the system up—knowledge and experience will prevail. Save the documentation on cards and drives you use most often. This will greatly enhance your productivity.

With all the VLSI (Very Large Scale Integration) SMT (Surface Mount Technology) chips being used, some with more than 100 pins as shown in Figure 1.2, documentation of just the pinouts is becoming extremely difficult to obtain, much less to stock for a number of different boards or devices. Add to that the fact that some have leads 0.05 inch on center,

**FIGURE 1.2    VLSI chip with 100 pins**

it is becoming extremely difficult even to see the leads and solder joints clearly, much less to desolder, replace, and resolder these surface mount devices neatly. These are only replaced by trained electronic technicians with specialized soldering equipment. Therefore, it is often more pragmatic to replace the failing board or device.

## 1.4 COMPUTER NUMBERING SYSTEMS

### THE SI SYSTEM

The system of measurements accepted all over the world is the International System (Le Système International d'Unités, SI). To keep large numbers from becoming cumbersome, SI prefixes are used. The twenty SI prefixes used to form multiples and submultiples of SI units follow:

| Decimal | Prefix | Abbreviation | English | Binary | Computerese |
|---|---|---|---|---|---|
| $10^{24}$ | yotta | Y | septillion | | |
| $10^{21}$ | zetta | Z | sextillion | | |
| $10^{18}$ | exa | E | quintillion | $2^{60}$ | Exabyte EB |
| $10^{15}$ | peta | P | quadrillion | $2^{50}$ | Petabyte PB |
| $10^{12}$ | tera | T | trillion | $2^{40}$ | Terabyte TB |
| $10^{9}$ | giga | G | billion | $2^{30}$ | Gigabyte GB |
| $10^{6}$ | mega | M | million | $2^{20}$ | Megabyte MB |
| $10^{3}$ | kilo | K | thousand | $2^{10}$ | Kilobyte KB |
| $10^{2}$ | hecto | h | hundred | | |
| $10^{1}$ | deka | da | ten | | |
| $10^{-1}$ | deci | d | tenth | | |
| $10^{-2}$ | centi | c | hundredth | | |
| $10^{-3}$ | milli | m | thousandth | | |
| $10^{-6}$ | micro | μ | millionth | | |
| $10^{-9}$ | nano | n | billionth | | |
| $10^{-12}$ | pico | p | trillionth | | |
| $10^{-15}$ | femto | f | quadrillionth | | |
| $10^{-18}$ | atto | a | quintillionth | | |
| $10^{-21}$ | zepto | z | sextillionth | | |
| $10^{-24}$ | yocto | y | septillionth | | |

Since 32 bits is 2 bits more than 30, and $2^2$ is 4, a 32-bit bus can handle numbers four times bigger than a 30-bit bus. Note that a 32-bit data bus will handle numbers up to 4 gigabytes. Since 64 bits is 4 bits more than 60 bits, and $2^4$ is 16, a 64-bit bus can handle 16 times a 60-bit bus, or 16 exabytes, which is 16 gigagigabytes, or 16 megamegamegabytes. Of course the last two prefixes are improper usage, but it shows just how *big* an exabyte is.

Regretfully, these are the highest (and lowest) values for which I can find prefixes. Also, note these are based on values of power of 10. Numbering systems for computers are based on powers of 2. Thus, $2^{10}$ is called a Kilobyte, although the real value is 1024, not 1000. So kilo, mega, and giga do not really mean the same thing to a computer user as they do to a scientist.

To overcome this discrepancy, some organizations have proposed prefixes for binary multiples that differ from the decimal multiples. They would be the first two letters of the SI prefix, and then "bi" (pronounced "bee") for binary. That would give:

| Value Prefix | Abbreviation | English |
|---|---|---|
| $2^{40}$ = tebi | TI | Terabyte |
| $2^{30}$ = gibi | GI | Gigabyte |
| $2^{20}$ = mebi | MI | Megabyte |
| $2^{10}$ = kibi | KI | Kilobyte |

Whether the "cuteness" of the sounds of "Keebee, Mebee, Gibee, and Tebee" will keep these from being taken seriously, or if it is just too late to change billions of computer users' hard-set habits, remains to be seen.

# THE BINARY AND HEXDECIMAL SYSTEMS

The numbering system you use every day is the **decimal** system. It uses a base of 10. Computers do not use the decimal system. Their digital electronic circuits use the **binary** numbering system, with a base of 2. It is extremely difficult for a person to read or write large numbers in binary, so we compromise and use **hexadecimal,** with a base of 16. Especially when considering addresses, it is essential to understand the hexadecimal system. To understand the hexadecimal system, it is necessary to understand the binary system.

First let's review the decimal system. In decimal, we use powers of 10. This gives value to each digit according to its position. This creates the 10s place, the 100s place, and so on. Since the digit 2 in the example below is in the 10s place, its value is 20 (two 10s). Because the base is 10, only ten numerals may be used: 0, 1, 2, 3, 4, 5, 6, 7, 8, and 9. Let's look at the number 3728 in base 10:

| Thousands | Hundreds | Tens | Ones |
|-----------|----------|------|------|
| $(10^3)$  | $(10^2)$ | $(10^1)$ | $(10^0)$ |
| 3         | 7        | 2    | 8    |

The above number is $3728_{10}$ (three thousand seven hundred twenty-eight) only because it is in base 10.

Electronic switches turn on and off. Therefore, a numeric counting system that gives only two choices is most compatible with electronic circuits. This is base 2, and is called the binary system, just as base 10 is called the decimal system. Because the base is 2, only two numerals may be used: 0 and 1. 0 stands for "off" or no voltage; 1 stands for "on" or voltage. The positional values of base 2 are shown below:

| Eights | Fours | Twos | Ones |
|--------|-------|------|------|
| $(2^3)$ | $(2^2)$ | $(2^1)$ | $(2^0)$ |
| 1      | 0     | 1    | 0    |

The above number $1010_2$ is a base 2, or binary, number with the decimal value 10. These binary numbers that are used with computers. Each positional value of 0 or 1 is called a **bit.** Four bits are called a **nibble.** Eight bits are called a **byte.** Two bytes (16 bits) are called a **word.** Two words (32 bits) are called a **longword** or a **double word.**

Two to the tenth power ($2^{10}$), 1024 decimal, is called a **kilobyte** (K). As in decimal, a thousand times a thousand is a million, so a kilobyte of kilobytes is a **megabyte** (MB).

The problem that occurs is in the human/machine interface. A 16-bit microprocessor needs a piece of data such as:

$$1011011001010010$$

As you can see, this is extremely difficult for a human to read, write, or even copy. Therefore higher bases are used. Regretfully, base 10 is not compatible with easy conversion. What is needed is a base that is a power of 2. The base 8 numbering system, octal, works. The base 16 numbering system, hexadecimal, works better. Being a larger base than 10, **hex** (for short) gives a shorter notation than decimal. Being base 16, sixteen numerals are needed. Since we run out of single-digit numerals after 0 through 9, letters are used to give a one-position numeral: A = 10, B = 11, C = 12, D = 13, E = 14, F = 15. An "H" is placed after a hex number for identification. A few simple examples follow:

| Hexadecimal | Decimal | Explanation |
|-------------|---------|-------------|
| 4H  | 4  | From 0 to 9, hex is the same as decimal. |
| AH  | 10 | The hex number A stands for decimal 10. |
| FH  | 15 | F is the largest hex digit. |
| 10H | 16 | The hex digit 1 is in the 16s place. |
| 15H | 21 | One in the 16s place and five 1s total 21. |

Here is the representation of the decimal number $50,986_{10}$ in hex:

| 4096s | 256s | 16s | 1s |
|-------|------|-----|-----|
| $16^3$ | $16^2$ | $16^1$ | $16^0$ |
| C     | 7    | 2   | A  |

Note that C72A$_{16}$ takes only four digits, compared to five in decimal. The following shows the conversions of binary, decimal, and hexadecimal from 0 through 15.

| Binary | Decimal | Hexadecimal |
| --- | --- | --- |
| 0000 0000 | 0 | 0 |
| 0000 0001 | 1 | 1 |
| 0000 0010 | 2 | 2 |

| Binary | Decimal | Hexadecimal |
| --- | --- | --- |
| 0000 0011 | 3 | 3 |
| 0000 0100 | 4 | 4 |
| 0000 0101 | 5 | 5 |
| 0000 0110 | 6 | 6 |
| 0000 0111 | 7 | 7 |
| 0000 1000 | 8 | 8 |
| 0000 1001 | 9 | 9 |
| 0000 1010 | 10 | A |
| 0000 1011 | 11 | B |
| 0000 1100 | 12 | C |
| 0000 1101 | 13 | D |
| 0000 1110 | 14 | E |
| 0000 1111 | 15 | F |

In the personal computer, 64 is a standard size jump or block of information. The higher numbers jump by 64 in decimal each time:

| Binary | Decimal | Hexadecimal |
| --- | --- | --- |
| 0100 0000 | 64 | 40 |
| 1000 0000 | 128 | 80 |
| 1100 0000 | 192 | C0 |
| 0001 0000 0000 | 256 | 100 |
| 0001 0100 0000 | 320 | 140 |
| 0001 1000 0000 | 384 | 180 |
| 0001 1100 0000 | 448 | 1C0 |
| 0010 0000 0000 | 512 | 200 |
| 0010 0100 0000 | 576 | 240 |
| 0010 1000 0000 | 640 | 280 |
| 0010 1100 0000 | 704 | 2C0 |
| 0011 0000 0000 | 768 | 300 |
| 0011 0100 0000 | 832 | 340 |
| 0011 1000 0000 | 896 | 380 |
| 0011 1100 0000 | 960 | 3C0 |
| 0100 0000 0000 | 1024 | 400 |

The microprocessor that the IBM clone PCs first used had twenty lines on which to place high (+5 V) or low (0 V) voltages to choose an address or location in memory. These twenty lines are called an **address bus.** Two to the twentieth power is 1,048,576. Starting at 0, you can give different combinations of 1s and 0s on the lines for addresses up to 1,048,575. If each location is an address in memory that is 8 bits (1 byte) wide, this is 1 megabyte (1 MB) of addressable memory you can have in a system with twenty address lines.

If the microprocessor's internal **registers** (locations that hold or store data) are only 16 bits wide, they will hold only $2^{16}$, which is 65,536 locations from 0 to 65,535. This is called 64 kilobytes (64KB). How can we get a 20-bit address? The microprocessor takes a 16-bit address and *shifts* it four places (in binary) to the left, filling in those four new places on the right with 0s, creating a 20-bit number. Now, in order to manipulate the four new digits, another 16-bit number is added in the microprocessor. The address shifted is called the **segment.** The address added to it is called the **offset.** Together they add to the real or **absolute address.**

An example follows in binary, then in hex. Four bits of binary = 1 bit of hex, so the binary is separated into 4-bit blocks just to make it easier to read.

|   |   | **Binary** |
|---|---|---|
| Segment = | 1100 0000 0000 0000 | 1100 0000 0000 0000 0000 |
| Offset = | 0000 1010 0001 0000 | + 0000 1010 0001 0000 |
| Absolute Address = | | 1100 0000 1010 0001 0000 |

|   | **Hexadecimal** | **Decimal** |
|---|---|---|
| Segment = C000 shifted | C0000 | 786,432 = 768KB |
| Offset = 0A10 added | + 0A10 | + 2,576 |
| Absolute Address = | C0A10 | 789,008 |

As you read this book, you will come across some chips and their bus widths. The width of the bus is the actual number of wires or copper strips it can use for highs and lows for binary 1s and 0s. In binary, this gives the total possible number of combinations and, therefore, the number of addresses or the largest number that the system can use. Setting two switches, A and B, to on (1) or off (0), gives the following possible combinations.

| **Binary** | | |
|---|---|---|
| **A** | **B** | **Decimal** |
| 0 | 0 | = 0 |
| 0 | 1 | = 1 |
| 1 | 0 | = 2 |
| 1 | 1 | = 3 |

This shows that two switches can represent four decimal numbers (0–3). By adding more switches, we can show that

$$3 \text{ switches} = 8 \text{ possibilities } (0\text{--}7)$$
$$4 \text{ switches} = 16 \text{ possibilities } (0\text{--}15)$$
$$5 \text{ switches} = 32 \text{ possibilities } (0\text{--}31)$$
$$6 \text{ switches} = 64 \text{ possibilities } (0\text{--}63)$$

Some of the more common addresses are shown in Table 1.3. The numbers under decimal and hexadecimal are actually **states** or numbers of addresses. The actual addresses start at **zero** and end one less than the states. For 256 states, the addresses would be from 0 to 255. In hex that would be 0 to FF.

The 8088 had 16-bit internal registers and could therefore accept only numbers up to $65,535_{10}$ or $FFFF_{16}$, or $1111111111111111_2$, which is 16 bits. Counting from 0 to the hex number FFFFH gives 65,536 combinations of data numbers. The 8088 had a 20-bit address bus, for 1 MB of total addressable memory. It had an 8-bit external data bus, allowing only numbers up to 255.

The 80286 has 16-bit internal registers also, but it has an external address bus of 24 bits and is therefore able to place addresses on the bus from 0 to 16 MB. The data bus of 16 bits allowed numbers up to 65,535.

**TABLE 1.3 Some standard bus widths and their addressing capability**

| Bus Widths | Bits | Decimal | Hexadecimal | K | MB | GB | TB | PB | EB |
|---|---|---|---|---|---|---|---|---|---|
| $2^8$ | 8 | 256 | 100 | 0.25 | | | | | |
| $2^{10}$ | 10 | 1024 | 400 | 1 | 0.001 | | | | |
| $2^{16}$ | 16 | 65,536 | 10,000 | 64 | | | | | |
| $2^{18}$ | 18 | 262,144 | 40,000 | 256 | 0.25 | | | | |
| $2^{20}$ | 20 | 1,048,576 | 100,000 | 1024 | 1 | | | | |
| $2^{24}$ | 24 | 16,777,216 | 1,000,000 | 16,384 | 16 | | | | |
| $2^{30}$ | 30 | 1,073,741,824 | 400,000,000 | | 1024 | 1 | | | |
| $2^{32}$ | 32 | 4,294,967,296 | 1,000,000,000 | | 4096 | 4 | | | |
| $2^{64}$ | 64 | | | | | | 16,777,216 | 16,384 | 16 |

The 80386 and 80486 have 32-bit internal registers and external data and address buses. This allows 4 GB of memory addresses and numbers up to 4 trillion.

Besides giving you a better understanding of the microprocessors you will read about in Chapter 3, this information will come in handy in Chapter 4 for addressing memory locations.

## 1.5 TESTING COMPONENTS

Although you can assemble and troubleshoot a personal computer system without a logic probe or a multimeter, using one can simplify the job. For those unfamiliar with them, a simple introduction to these tools to test electrical components follows. All mention of oscilloscopes in this book is meant only for those who already have and know how to use one.

### METERS

First, the multimeter should have at least the two functions you will use most often, voltage and resistance. Second, it should have appropriate scales, voltage going above 300 volts direct current (V DC) and above 300 volts alternating current (V AC). Read the operator's manual for the meter and familiarize yourself with the meter's function switch and scales. Figure 1.3 shows an analog and a digital meter (DCV = V DC; ACV = V AC).

### Analog Meters

Note the combination function and range switch on the analog meter. It is shown in the off position for transporting. The small, slotted circle just above the word OFF is the mechanical zero adjustment. With a screwdriver, adjust carefully for exactly zero reading with the pointer. Note the top right quadrant of settings is OHMS (abbreviated by the Greek letter omega, $\Omega$). Set the rotary switch on one of these scales. Touch the red (positive) and black (negative) leads together and adjust the Ohms Adjust thumbwheel (bottom left corner) for a zero reading on the top scale. Note that the top scale is reversed from the bottom scale. Note also that the bottom scale is symmetrical and linear. The top is logarithmic. The top scale is for resistance readings, the bottom for voltage or current. If you change the function switch to a different resistance scale, you must rezero by touching leads and using the thumbwheel.

The resistance scale is read at the pointer and multiplied by the value the function switch is pointing at in the top right quadrant. Note that the needle is past a mark for 40 (though 20 was the last number from the right). When the function/range switch points at R × 100, the reading is 42 × 100 or 4200 ohms ($\Omega$), or 4.2 kilo-ohms (k$\Omega$) of resistance.

**FIGURE 1.3  Analog and digital meters**

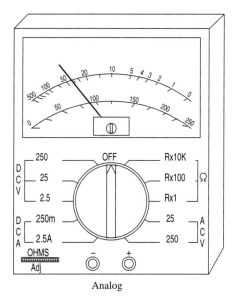
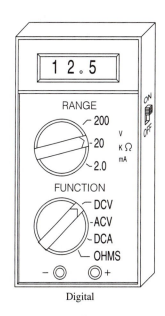

Analog                                                Digital

The bottom scale is for voltage or current. The needle is pointing at about 90 on the bottom scale. When the function/range switch points at 250 in the top left quadrant (DCV) the reading is 90 V DC. However, when the switch is at 25 the reading is 9.0 V DC, and when the switch is at 2.5, the reading is 0.9 volts.

### Digital Meters

Digital meters are easier to use because they do not require much interpretation when read. Note the separate function switch (bottom) and range switch (top). At the current settings, DC volts and 20, the 20 means that is the highest voltage that can be displayed. This makes the 12.5 displayed just that, 12.5 V DC. There are no user adjustments to the resistance scales on a digital meter.

## RESISTANCE

Resistance is the measurement of the amount the device being measured *resists* the flow of electric current. It is measured in ohms, kilo-ohms, and mega-ohms where kilo means 1000 and mega means 1,000,000. So 6000 ohms is 6 kilo-ohms (k$\Omega$). Some (mainly analog) meters multiply the reading of the pointer needle by the number the scale switch (or combination scale/function switch) is turned to. For example, if the pointer points to 4.7 and the scale switch is set to k$\Omega$, the value is 4.7 k$\Omega$, or 4700 $\Omega$. Others (mainly digital meters that show their readings like a digital watch) give the reading directly, printing out the ohms or kilo-ohms on the display.

## VOLTAGE

Voltage is electrical potential. Voltage DC is direct current, which is constant. Voltage AC is alternating current, which alternates between negative and positive peak values. For the voltage function on analog meters, the scale readings are usually *not* multiplied as the resistance scale is, but instead show a *maximum* value of the largest (last) number on the scale. For example, if the scale switch is set at 50 V DC, find a scale on the display that ends in 5, 5.0, 50, or 500. You will take your reading from this scale, considering the last number to be 50 (even though it is one of the above). Of course, lower numbers are relative.

   **Caution:** Direction (polarity) is important in voltage reading. Voltage DC is positive or negative, and connecting your test leads incorrectly could damage the meter and the device. Positive voltage is measured with the positive (red) lead connected to the point to be measured and the negative (black) lead connected to ground. Sometimes voltage *across* a device is measured; the negative lead then goes to the *least* positive device lead.

## TESTING

We shall concern ourselves with measuring six different measuring situations. Typical packages of the electrical components covered are pictured in Figure 1.4. There are hundreds of variations, but this figure should give you the general idea and allow you to identify the necessary component.

### Continuity

On wires, cables, and printed circuit board traces, the copper wires are conductors and should have very little resistance, such as 0 $\Omega$, or less than 1 $\Omega$. Set your meter for resistance function and the "times 1" scale (R $\times$ 1). On this scale (the lowest resistance scale) you multiply the pointer reading by 1, which does not change it. Set your meter's Ohms Adjust and Zero Adjust before using. Now by placing the positive (red) lead on one end of the wire and the negative (black) lead on the other end, you are testing whether the wire does indeed conduct (has continuity) or does not conduct (open, broken). A reading of below 1 $\Omega$ is

**FIGURE 1.4 Typical electronic component packages**

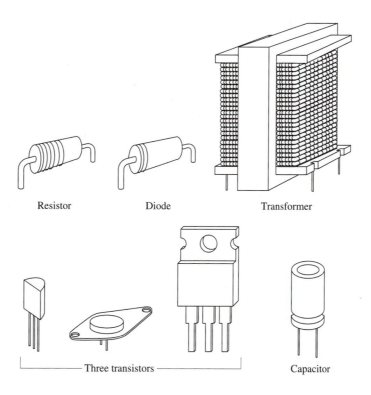

Resistor          Diode          Transformer

Three transistors          Capacitor

required to be good. Direction or polarity (which end each lead goes to) is not a concern. Power to the system must be *off!*

## Resistors

Resistors are components that resist the flow of electric current. The more they resist, the higher their value. Their physical appearance is usually a cylinder about 1/8 inch in diameter and 1/2 inch long, although some are much larger. These usually have a brown color overall, with different colored bands around them to indicate value by a color code. On newer technology boards, resistors may be surface mount (SM) and appear as tiny black squares about 3/32 inch square. While entire courses are taught on properly using a multimeter to read resistance, voltage, and current, the following should suffice.

Set the meter at a scale consistent with the expected value of the resistor. This may be found from the resistor color code (if you are going to use a meter, get information on the code and learn it), from writing on the device, or from a schematic diagram. So, set your meter on the scale and connect your leads. Again, direction (polarity) is unimportant. What is important is whether the resistor is connected to other devices, because they may change the expected reading. Therefore, if the reading is very much different than expected, if possible, disconnect (unsolder) one lead of the resistor and measure again. Power to the system must be *off!*

## Transformers

Transformers transform alternating current from one voltage to another. Their physical appearance is a block or toroid (donut) of iron (laminated sheets or solid ferrite) wrapped in copper wires. Testing the transformer is accomplished by taking continuity checks of the separate windings (the wrappings of wire) and by a voltage (V AC) measurement. When taking the voltage measurement, the power to the system obviously must be on. When taking the resistance measurement, the power to the system must be *off!*

It is possible that a winding became shorted (electrically connected) to another winding. This may be checked by an **isolation test,** which is the opposite of the continuity test. Connect either lead of the meter to one lead of one winding, and connect the other meter lead to one lead of the other winding. The reading should be very high, showing the windings are isolated from each other.

## Diodes

Diodes are one-way valves. They allow electrical current to flow in one direction but not in the other. Their physical appearance is usually a small cylinder about 1/8 inch in diameter and 1/2 inch long. They are usually black overall with a white or silver band around one end. That end is the *cathode;* it has *positive* current coming from it.

Testing a diode (also called a rectifier) requires confirming the one-way function. That means there should be a high resistance one way (polarity) and a low resistance the other. The ratio of the resistances should be at least 1:10, usually much higher. Set the meter on the resistance function, on a scale of ×10 or ×100. Measure the diode in one direction (polarity) and then in the other. You can usually get a good reading this way, but if the reading is questionable, you may need to remove (desolder) one end of the diode and test again. Power to the system must be *off!*

## Transistors

Transistors are made of two diodes. They appear as cylinders 1/8 inch in diameter and 1/4 inch high, with three leads from the flat bottom, or as a rectangular solid (cigarette pack) with three leads from the bottom, or as a short "top hat" with two holes in the "brim" and two leads out the bottom, or finally, for SMT, as a tiny square with three tiny leads.

To test a transistor, you test two diodes. The three leads are named the *emitter, collector,* and *base.* Often these are labeled in silk screen on the printed circuit board. The local electronics store may have a data book giving pictures and pinouts. If you can identify them, great. If not, try all combinations.

Set the meter on the resistance function on a scale of ×10 or ×100. Connect the positive (red) lead to the base and the negative (black) lead to each of the other two leads, one at a time. The readings should be similar. Note the reading. Now connect the negative lead to the base and the positive lead to each of the other two leads, one at a time. The readings should be similar to each other and at least ten times lower or higher than the first readings. This reading shows the transistor is functioning properly. When tested in the circuit, other components can change these readings. Therefore, it may be necessary to remove the transistor for a definite reading. Power to the system must be *off!*

## Capacitors

Capacitors (or caps) are devices that allow alternating current (AC) to pass through them but not direct current (DC). Because the meter uses DC to test resistance, the capacitor shows an infinite resistance (too much to read). Of course in a circuit, other components affect measurements, so a *low* resistance reading across a capacitor is what needs to be checked out further. If you cannot identify any other component giving the low resistance, you may have to desolder one lead of the capacitor and test it alone.

## Voltages

Voltages must be measured with power to the system *on,* so use caution. In a personal computer system, outside of the power supply, there is no voltage present over 12 V, and that won't harm you under normal circumstances. However, *you* may harm the system. Try not to touch circuit board contacts (edge connectors) with your fingers. Sweat and grease ruin electrical contacts. Never let a tool or screw fall into a system. To prevent that, do not insert or replace a screw with the power on. In fact, never connect or disconnect anything with power turned on.

Now, set the function scale to DC or AC as desired. AC is used exclusively in power supplies or display monitors. Choose a scale that is higher than the voltage you expect. The most common DC voltages in a system are 5 and 12 V. Meter scales of 15 and 20 are common scales. Be careful to connect the positive (red) lead to the most positive (highest) point to be tested, and connect the negative (black) lead to ground or to the most negative (lowest) point to be tested.

If you plan to troubleshoot at a deeper level, you must become proficient with at least a meter and, if possible, an oscilloscope.

## DIGITAL LOGIC PROBE

The digital logic probe is a simple test device that alone can show limited information. It can show a low voltage, below 0.8 V. It can show a high voltage, above 2.0 V. Finally, it can show a pulsing signal, either by a third LED or by having the low and high LEDs on at the same time. You can tell whether a pin or contact is on, off, or pulsing. Some documentation gives the logic levels at certain points under certain conditions, allowing the logic probe to be used to better advantage. The probe will have a red lead to connect to +5 V, and a black lead to connect to ground. These borrow power from the system to run the probe. Figure 1.5 shows a typical digital logic probe.

With an additional piece of equipment, the **logic pulser,** you can insert pulses and, with the logic probe, see if they go through a device properly. This type of deeper troubleshooting takes more electronics training.

## 1.6 PURCHASING SUBASSEMBLIES

Acquiring the subassemblies needed to build a computer system is no small feat. You need to start with a clear understanding of the quality, capabilities, and desired cost of the finished system. Appendix B gives some sources for subassemblies and parts; however, no warranty is given here of their quality, or even their honesty. Also included are some computer magazines that have ads for parts. *The Computer Shopper* is the most comprehensive.

As good as mail-order prices are, there is much comfort and wisdom in becoming a known customer at local computer stores. Local computer hobby groups are also valuable contacts in locating parts at reasonable prices in a reasonable amount of time.

The case usually has a power supply already installed. Personal taste plays a part in choosing the case. Figure 1.6 shows a few typical cases: the AT desktop case (above), mini-tower case (lower left), and tower case (lower right). Towers are becoming popular because they can be on the floor by your desk, leaving an uncluttered desktop.

The power supply is the foundation of your system. It delivers all the voltage and current to run everything in the computer case. The power supply must be powerful enough to handle all the devices you plan to install in your system and still have leftover power for future expansion. Power is rated in watts, just like a lightbulb. Typical ratings are 200, 230, and 250 watts. See

**FIGURE 1.5   Digital logic probe**

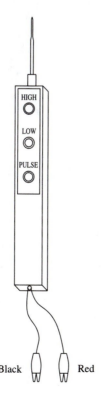

Black          Red

**FIGURE 1.6  Typical cases**

Desktop

Minitower

Tower

Chapter 2 for help selecting the power supply you need. If the power supply does not come with your case, the form factor, or physical size and shape, must be chosen to fit your case. If you are inexperienced, it is wise to buy a case with the power supply already installed.

Keyboards come in only three types, but have many styles. The types are PC/XT, AT, and Windows. They are not interchangeable, although many are switchable to work on any type of system. Jacks are either 5-pin DIM or 6-pin mini-DIM(PS/2) and must match the motherboard plug or use an adapter. The styles are just a matter of personal choice.

The motherboard is the heart of the computer system, and the microprocessor (CPU) is the brains of the motherboard. While you may want the very latest, fastest CPU, you may save hundreds of dollars by choosing one that meets your needs, with some possibility of future upgrading. See Chapter 3 (Motherboards) to help you understand your choices better.

The video display is what you will look at all the time you are using the system, so it also must be chosen wisely. An inexpensive monochrome monitor, which shows only shades of gray, black, and white, may meet all your needs for less than $100. If you use graphics-oriented programs, however, that may compel you to spend about $300 for a color display. VGA is the lowest quality sold today. The display needs a display adapter card to drive it. This adapter must be compatible with the display. See Chapter 6 on displays for selecting adapters and displays.

All the physical parts we have been talking about assembling are called the *hardware* of the system. The instructions that tell your computer what to do are called programs. Programs are called *software;* they are stored on floppy disks, CD-ROMs, or the hard drive. Floppy-disk drives are needed to get information into your computer system. The standard drives used today are the 1.2 MB 5.25" and the 1.44 MB 3.5". Although the 1.2 MB 5.25" is fading from the scene, many systems have one of each, giving you the ability to read any floppy disk. A 2.88 MB 3.5" is available but is not as popular yet. A floppy controller card is needed to drive the floppy drives. See Chapter 7 for more information on floppy drives.

While floppy disks hold much more information than they used to, it is irritating to have to continually shuffle them in and out of the drives. Hard drives are very similar to the floppy drives, but the media is on a solid metal platter that is never removed. A stack of platters increases the amount of information that can be held, so that the typical hard drive today is 200 MB or larger. The hard drive is much faster than floppies and much more convenient. After you first install the software, you hardly ever touch a floppy again, except to load new programs or copy data from the hard drive to a floppy disk for safety against losing that data if the hard drive fails. These are called your backup disks.

The final hardware component in a basic system is the mouse or trackball. These are called pointing devices, because you use them to move a little arrow called a pointer around on the screen, to point at things. Mice and trackballs are discussed more in Chapter 9, Peripherals.

Table 1.4 is a list of the subassemblies you will need to build a basic system. Note that many are optional. Of course, the adapter cards must match the type of I/O expansion connectors available, ISA, EISA, PCI, or VESA. This is discussed in the chapter on motherboards.

Without software, the system hardware won't do anything. DOS is the first program needed; it tells the system how to use the floppy and hard drives. DOS stands for Disk Operating System. The next most commonly used software is Windows, Windows 95, or Windows 98. Windows uses pictures called icons to make it easier, in theory, to use your system. These programs alone allow you to do many things with your system.

**TABLE 1.4   Subassemblies needed to assemble a system**

| Subassembly | Cost |
| --- | --- |
| Case (no power supply) | $40 |
| Power supply | $40 |
| Keyboard | $30 |
| Motherboard (with processor and RAM) | $100–450 |
| Video adapter card | $30–100 |
| Floppy drive controller/cables | $30 |
| Floppy drive | $55–90 |
| Parallel port | $15 |
| Video display | $250–450 |
| **Optional Subassemblies** | **Cost** |
| Hard drive controller/cables (This can be a multi-I/O card with floppy, hard, parallel, serial, and game ports.) | $30–50 |
| Hard drive | $150–200 |
| Mouse or trackball | $15–50 |
| Second floppy | $55–90 |

The next most commonly used programs are word processors, databases, and spreadsheets. A word processor allows you to write letters, documents, books, and so on. Word processors are the most-used type of software. A database lets you keep information in an organized format, like keeping a 3-by-5 card file. A database may be used for your Sunday school class members' information, your gardening information, or a store's inventory. A spreadsheet manipulates numbers in rows and columns, adding, subtracting, and performing varied calculations. Spreadsheets are used to keep a budget, a bank account record, or other data that requires calculations.

Selecting, purchasing, and installing software is covered in this book only to the extent needed to set up an operating system. Software required to install a device is covered in the section on that device. DOS and Windows are covered in Chapter 10.

## 1.7 ASSEMBLING SYSTEMS

### SAFETY FIRST

#### Personal Safety

Working with any electronic or electrical equipment exposes the technician to shock or electrocution dangers. In personal computer systems, there are only three places where dangerous voltages are present. First, the only source you could accidentally touch is the wires from the sealed power supply go to a power switch on the front of the case. These have 120 V AC present, and, although they are usually insulated with shrink-wrap, or "spaghetti," they are still sources of danger. Always unplug the computer before connecting or disconnecting these wires. Second, if you remove the case of the power supply, 120 V AC and 330 V DC are present. Third, if you open the monitor or display, it has 120 V AC and various voltages to 35,000 V DC. Capacitors in the power supply and display may hold dangerous charges *even when unplugged.* Do not open the power supply or display unless you are a trained technician.

The amount of damage to the human body from electrical current depends upon the type of current (AC or DC), the magnitude of the current, the skin resistance, the time the current flows, and the path through the body. Low frequency AC (60 Hz) is more damaging than DC or high-frequency AC, such as radio waves. Higher voltage or lower skin resistance means higher current. Power is current times voltage, so with higher current there is more power to burn skin or internal body parts. There is the danger the body will convulse and stay in contact with the current, increasing the time of contact and the damage. Obviously, a path in the knee and out the foot exposes fewer critical body parts to damage, compared with one into the head, through the chest, and out the foot. You can feel about 10 milliamps of DC and about 1 milliamp of 60 Hz AC; 100 milliamps of AC may take over control of the heart. Dry skin has an average resistance of 30,000 ohms/sq. cm, but wet skin is about 500 ohms/sq. cm.

Considering the above, never work with electricity around dampness or water. Never ground your body with anything other than a proper grounding strap. This will include a resistance to limit current. If you witness a person being shocked, first attempt to turn off the current. If that is not possible, be careful to use a nonconductor, such as a wood stick or a dry cloth, to pull the person away from the current. When working with electrical equipment, always be aware of the presence of dangerous voltages.

#### Equipment Safety

Electrostatic Discharge (ESD)—Electricity may also damage your equipment. Friction between different materials can cause the buildup of static charge. When this voltage is high enough, it will find a path to dissipate to ground. This is the discharge or arc you may see in the dark between your hand and the doorknob, or your shirt and your nylon windbreaker when you pull it off. As few as 100 V of static discharge may damage a component in a computer. Humans may not feel anything less than 3000 V. A good jolt when you touch the doorknob after dragging your feet across the carpet may reach 30,000 V. Just working at a bench may develop 6000 V of static electricity on your body, and you will be unaware of it until you touch a path to ground.

Many electronic devices are considered to be electrostatic discharge sensitive (ESDS). Damage may occur in Vmos (Vertical metal oxide semiconductor) devices at 30 V, in Mosfet (Metal oxide semiconductor field effect transistor) devices such as EPROMs at 100 V, in Jfet (Junction field effect transistor) devices at 150 V, in op-amps at 190 V, and in TTL (transistor–transistor logic) devices at 1000 V. On semiconductor devices, such as a chip used in computers, ESD causing damage fatal to the circuit may be the best that can happen. What could be worse? Latent damage that allows the equipment to pass your tests but fail when the customer is using it. This can hurt your reputation and your business.

To protect equipment from damage, use antistatic packaging and ground paths such as wrist straps, workbench pads and floor pads to dissipate the charge so there is no buildup and, therefore, no discharge. For safety for the worker, these protective devices are not hard-wired directly to ground but through a current limiting resistor. Always use static protection when working on computer equipment.

Corrosion—The human hand carries dampness, salts, and oils that can harm electronic circuitry. Touching the contacts of an edge card connector or a header connector deposits residue from the hand and may lead to corrosion or oxidation of the surface of the metal over time. This can result in a faulty contact leading to system failure. Never touch contacts of any kind. Hold printed circuit boards by the edges that do not have contacts.

If you ignore ESD and handling precautions, you may harm the circuitry. Do not destroy the system you set out to build or repair.

## Fire

Any time we use electrical power it produces heat, maybe sparks, and the possibility of fire. It is therefore important to know the proper technique for putting out fires. Proper extinguishers are chosen by the type of fire or materials involved. Class A means a fire involving ordinary combustibles such as wood, rubber, paper, plastics, and fabrics. The proper fire extinguisher for this would be a type A extinguisher, and may be water or foam. An ABC dry chemical extinguisher will also work, because it is made for Class A, B, or C fires. A Class B fire involves flammable liquids such as gasoline, paint, oil, or grease. The extinguisher to use on this type of fire is a foam, ABC dry chemical, or carbon dioxide extinguisher, all of which smother the flames. Class B also includes flammable gases for which a dry chemical extinguisher is used. A Class C fire involves energized electrical equipment. This would be a computer wiring, motors, generators, or appliances burning. The extinguisher to use is a regular dry chemical, ABC dry chemical, or carbon dioxide extinguisher.

As neat as that sounds—A, B, C—many fires involve more than one of those types of materials. A Class AB fire involves ordinary combustibles and flammable liquids and requires a foam or ABC dry chemical extinguisher. A Class BC fire involves flammable liquids/gases and energized electrical equipment and requires an ABC dry chemical or carbon dioxide extinguisher. A fire department representative told me the best choice in a computer shop would be a Haylon 1211 or Haylon 1301 gas extinguisher. These are Class BC. If these are heavier than 9 pounds, they have a small Class A rating also. Of course you would disconnect current first if at all possible. Always warn others in case of any fire.

## PUTTING IT ALL TOGETHER

From the general system block diagram in Figure 1.1, you can see most of the subsystems that are needed to assemble a computer system. A list of steps follows that will aid you on your first few systems. The list is just a general guideline. You should certainly read through the "Assembling" sections of all chapters before starting to build a system. If special situations or problems arise, refer to the "Basic Troubleshooting" section in the relevant chapter. There you will find hints, aids, and suggestions for each specific subassembly.

FIGURE 1.7    Bare case

In step 1 below, you prepare the case if necessary (Figure 1.7). Most XT, AT, and ATX cases need no preparation, with the rare exceptions of adding self-adhesive feet on the bottom, to stop vibration and protect furniture, or snapping or screwing in the speaker. The towers may need the four operations in step 1.

Figure 1.8 shows a typical front panel system speed display jumper arrangement. It is mounted in the front of the case, where the LED numerals show through. The jumper is a cheap switch that shorts two pins together. It may be necessary to use many jumpers to set the speed. It should be set to a value closest to (or just below) the CPU speed. This speed is usually shown on top of the CPU chip itself. On the speed display shown, documentation was poor, but absolutely necessary. Remember the speed displayed has nothing to do with the speed of the CPU; it only shows what someone has set the speed display to.

**FIGURE 1.8    Speed display jumpers**

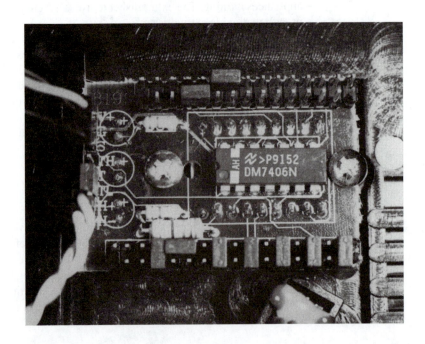

Because the case manufacturer does not know how many drives you will install, plastic bezels are included, covering the openings of the drive bays where the drives may be installed. The bezels usually just snap over the front of the bays, but sometimes they are screwed in. These should be removed or added as needed. Finally, the case has front panel switches and a speaker with wires that need to be connected to the system motherboard.

## STEPS TO ASSEMBLE A PERSONAL COMPUTER SYSTEM

1. If case has digital clock speed display, set jumpers to show speed closest to CPU speed, as shown on CPU chip. Install self-stick rubber feet on bottom of case. Install power switch in case, if necessary. Install speaker in case, if necessary.
2. Install RAM into system board.
3. If any exsist, set system board jumpers for RAM and video. In most modern motherboards, these are automatically recognized.
4. Install system board in case (yes, before power supply, if supply is not in case already) (Figure 1.9). This includes connecting wires to the case front panel switches, and the speaker. The board often has cryptic names silk-screened close to the connectors, such as "rst" for the reset switch and "turb" for the turbo switch. The key switch and 4-pin speaker connector are usually well labeled.
5. Install power supply (do this after system board because it gives you more room when installing system board). Connect power supply leads to case front panel power switch on a style case.
6. Connect power supply to system board (Figure 1.10).
7. Connect keyboard to system board (Figure 1.11). The "indent" in the metal cylinder of the keyboard plug faces the "top" of the expansion slot where the adapter cards are screwed down.
8. Insert video display adapter card in expansion slot (Figure 1.12). It may be an ISA, PCI, or AGP card. If it is an AGP card, it will be in the only AGP slot on the board. The others typically are placed in the slot farthest from the power supply. Check for a mono/color motherboard jumper, and set for your display.

**FIGURE 1.9   Case with motherboard**

**FIGURE 1.10 Case, motherboard, and power supply**

9. Connect monitor signal cable to video adapter card (Figure 1.13).
10. Plug system board power cord in wall.
11. Plug display monitor power cord in wall.
12. Turn on display power.
13. Turn on system power. Turning on the power is **cold booting** the system. At this point, the system should start its power-on self-test (POST). This is an automatic system test

**FIGURE 1.11 System with keyboard**

FIGURE 1.12   System with
video card

run by the BIOS (basic input/output system, a program in ROM). The POST should display some message on the display monitor (if not, vary contrast and brightness to be sure it is not just too dim to see). A "beep" should be heard from the speaker. If you hear any beep sequence or see any message on the display, that's good. Do not be overly concerned at this point about the error messages on the display. The system expects a boot device (hard or floppy drive) to be present and may warn that none is detected. The system may warn that installed hardware does not match CMOS configuration and that we should run SETUP. Any intelligible message on the display proves the power

FIGURE 1.13   System with
display

supply, system board, display adapter, and display are functioning well. The cold boot process causes the POST to test the RAM, which can take considerable time. Press Ctrl, Alt, and Delete at the same time on the keyboard; the system should reboot. This is considered a **warm boot,** because the system was already on. A warm boot skips the memory test. This proves the keyboard is functioning well. If any of the above do *not* occur, consult the troubleshooting guide in Chapter 11 or in the chapter relevant to the suspected subassembly. If the above goes smoothly, turn off the power to the display and the system and continue assembling the system.

14. ATX motherboards include hard and floppy drive controllers on board. If you have an older motherboard, install the floppy drive controller card (Figure 1.14). Usually this is in the I/O slot closest to the power supply. This may be either a
    a. Dual floppy controller card
    b. Combined floppy and hard controller card
    c. A multi-I/O card with both controllers and parallel, serial, and game ports

15. Most cases have plastic bezels or covers over the drive openings. You must remove the bezels before installing the drives. Some snap in; some are screwed in. Install floppy drive(s) and connect to power supply and drive controller adapter (Figure 1.15). Some system boards have the connectors for floppies and IDE (integrated drive electronics) hard drives built in. Watch for the red or blue stripes on one side of a ribbon cable to show pin 1 and check the card for same. Some older cases did not have enough room between the power supply and the drives, and the supply had to be removed to install or remove the drives.

16. If step 14 did not include a hard drive controller, install one. Install the hard drive(s) and connect to power supply and drive controller adapter.

17. Power up the display, then the system. At this point, the AT system should display a warning and request to run SETUP. Enter setup by obeying the commands on the display. For running the setup program, see Chapter 3 under 80286 Motherboards and Chapter 8.

18. Partition and format the hard drive. Now that the entire basic system is functioning, you need to prepare your hard drive. Most drives today are low level formatted when you receive them, so in most cases this will be only partitioning and high level formatting (see Chapter 8).

19. Install printer adapter cards, mice, trackballs, modems, fax cards, joysticks, optical scanners, network cards, voice recognition cards, speech synthesizers, and any other peripherals. These often conflict with each other. To resolve interrupt conflicts, see Chapter 9.

20. To optimize the system, create the AUTOEXEC.BAT file and the CONFIG.SYS file according to the installation directions of the software you are going to install. See Chapter 10 under AUTOEXEC.BAT and CONFIG.SYS. Windows 95 and 98 take care of this for you.

**FIGURE 1.14   System with drive controller card**

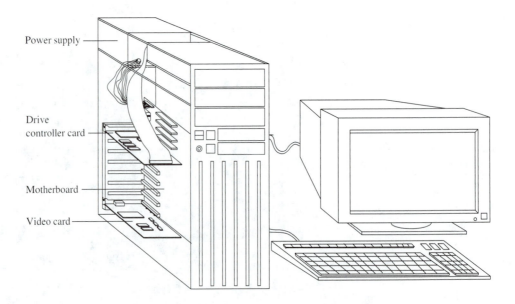

**FIGURE 1.15  System with floppy drives**

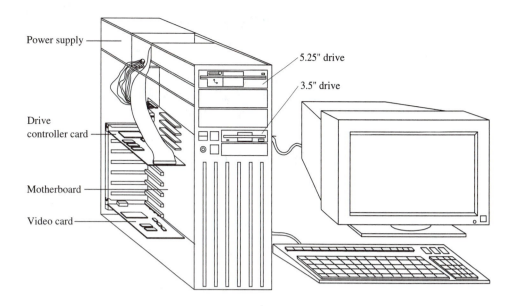

Power supply

5.25" drive

3.5" drive

Drive controller card

Motherboard

Video card

Remember, this is a general list. If the system fails at any step, undo what you did in the previous step and check the system again. If it works, reinstall the next subassembly. If it fails, it may not be the system or the subassembly but an interrupt conflict. Documentation supplied with the adapter card should give interrupt choices.

## 1.8  INITIAL TROUBLESHOOTING

### BEFORE YOU BEGIN ANY JOB

#### Protect the Customer

If the customer contacts you before she brings the system to you or you go to the site to work on the system, ask if she has performed a backup. Copying data or work files either to floppies, another hard drive, a magnetic tape, a writable CD, or another network drive protects the user against catastrophic loss of valuable business information. This should be done on a regular basis. If the customer has not, do not denigrate her for failing to do so. When the system is fixed, teach the customer how to do a backup and help her design a backup schedule. If the system is functioning enough to perform a backup, ask her if she wants to do it before it is worked on, or if you should do it. This will protect your customer from loss.

#### Protect Yourself

Before working on a system:
1. Make sure the customer knows your fees so there will be no misunderstanding.

If the system functions enough to allow it, you should do the following:

2. Do a virus scan of all system hard drives. You don't want to have your disk infected by a customer's system.
3. Run a full diagnostic check, record the results, and inform the owner of the results before you open the case. This will protect you if the owner says that you caused some major problem when you worked on it. DOS 6.0 and later versions have a program called Microsoft Diagnostics (MSD) that shows information about the system and saves a log of this. Windows 95 and 98 do this through the Control Panel using the Device Manager (see Section 1.2 or Chapter 10).
4. Back up your customer's data before you work on the system. This ensures that you will not be liable for loss of their business files.
5. Keep a log of all work you performed. This helps in troubleshooting a complex problem.

## Protect the Equipment

1. Be sure all software you use is virus scanned, so you don't infect your customer's system.
2. Be sure to follow all ESD protective measures to prevent damaging his system.
3. Keep an organized work area, labeling the customer's system components so they are not mixed with another customer's.

## TROUBLESHOOTING

When a system is first assembled, you test the system at various steps of assembly described in the previous section. This helps track the proper operation and identify just which component started causing problems. When a previously operating system has failed, you must take into account more than hardware. The software or programs may be failing or interacting with each other to cause a more confusing set of symptoms than just hardware.

The following set of steps is given here mainly to show the depth of thought sometimes needed to identify the cause of failure. Some mention will be made of things you have not yet read about. After reading the rest of the book, refer to these steps when problems arise and to Chapter 11, Symptoms and Repairs.

1. From a distance, you can make some observations of a "dead" system that is turned on while you are watching it. Is the display pilot light on? If so, there is 110 V AC to the system. If not, you probably do not have 110 V AC at the wall. Less likely, you may have a dead display.
2. If the display pilot is on, can you hear the system power supply fan? It is the only noisy thing in the system, with the possible exception of the hard drive. If you hear a motor noise, the power supply is probably good, because it is designed to shut down and turn off the fan if it is bad.
3. If the display pilot is on and you hear the fan, do you hear one "beep" from the speaker as the system is booting? If so, the motherboard is probably OK. That is what one "beep" shows (see Appendix A, POST Error Codes).
4. Is there any display at all on the screen? If the display is proper, fine! If it is not, note exact symptoms to try to isolate a video adapter problem from a display problem.
5. If the display is OK, does the system now boot its floppy or hard drive? This is indicated by a noise and pilot light on the floppy or hard drive. You should have a bootable DOS disk in floppy drive A: when booting. If you hear and see these clues, the system is attempting to boot. This means the motherboard is doing its job.
6. Does action in either drive end with a new display of some type on the display screen? The display should be the DOS prompt, the date and time request, Windows 95 or 98 desktop, or another program that was loaded by the AUTOEXEC.BAT on the hard drive.

Steps 1 through 6 show that the basic hardware is operating. Now:

7. Is it really a hardware problem, or could it be an operator or a software problem? Observe what the operator is doing when he thinks the system is failing. Is there an obvious operator error?
8. Does the error occur in all programs that do the same thing? For example, does it not print from a word processor but print from the DOS PRINT command or the DOS COPY TEXT LPT1: command? This would seem to indicate software setup problems or a printer driver problem.
9. Rename AUTOEXEC.BAT and CONFIG.SYS (don't erase them!). Power down, wait 30 seconds, then power back up. This has eliminated TSR (Terminate Stay Resident) programs and exotic configurations. Use a tried-and-true diagnostic program or DOS to perform the same operation the operator was having problems with. If the hardware functions properly now, backtrack and find the memory or interrupt collision that caused the problem.

10. *Document everything you do.* Has the user knowingly changed anything about the system? If he has added a card or memory, check the installation, or remove the item and do a cold boot.
11. Strip all boards from the expansion slots. Disconnect the keyboard. That's right: no display, no cards, no keyboard, just power supply, speaker, and motherboard. Power up. Does the fan turn at full speed? (Yes = power supply probably OK. No = see Chapter 2.) Do you hear any beeps from the speaker? (Yes = motherboard probably OK [see audio error codes in Appendix A]. No = see Chapter 3.)
12. Power down. Connect keyboard, video card, and display. Check motherboard video switch setting. Power up. Any video at all? (Yes, perfect = card and display OK. Yes, marginal = check video switches any. Perform CMOS setup for video. Windows 95 or 98 click on Conton Panel, display Settings. No = see Chapter 6.)
13. Audio or video error codes? (Yes = see chapter indicated by error codes in Appendix A. No = boot DOS from hard drive or floppy.) Keyboard problems, see Chapter 5. Memory problems, see Chapter 4. In general: Use a pencil eraser on the edge connector of a card causing trouble (not over the slots); power down; firmly seat all RAM, ROM, and other socketed chips; check all cables and connectors; check all setup switches and the CMOS SETUP.
14. SCANDISK or CHKDSK/F command on drives A:, B:, and C:. OK?
15. MEM command. OK? Run MSD and a full diagnostic program.

## 1.9  CERTIFICATION

The computer technician needs an understanding of business philosophy to succeed. He needs an understanding of interpersonal relationships to smoothly interface with customers. The technician needs to be fluent, communicate well, and present an organized, positive impression. He also needs to be many things to many people, but one thing overrides all of these: The computer technician needs to be knowledgeable about the field.

There are many ways to obtain this knowledge. Most localities have a community college nearby. Community colleges usually have excellent two-year computer degree programs at very reasonable costs. There are many good private technical schools that have good programs, although the cost is usually higher. With luck or a contact person in the industry, a person can start in the business with no knowledge. An experienced technician may take you on as an apprentice and train you as you help out in the shop. Regardless of where you learn about computers, the person hiring you can only make assumptions about your level of expertise. There is one way to prove to a potential employer that you are a professional. That is certification. There are a number of organizations that certify technicians of different types for different purposes.

### A+ CERTIFICATION

The most basic certification that is directly related to the computer technician is from the Computing Technology Industry Association (CompTIA), which was formed in 1982 and has more than 7000 member businesses. Their two main certifications are A+ certification for computer technicians and (soon) Network+ certification for network technicians. CompTIA has a number of test sites available nationally, and charges are approximately $180 for the combination of a core test of basic computer knowledge, and a specialty of DOS/Windows or Mac. Given as two tests of about one hour each, passing will show you as A+ certified. Just look in the classified ads under "Computer" and you will notice a number of employers requiring A+ certification. Information about CompTIA is available at www.comptia.org or (630) 268-1818 ext 305 or 359. Training or test prep materials are available from www.heathkit.com, www.aplusexam.com, and www.aplushelp.com.

### MICROSOFT CERTIFICATION

The Microsoft Corporation offers Microsoft Certified Systems Engineer (MCSE) certification. This shows that you are quite familiar with Microsoft Windows NT Server and other Microsoft products. A certification of "MCSE + Internet" means you are a pro on the Net and Web pages. The Microsoft Certified Professional (MCP) certification means you have in-depth knowledge of a Microsoft operating system. The Microsoft Certified Trainer (MCT) certification shows you are qualified to teach courses at Microsoft Authorized Technical Education centers and Microsoft Authorized Academic Training Program schools. Information is available from www.microsoft.com/train_cert. Study guides for the MCSE exams are sold at www.troytec.com.

## CET CERTIFICATION

The International Society of Certified Electronics Technicians (ISCET) tests and certifies thousands of electronic technicians. They have a number of tests, including the Associate-Level CET Exam, the Journeyman-Level CET Exam, the CAT Exam, the ISCET Exam, and the FCC (Federal Communication Commission) Exam. ISCET is available at www.iscet.org.

## NASTEC CERTIFICATION

The National Appliance Service Technician Certification (NASTEC) is for technicians who work on large home appliances. Information is available at www.nastecnet.org.

## NICET CERTIFICATION

The National Institute for Certification in Engineering Technologies tests and validates engineering technicians and engineering technologists. More information is available at www.nicet.org.

## NARTE CERTIFICATION

The National Association of Radio and Telecommunications Engineers (NARTE) also has a number of tests and certifications, including Telecommunications Certification, Electromagnetic Compatibility (EMC) Certification, Electrostatic Discharge (ESD) Control Certification, FCC Commercial License Testing, and FCC Commercial License Examinations. You may find out more about NARTE at www.narte.org.

There may be others, but you should check into the certifications above that apply to your goals, particularly the A+ certification.

## 1.10 CHAPTER REVIEW/QUIZ QUESTIONS

1. Before opening a customer's system,
   a. Clean it up to look better.
   b. Label all parts.
   c. Write down the serial numbers of the components.
   d. Run diagnostics to document its condition when received.

2. Tools needed to assemble or repair a personal computer include
   a. Oscilloscope, multimeter, Phillips screwdriver.
   b. Multimeter, Phillips screwdriver, nut drivers.
   c. Numerous hand tools, general electronic test equipment.
   d. One screwdriver and a digital logic probe.

3. On some cases, installing the floppy or hard drive may require removing the
   a. Keyboard.
   b. Motherboard.
   c. Power supply.
   d. Video adapter card.

4. Cases may need
   a. Painting.
   b. Rubber feet installed, front panel clock speed LED display set, speaker and front panel power switch installed.
   c. Holes drilled for screws to hold hard drives.
   d. Electrical insulation.

5. You perform initial basic system tests
   a. After the system is completely assembled.
   b. At stages of assembly.
   c. Before assembling any parts.
   d. After all software is installed.

6. SMT devices are replaced
   a. When the system fails.
   b. After double-checking with diagnostic software.
   c. For quick repairs only.
   d. By trained electronics technicians with specialized equipment.

7. When handling PC boards for assembly,
   a. Touch the power supply or wear an antistatic wristband.
   b. Wear gloves.
   c. Use a carpeted worktable for cushioning boards.
   d. Ground each board to the power supply.

8. VLSI are initials for
   a. Visible Leading Signal Indicator.
   b. Variable Low Signal Indicator.
   c. Very Large Scale Integration.
   d. Vertical Local Storage Integrator.

9. The foundation all other circuits rest upon is the
   a. CPU.
   b. Power supply.
   c. Keyboard.
   d. RAM memory.

10. The only I/O device directly connected to the motherboard is the
    a. Video display.
    b. Hard drive.
    c. Floppy drive.
    d. Keyboard.

11. For troubleshooting, the oscilloscope is
    a. Absolutely required.
    b. Almost a necessity.
    c. Often required.
    d. Optional.

12. The binary number system requires numerals
    a. 0 through 9.
    b. 0 and 1.
    c. 0 through 7.
    d. 0 through 9 and A through F.

13. The hexadecimal number system requires numerals
    a. 0 through 9.
    b. 0 and 1.
    c. 0 through 7.
    d. 0 through 9 and A through F.

14. One of the first steps in assembling a system is installing
    a. The floppy drive.
    b. A battery for the CMOS RAM.
    c. The video adapter card.
    d. The RAM in the motherboard.

15. The front panel clock speed LED display must be set to show
    a. The exact speed of the system clock.
    b. A choice as close to the CPU speed as possible.
    c. Twice the system speed.
    d. The RAM speed.

16. The POST is the
    a. Small metal standoff under the motherboard to ground it.
    b. Center wall of the case to mount the drives to.
    c. Test of the system the BIOS performs on cold booting.
    d. Practical Operating System Technique.

17. Voltages in the circuitry exposed when removing the outer computer case may be
    a. Lethal 110 AC at the front cover power switch.
    b. Shocking high DC in the power supply.
    c. Nothing dangerous.
    d. Lethal AC and DC.

18. Voltages in the opened power supply may be
    a. Shocking high DC only.
    b. Shocking high AC only.
    c. Lethal DC only.
    d. Lethal AC and DC.

19. The highest voltage in the entire computer system is in the
    a. Main case on the motherboard.
    b. Power supply.
    c. Monitor.
    d. Drives.

20. The right type of fire extinguisher to use on an energized computer system is Class
    a. A.
    b. B.
    c. AB.
    d. ABC.

21. Electrostatic discharge may destroy electronic circuitry at voltages as low as
    a. 10 V.
    b. 100 V.
    c. 1000 V.
    d. 10,000 V.

22. For you to even feel a discharge (spark), it must be at least
    a. 30 V.
    b. 300 V.
    c. 3000 V.
    d. 30,000 V.

23. Current as low as this passing through the chest may take over control of the heart muscles:
    a. 1 milliamp.
    b. 10 milliamps.
    c. 100 milliamps.
    d. 1000 milliamps.

24. To protect the circuitry you work on from ESD, always
    a. Wear nonconductive clothing.
    b. Ground yourself to an earth ground.
    c. Wear an antistatic wristband, properly grounded.
    d. Wear rubber or cloth gloves.

25. To protect against ESD, your workbench should have
    a. A soft nylon carpet surface.
    b. An antistatic mat, properly grounded.
    c. A smooth plastic surface.
    d. A rubber mat.

26. To protect your customers' data
    a. Always backup their data.
    b. Always be sure their data is backed up.
    c. Chastise them for not making backups.
    d. Install circuit protectors.

# 2

# Power Supplies

**OBJECTIVES**

After completing this chapter you should be able to:

- Describe the difference between linear and switching power supplies
- List the advantages and disadvantages of the linear supply
- List the advantages and disadvantages of the switching supply
- Estimate the size supply needed for a system
- Quickly diagnose a power supply
- Thoroughly test a power supply (user)
- Locate and replace failing internal components (tech)
- Replace a power supply

## 2.1 LINEAR SUPPLIES

The old-fashioned linear power supply is easy to design and repair by even the least experienced technician. It consists of a switch, a fuse, and a pilot lamp on the 110 V AC line, which then goes across the primary or input of a transformer. The transformer has a heavy iron core to be efficient with the low frequency (60 Hz) of the AC line. The core must be quite large to dissipate the heat developed.

A typical linear supply transformer may be 4 inches in height, length, and width, weighing up to 10 pounds. The voltage from the secondary (output) or secondaries of this transformer are rectified, filtered, and regulated. These linear supplies are usually considered too massive to be designed into a personal computer.

## 2.2 SWITCHING SUPPLIES

Almost all computer power supplies are switching power supplies. To lower the size, bulk, and cost of the large power transformer required for the linear supply, the switching supply uses an oscillator that "switches" voltage on and off to raise the frequency of the input voltage. This higher frequency has higher efficiency through a transformer, allowing the iron core to be smaller and lighter.

The switching supply's main power transformer may be $1.5 \times 1.5 \times 1.0$ inches and have a weight measured in ounces instead of pounds. Because of their smaller size and lighter weight, switching power supplies are used for today's computers.

## POWER SUPPLIES

The power supply is the foundation of the computer system. It is one of the few subsystems for which you can still find generic parts for repair. You don't want a house built on sand, so the power supply's power rating must equal or exceed the requirements of the system. This power is rated in watts, just like a standard lightbulb. A 100-watt power supply can deliver the same amount of power used by a 100-watt lightbulb.

Adapter cards and drives usually give their electrical requirements in amperes (amps) and volts, so we will need to convert the amps and volts to watts. This will allow us to compare the system's power requirements with the power supply's capabilities. Table 2.1(a) gives an estimate of some power requirements of various devices and adapters. Table 2.1(b) gives the power breakdown and capabilities of some standard supplies.

Only two equations are needed for the calculations. First, $P = I \times E$ where $P$ is power in watts, $I$ is current in amperes, and $E$ is voltage in volts. Second, $R = E/I$ where $R$ is resistance in ohms, and $E$ and $I$ are the same as in the first equation. Because cards and drives from each manufacturer may be very different, the following are rough estimates. You can add up the power requirements of the cards and devices you want to install to arrive at an estimate of the minimum power supply needed to support your system.

From Table 2.1b, you can see that a standard 200-watt supply can deliver approximately 20 amps of 5 V DC (100 watts), 8 amps of 12 V DC (96 watts), 1/2 amp of –5 V DC (2.5 watts), and 1/2 amp of –12 V DC (6 watts), which adds up to slightly less than 200 watts total. This is the number of watts it can deliver. This total must be greater than the number of watts your system will require.

The form factor of a device is the shape and size. In general, there are four main types of power supplies. The PC/XT supply, the full-sized AT supply, the mini AT supply, and the ATX supply. The mini usually has a cable out to a front panel switch for power on/off.

The standard AT supply has two 6-pin plugs to connect to the motherboard. These are designated P8 and P9. **Caution:** These must be plugged into the correct connector or damage to

**TABLE 2.1   Rough current requirements and capabilities**

| (a) Power Requirements of Standard Cards and Drives | | | |
|---|---|---|---|
| Device | Current Required | 5 V Power | 12 V Power |
| Motherboard, no cards | 1 A from +5 V | 1 A × 5 V = 5 W | 0 W |
| Video adapter | 1 A from +5 V | 1 A × 5 V = 5 W | 0 W |
| Floppy drive | 0.2 A from +5 V | 0.2 A × 5 V = 1 W | |
|  | 0.3 A from +12 V | | 0.3 A × 12 V = 3.6 W |
| Hard drive | 0.3 A from +5 V | 0.3 A × 5 V = 1.5 W | |
|  | 0.7 A from +12 V | | 0.7 A × 12 V = 8.4 W |
| CD-ROM drive | 0.3 A from +5 V | 0.3 A × 5 V = 1.5 W | |
|  | 0.6 A from +12 V | | 0.6 A × 12 V = 7.2 W |
| Most adapter cards | 1 A max +5 V | 1 A × 5 V = 5 W | 0 W |

| (b) Power Breakdown of Standard Power Supplies | | | | |
|---|---|---|---|---|
|  | PC | XT | AT | |
| Supply Volts | 64 W | 130 W | 200 W | 230 W | 250 W |
| +5 | 7.0 A × 5 V = 35.0 W | 15.0 A × 5 V = 75.0 W | 20.0 A × 5 V = 100.0 W | 23.0 A × 5 V = 115.0 W | 25.0 A × 5 V = 125.0 W |
| –5 | 0.3 A × 5 V = 1.5 W | 0.3 A × 5 V = 1.5 W | 0.5 A × 5 V = 2.5 W | 0.5 A × 5 V = 2.5 W | 0.5 A × 5 V = 2.5 W |
| +12 | 2.0 A × 12 V = 24.0 W | 4.2 A × 12 V = 50.4 W | 7.5 A × 12 V = 90.0 W | 9.0 A × 12 V = 108.0 W | 9.5 A × 12 V = 114.0 W |
| –12 | 0.3 A × 12 V = 3.6 W | 0.3 A × 12 V = 3.6 W | 0.5 A × 12 V = 6.0 W | 0.5 A × 12 V = 6.0 W | 0.5 A × 12 V = 6.0 W |
| Total | 64.1 W | 130.5 W | 198.5 W | 231.5 W | 247.5 W |

the motherboard may result. An orienting "lip" on the motherboard connector allows the plugs to be connected in one direction only. The problem is which is on the left and which is on the right. The color of the wires is the key to the answer. Always place the three red wires on one end, all the black wires on both plugs together in the middle, and the orange, red, and yellow wires at the opposite end. This will ensure correct placement (Figure 2.1a).

The supply will also have a number of 4-pin plugs, each with a yellow, two black, and a red wire. These are the drive power connectors for floppy, hard, and CD-ROM drives. The plugs may be one of two types: a standard large connector or a miniconnector. They are both physically keyed to ensure proper connection. You cannot plug them in backward.

See all power supply plugs in Figure 2.1.

The ATX motherboards (Chapter 3) require the ATX power supply form factor. There are numerous changes incorporated in this new design. The fan pulls air into the case and exhausts it out over the microprocessor to aid in cooling the CPU. This eliminates the need for a second CPU fan, although all processors for the ATX systems have cooling fans installed. There is only one (main) motherboard power connector, a typical 20-pin (2 rows x 10 pins) Molex™

**FIGURE 2.1   Power supply pinout**

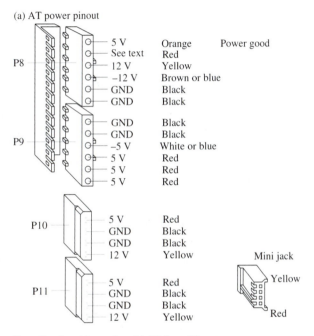

(a) AT power pinout

| P8 | | |
|---|---|---|
| | 5 V | Orange        Power good |
| | See text | Red |
| | 12 V | Yellow |
| | −12 V | Brown or blue |
| | GND | Black |
| | GND | Black |

| P9 | | |
|---|---|---|
| | GND | Black |
| | GND | Black |
| | −5 V | White or blue |
| | 5 V | Red |
| | 5 V | Red |
| | 5 V | Red |

| P10 | | |
|---|---|---|
| | 5 V | Red |
| | GND | Black |
| | GND | Black |
| | 12 V | Yellow |

| P11 | | |
|---|---|---|
| | 5 V | Red |
| | GND | Black |
| | GND | Black |
| | 12 V | Yellow |

Mini jack
Yellow
Red

*Note:* Usual position on board is P9 front, P8 rear.

(b) ATX power pinout
(Sense)                    11      1

| | | |
|---|---|---|
| (Violet) Orange  3.3 V | | 3.3 V  Orange (Violet) |
| Blue  −12 V | | 3.3 V  Orange (Violet) |
| Black  COM | | COM  Black |
| (Gray) Green  PS-ON | | 5 V  Red |
| Black  COM | | COM  Black |
| Black  COM | | 5 V  Red |
| Black  COM | | COM  Black |
| White  −5 V | | PW-OK  Gray (Orange) |
| Red  5 V | | 5 V SB  Purple (Brown) |
| Red  5 V | | 12 V  Yellow |

20      10

Optional
4      1

| | | |
|---|---|---|
| Black/White  1394R | | FANM  White |
| Red/White  1394V | | FANC  Blue/White |
| RASVD | | 3.3 V SNS  Brown/White |

6      3

connector. It is strongly keyed (Figure 2.1b) Like I/O connectors, the pins are numbered down one row and continue in the same direction for the second row. This is the opposite of chip pin numbering, which wraps around on the second row and reverses direction. This difference in numbering has always been an irritation, but at least remains consistent with the past. The following shows the pinout:

ATX Power Supply Pinout

| Color | Function | Pin | Pin | Function | Color |
|---|---|---|---|---|---|
| Orange (violet) | 3.3 V | 11 | 1 | 3.3 V | Orange (violet) |
| Blue | −12 V | 12 | 2 | 3.3 V | Orange (violet) |
| Black | COM | 13 | 3 | COM | Black |
| Green | PS-ON | 14 | 4 | 5 V | Red |
| Black | COM | 15 | 5 | COM | Black |
| Black | COM | 16 | 6 | 5 V | Red |
| Black | COM | 17 | 7 | COM | Black |
| White | −5 V | 18 | 8 | PW-OK | Gray |
| Red | 5 V | 19 | 9 | 5 V SB | Purple |
| Red | 5 V | 20 | 10 | 12 V | Yellow |

Pins 1, 2, and 11 supply 3.3 V DC for the new lower power CPUs. Note that pin 11 may also have a brown wire for 3.3 V sensing by the main board. Pins 4, 6, 19, and 20 supply the 5 V DC for standard chips. Pins 3, 5, 7, 13, 15, 16, and 17 are the commons or grounds. Pin 8 is the power OK signal from the power supply to the motherboard. A high indicates the 3.3 V and 5 V outputs are OK. Pin 9 is a 5 V DC standby voltage, staying at 5 V even when the rest of the system goes into low power or sleep mode. Pin 10 is the main 12 V DC supply for drive motor power, though many drives are using 5-V motors now. Pin 12 is the negative 12-V supply, used for dual-sided signals such as the standard serial port needs. Pin 14 is the "power on" signal from the motherboard to the power supply. Normally held high by the power supply itself; when the motherboard pulls it low, it informs the supply it is OK to turn on main power voltages.

An optional 6-pin (2 rows × 3 pins) connector may exist. Pins 1 and 2 are fan signals. Pin 1 is the fan monitor (FANM) and monitors that the fan is indeed moving by sending two pulses per revolution to the motherboard. This would give an alarm if the fan fails. Pin 2 is the fan control (FANC) signal from the motherboard, going from 12 to 0 volts to control the fan speed for low power periods. Pin 3 is a 3.3-V sensing line to regulate that voltage on the board. Pin 4 is an isolated ground for IEEE-1394 communication circuitry; pin 5 is the voltage for that circuitry, which may be from 8 to 40 volts. The pinout follows.

ATX Optional Power Connector Pinout

| | | Pin | Pin | | | |
|---|---|---|---|---|---|---|
| Black/white | 1394R | 4 | 1 | FANM | (monitor) | White |
| Red/white | 1394V | 5 | 2 | FANC | (control) | Blue/white |
| | Reserved | 6 | 3 | 3.3 V | Sense | Brown/white |

## ENERGY-EFFICIENT "GREEN" SYSTEMS

New systems have power-down features to lower power usage. These are called "green" features because they are environmentally friendly. These make the system and display go into a very low power mode, and the display screen is black. Some try to make you aware of this by flashing the power LED on the case front. This can be very irritating, if you think your system is off, and hit the power button to turn it on, only to find you just turned it off. Windows then runs SCANDISK when you turn it back on, to secure that the hard drive information was not corrupted by an improper shutdown.

To lower the contamination created by the production of power, the U.S. Environmental Protection Agency created the "Energy Star Office Equipment Program." This program has some memorandums of understanding for compliance to low power consumption standards. These are available at www.epa.gov/energystar.html.

The standards describe a low-power mode of equipment as the lowest power state the equipment is designed to enter after some period of inactivity, without actually turning off. That period is the default time, which is the time period set by the factory that determines how long the equipment sits idle before it enters the low-power mode.

Computers may be set in the CMOS SETUP (Chapter 3) to enter a low-power 30-watt "sleep" mode after a period of 15 to 30 minutes. Monitors may have two low-power modes: one of 15 watts and one of 8 watts for 15 to 30 minutes of inactivity. With Windows 95, Click on Start, Settings, Control Panel, System, Device Manager, System Devices, Advanced Power Management Support, and Settings. Windows 98 has a Power Management icon in the Control Panel for setting modes. Neither Windows 95 or 98 will allow hibernation if the 32-bit file allocation (FAT32) has been utilized. See Chapter 3 for more on power management.

Printers may power-down to 15 to 45 watts, depending on print speed. Personal printers that print up to seven pages per minute (ppm) should go to 15 watt mode after 15 minutes. Small-business printers giving more than 7 but less than 14 ppm should go to 30 watts after 30 minutes. Industrial or commercial printers that print more than 14 ppm should go to 45 watts in 60 minutes.

Optical scanners should power-down to 12 watts when idle for 15 minutes.

## SURGE SUPPRESSORS

A surge suppressor is designed to prevent momentary high voltage spikes—introduced in the line by lightning, motors, or other sources—from damaging your computer system. Typical power strip surge protectors have outlets rated at 125 V AC, 15 amps. Since power is the product of voltage and current, the power is $125 \times 15 = 1875$ watts.

Protection in inexpensive units is given by metal oxide varistors (MOVs) alone. The metal oxide varistor temporarily shorts voltages higher than its clamping voltage. If the overall power dissipated is lower than its maximum surge power dissipation, it will not be destroyed. If it is higher, the MOV will be destroyed, and usually there is no indicator of this to the owner of the device. More expensive units include toroidial coils and perhaps high-frequency capacitors. Depending on model and cost, typical ratings are:

> Maximum spike current: 4000 amps to 26,000 amps
> Maximum surge voltage: 6000 volts
> Maximum surge power dissipation: 50 to 480 joules
> Clamping voltage: 330 to 400 volts
> Clamping response time: 1 nanosecond

## UNINTERRUPTIBLE POWER SUPPLIES

Uninterruptible power supplies (UPSs) are devices designed to save your computer system from crashing when the wall AC line current fails. They also act as line filters, smoothing out surges and bypassing electrical noise on the AC line. A differentiation is made between two kinds of uninterruptible supplies that have different strengths and weaknesses.

## STANDBY POWER SUPPLIES

The standby power supply (SPS) passively waits until wall line power fails and then switches on instantly to maintain continuous operation of the computer system. Switching time is critical to resist glitches and lost data or crashes. The AC produced by these supplies is best if it is a pure sine wave, as opposed to a square wave that the typical inexpensive inverter would produce. Switching time is about 3 milliseconds and sine wave compliance about 95%. Because this kind of system is in standby mode most of the time, there is less circuitry and less strain in the batteries. This system therefore costs less. However, because the system has to "switch" on when the power fails, there may be the problems mentioned above. More information on this type of supply is available at www.apcc.com.

## TRUE UNINTERRUPTIBLE POWER SUPPLIES

The second kind of supply is a true UPS. This supply converts wall line current to DC to keep batteries charged, while the batteries continuously drive an inverter to deliver AC to a standard AC receptacle into which the computer system is plugged. This type does not have to switch on during a power failure; it is always on, so there is no danger of a "switchover" spike or lapse that causes problems. These are sometimes described as "in-line" supplies. These are the most fail-safe type of system you can buy, but they cause more wear on the batteries, which shortens their life. Thus, costs for battery replacement, added to the initially higher price, make this a more expensive option.

A new type of internal UPS is actually installed inside the computer case close to the ATX power supply. It is charged by DC from the supply, and in case of power line failure, supplies DC directly to the supply from the nicad batteries. This saves the AC-to-DC conversion to charge the batteries, and then the DC-to-AC conversion for the standard power plug of the PC. These supplies are designed to give about 5 minutes to shut the system down. More information is available at www.silverline-power.com.

Whatever the type, UPS has a rechargeable battery and uses the battery to supply power to your computer system. These batteries are usually meant to give you only a few minutes of power, allowing you to shut your system down properly and turn it off.

The longer the run time, the larger and more expensive the battery. A typical color display may pull 1.8 amps at 120 volts, or 216 volt-amps (VA). Your computer is probably not drawing the maximum the computer power supply is designed to deliver, but use that as a safe high end.

Volt-amps are not exactly the same as watts, because the current and voltage in an AC system can be out of phase. However, assuming they are the same, a 250-watt supply needs 250 VA. So the display monitor requires 216 VA, and the computer with all that is in the case requires 250 VA, for a total of 465 VA.

Choosing a UPS requires some calculation. A UPS may say it has 500 VA capability. To match this to your system, take your combined volt-amps of 465 in our example, and divide by 120 volts for 3.9 or 4 amps. Divide the UPS volt-amps by 120 volts for 4.2 amps.

This UPS has enough power to run your system. The only question left is "for how long?" The physically larger battery (pack) can deliver the power for a longer time. Batteries are rated in volts (V), amps (A), and amp-hours (A-h). A typical small UPS may have 12 V, 4 A-h batteries. Four amp-hours is a product. This battery will deliver 4 A for 1 hour, or 1 A for 4 hours, all at 12 V. Power is voltage times current, so a 12 V, 4 A-h battery has 48 VA-h. When the circuitry of the UPS raises the voltage ten times to 120 V, it must lower the amp-hours by ten to 0.48 A-h, because energy cannot be created or destroyed. Dividing the 0.48 A-h by the 4 A our system requires (if loaded) gives 0.12 hours or 7.2 minutes. This is more than enough to save all data, and properly shut the system down.

## 2.4 GENERIC SYSTEM OPERATION

## OPERATING VOLTAGES

Proper operating voltages are the foundation on which the entire system rests, so it is imperative to confirm that the power supply is operating properly. A multimeter may be used to confirm voltages as indicated in Figure 2.1.

Note that plug 8 pin 2 is not connected on the PC or XT, but it is connected on newer units and should read +5 V. The type of chips used in the system are TTL, or transistor–transistor logic. This type of electronic chip will not run if the voltage supplied to it is less than 4.75 V, and it may be destroyed if the supply voltage is more than 5.25 V. Since the 5 V supply feeds transistor–transistor logic (TTL), it should be between 4.75 and

5.25 V. The +12 runs motors for drives, and the RS-232 card (with the –12 V), so it may be as low as 11.0 V DC. The ATX supply voltages should all be within ±5%, with the exception of 3.3 V, in which must be within ±3%.

If the voltages are off, is the supply bad or is a load dragging it down? You may find this out by disconnecting the supply. (**Warning:** Always turn the supply off when connecting or disconnecting anything.) Most supplies are switching supplies, and many will not operate without a load. They need to deliver power to something, so disconnect all plugs coming from the power supply to the motherboard and to all drives and connect a load resistor across the 5 V supply. A 2-watt 22-ohm resistor draws about 0.227 amps and dissipates about 1.14 watts.

You need to draw enough current to load the supply but less than the maximum power rating. A handy place to insert the load resistor is at one of the drive power connectors between the red and black lines. Calculations for building a load that would test each supply at maximum are given in Table 2.2. Resistors, plugs, and wire are available at electronics parts stores. The ATX supply will also need pin 14, PS-ON, pulled low to tell the supply to turn on.

Measure all voltages of the disconnected but loaded supply. If the voltages are now good, something in the system was loading the supply down. If the voltages are wrong, you must repair or replace the supply.

## POWER SUPPLY MECHANICS

Power supplies are usually removed by removing four screws from the rear of the case (right-hand side except in towers), then sliding the supply toward the front of the case to undo two tabs in slots on the bottom of the supply. In some cases, the drives in front of the supply must be removed to give more room to slide the supply forward. Connect a new supply electrically, and test the system before actually replacing the supply.

Some of the new tower and slimline cases use a power supply with the main power switch externally mounted on the front of the system case instead of on the rear right of the supply case itself. This switch is a possible cause of total supply failure.

Five lines go to the switch. The green, or thickest, wire having a ground lug is the safety ground. Both sides of the AC line are broken; these are almost always white (neutral) and black (hot).

**TABLE 2.2 Calculations for building loads to test power supplies**

| 64-W Supply | 130-W Supply |
|---|---|
| +5 V/7.0 A = 0.71 Ω at 35.0 W | +5 V/15.0 A = 0.33 Ω at 75.0 W |
| –5 V/0.3 A = 17.0 Ω at 1.5 W | –5 V/0.3 A = 17.0 Ω at 1.5 W |
| +12 V/2.0 A = 6.0 Ω at 24.0 W | +12 V/4.2 A = 2.8 Ω at 50.4 W |
| –12 V/0.3 A = 40.0 Ω at 3.6 W | –12 V/0.3 A = 40.0 Ω at 3.6 W |
| **200-W Supply** | **230-W Supply** |
| +5 V/20.0 A = 0.25 Ω at 100.0 W | +5 V/23.0 A = 0.22 Ω at 115.0 W |
| –5 V/0.5 A = 10.0 Ω at 2.5 W | –5 V/0.5 A = 10.0 Ω at 2.5 W |
| +12 V/7.5 A = 1.6 Ω at 90.0 W | +12 V/9.0 A = 1.3 Ω at 108.0 W |
| –12 V/0.5 A = 24.0 Ω at 6.0 W | –12 V/0.5 A = 24.0 Ω at 6.0 W |
| **250-W Supply** | |
| +5 V/25.0 A = 0.20 Ω at 125.0 W | |
| –5 V/0.5 A = 10.0 Ω at 2.5 W | |
| +12 V/9.5 A = 1.2 Ω at 114.0 W | |
| –12 V/0.5 A = 24.0 Ω at 6.0 W | |

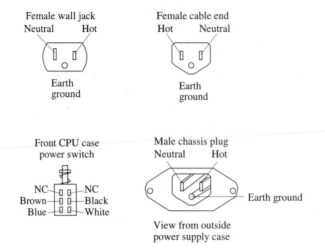

**FIGURE 2.2   Power supply external switch wiring**

Female wall jack
Neutral    Hot

Earth
ground

Female cable end
Hot    Neutral

Earth
ground

Front CPU case
power switch

NC —         — NC
Brown —      — Black
Blue —       — White

Male chassis plug
Neutral    Hot

Earth ground

View from outside
power supply case

You can make a continuity check of the DPST (Double Pole Single Throw) or DPDT (Double Pole Double Throw) switch. With power off, one position of the switch should give 0 Ω resistance and the other position should give infinite resistance. Write down the color of the wire going to each pin. Be careful to reconnect in the correct positions, so you don't reverse hot and neutral. See Figure 2.2 for typical wiring.

**Note:** The rest of this section is recommended for technicians, *not* for nontechnician users.

**Warning:** The following discussion is for background only. Do not open the metal case of the power supply unless you are a trained technician.

The switching supply takes the input line of 110 V AC 60 Hz, or cycles per second, through a switch and a fuse and immediately rectifies it in a monolithic bridge or four separate diodes. This DC voltage of approximately 155 to 170 V DC is usually doubled to 330 V DC. It runs one section of circuitry directly connected to the AC line and must be treated with care. All test equipment should use an isolation transformer.

This DC high voltage is divided down to run an oscillator at up to 50 kHz. This is fed to a power transistor that oscillates the 330 V DC, creating 330 V AC 50 kHz. This high-frequency signal goes to the primary of the main power transformer.

In ISA or EISA computers, this transformer typically has three secondaries, one each (with two terminals) for –12 V, +12 V, and +5 V. The –12 V is used to develop the –5 V. These outputs are high frequency AC, which is easier to filter than the low frequency of the linear system, so the electrolytic filter capacitors are much smaller in a switcher also. They are filtered, regulated, and output.

Part of the filtering is to feed back the outputs to comparators to adjust the oscillator's output, to raise or lower the voltage. Part of the feedback is used to shut down the supply should a potentially damaging overvoltage occur.

The oscillator is in the part of the circuit before the main transformer, so it is tied directly to the AC line and must be kept isolated from the output. This is done through an optocoupler or feedback transformer. Due to design, the switching supply usually needs a load on the output to operate.

## POWER SUPPLY BLOCK DIAGRAM

Since a schematic is specific for a unique supply, a relatively detailed block diagram shows enough to help you troubleshoot a power supply. Figure 2.3 shows a block diagram of a typical supply.

Beyond the expected on/off switch and a fuse, the 110/220 switch must be in the correct position of 110 V. A line choke follows to smooth current and act as an EMI (Electromagnetic Interference) filter. A typical size for this is a dual toroidal (doughnut-shaped) inductor about 3/4 inch in diameter.

The incoming 110 V AC is a RMS (Root Mean Square) measure; peak to peak it is 339.36 V. This is directly rectified in a monolithic bridge or four high current diodes (or in a voltage

**FIGURE 2.3 AT power supply block diagram**

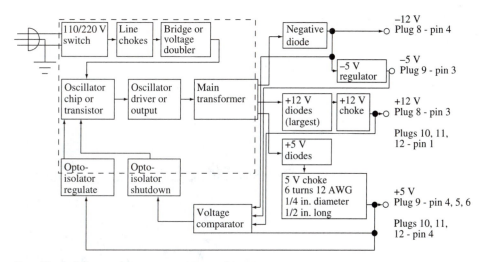

*Note:* Circuits in box use common ground. Be careful with scope or meter.

doubler circuit) without going through a large, heavy step-down transformer as in a linear supply. This is filtered by two large electrolytic caps, often 220 or 330 μF/200 V DC. The actual direct current voltage after losses in the inductors used to smooth current is around 325 V DC.

**Warning:** Remember to carefully discharge these two caps across their own terminals when the supply is *off* before poking around much, and watch your fingers when the supply is *on*.

This high DC voltage is then oscillated (usually the first transistor) at about 20 to 90 kHz, depending on the particular manufacturer. This new high-frequency AC is then sent (sometimes through a small transformer) through a power amp driver (usually the second transistor) to the primary of the main transformer.

Note that up to this point both sides of the 110 V line have been used without the benefit of a transformer to isolate the circuit from the line. So far then, all "grounds" are really a common return, and your oscilloscope (with its ground on the chassis) will not properly display the signals. A multimeter running with a battery will work fine, using the "common" for the negative lead. On the secondary of this main transformer and thereafter, the scope works fine. This is also why optocouplers (usually two) are needed to feed back the oscillator regulation reference signal and the oscillator shut-down signal from beyond the secondary back to before the primary. Photocouplers or optoisolators can be used in place of the optocouplers.

At this higher frequency, the main transformer can be much smaller. The primary winding is about 0.3 to 0.4 Ω. The secondary windings are from 0.06 to 0.1 Ω. In almost every case they should measure less than 1 Ω on primary and secondary. Obviously, these secondary windings use heavy gauge wire for high current capability and therefore have lower resistance.

Figure 2.4 shows the line filter cap and inductor, the monolithic bridge rectifier, and the two large DC filter/doubler caps. These are the ones that may be charged to 165 V DC each.

The outputs from the secondary of the main (and often only) transformer are three or more taps at 20 to 40 V AC (Figure 2.5). These are then rectified for the final DC outputs of +5, –5, +12, and –12 V DC.

The –5 V and –12 V rectification is usually through one small diode (similar to 1N4001). Some supplies use the common 7905 to regulate the –12 V down to get the –5 V. The +12 V uses the two largest diodes (with cathodes connected, mounted on a large 1.5" × 1.5" heat sink). You can simply identify them by a resistance continuity check from plug 8, pin 3 to the cathodes of a pair of diodes. Either this +12 V or the –12 V is used to run a fan (usually brushless), which aids cooling of the supply and the system. The +5 V uses two small diodes. They may be simply identified by a continuity check from plug 9, pin 4 or 5 or 6 to the cathodes of a small pair of diodes. The +5 V supply is the most heavily loaded, so it usually has an extra choke. Typically, this is about six turns of number 12 awg (American wire gauge) on a ferrite core 1/4 inch in diameter by 1/2 inch long. Check –5 V anodes to plug 9, pin 3 and –12 V anodes to plug 8, pin 4.

**FIGURE 2.4  Closeup of power supply components**

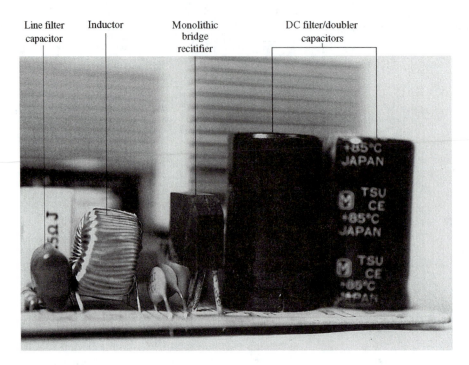

Line filter capacitor · Inductor · Monolithic bridge rectifier · DC filter/doubler capacitors

**FIGURE 2.5  Power supply main transformer secondary winding signals**

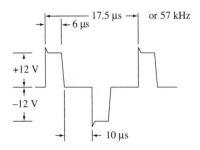

*Note:* Signal on one tap of secondary of main transformer. Another tap has double this or 25 peak.

Regulation occurs as one of the outputs, 15.48 V DC or the 5 V DC, is fed back through an optocoupler to the oscillator, which adjusts the level of the high frequency on the primary, raising or lowering overall output. Also there is usually another toroidal coil choke (the largest in the system, about 1 1/4 inch in diameter with multiple windings) to smooth the currents. Resistance on all windings should be well below 1 Ω.

All four DC voltages are run to a quad comparator, often an LM339 or HA7339, that checks them. See Figure 2.6 for the pinout of the LM339. If any are below tolerance, the comparator output, passing through two small transistors as an amp and then a driver, turns the power good signal off. Without this signal on power plug 8, pin 1, the oscillator for the CPU will not function and the computer is dead. This line also goes to a second optocoupler and to the oscillator to shut it down if one of the voltages is too high or low.

## REPAIRING THE POWER SUPPLY

To begin repairing the supply, totally disconnect it, and disassemble, baring the circuit board. (**Warning:** Short the two large caps before they zap you.) Look the board over and identify major components from the block diagram in Figure 2.3. With a marker, mark the bottom of the circuit board, boxing in transistor leads and transformer leads, marking polarity of caps, and the like.

Switching supplies are usually "dead" because the comparator circuit shuts down the oscillator if any little thing goes wrong, killing all outputs. This kind of feedback makes it a little

**FIGURE 2.6   LM339 pinouts**

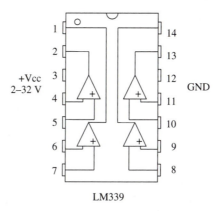

more difficult to work on a switching supply than on a linear supply. The oscillator does not control the high-voltage DC from the voltage doubler or large bridge close to the input, so find the two large caps and use an isolated meter to check for about 160 V DC across each.

If your meter is battery driven, that is fine. If it plugs into the wall, you must use an isolation transformer. These are available at electronic parts stores. This 320 V DC runs everything. If it's not there, just follow from the AC input with your meter through the small line input choke, the power switch, the fuse, another small input choke, and finally the bridge or doubler diodes and large filter caps.

Now that you have the high-voltage DC, check the outputs of the secondary of the main transformer. If you have the waveform signal shown in Figure 2.5, or anything similar, you are home free. If not, you need to get the oscillator going.

You must separate the oscillator from the shutdown signal from the comparators. If there is a bad short somewhere past the secondary, this will pump current into it and could heat up some component to smoking or worse. Therefore, prepare for measurements with the power off, then apply power, take quick measurements, and turn off the power at the first sign of trouble.

Pins 5, 7, 9, and 11 are the positive voltage inputs, and pins 4, 6, 8, and 10 are the negative inputs. If the positive input voltage is above the negative input voltage, the output is high, or 5 V. If the positive input voltage falls below the negative input voltage, the output is low, or 0 V. One of each pair of inputs is the voltage being checked against the reference and the other is the reference.

Two of the four 339 pins 1, 2, 13, and 14 may be tied together and go out to power good (either directly or through two small driving transistors) and to a photocoupler back to shut down the oscillator. On some, pin 1 or 2 goes directly to power good but *not* to shutdown, leaving 13 and/or 14 to go to shutdown. In those cases, the proper state of 13 and 14 is *low*, not high.

On a working supply, you may use a 1 kΩ to 2.7 kΩ resistor (4.7K may be too large to work) and pull down (lower the voltage below 0.8 V) pins 1 and 2 or 13 and 14 (whichever outputs are high) of the 339 and watch the system shut down, or you can pull up 13 or 14 (whichever is down).

**FIGURE 2.7   TL494 pinouts**

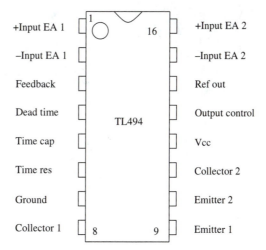

The TL494 with pinouts, shown in Figure 2.7, is a **pulse width modulator.** The inputs are on pins 1 and 2, and the outputs are on pins 8 and 11. They are sometimes pulled up to Vcc by a 60 to 150 Ω 2-watt resistor each. The timing resistor and cap are pins 6 and 5. Using this device gives system feedback control over the oscillator in the circuit.

In the total printed circuit board in Figure 2.8, note the two large filter/doubler caps. Just below them are four power rectifier diodes. Just below the diodes are the line choke inductive filter and, finally, the fuse. At the top center is the single heat sink with three major power transistors. Just below is the main transformer. The transformer is unbelievably small for delivering 200 W, but that is what a switcher is all about. To the right of the transformer are the output inductive filter, and to the right of the filter, the output filter caps. Just below the output inductor are two heat sinks for the output regulators.

The power supply of the system and the power supply of the display (monitor) have components that run hot. They may have large aluminum fins called heat sinks attached. Expect heat sinks to be very warm. This heat can cause intermittent failures of components. Often, a component fails after a certain time period then corrects itself automatically because its failure stopped the current that was heating it.

When you have this kind of symptom, **freeze spray** may be very useful. Spray the suspected component. If the component starts working again, then heats up and fails again, you have confirmed it as the problem component.

These cooling sprays are often misused and can themselves cause component failure. Again, the proper use is to identify the component or group of components *first,* allow the system to heat up and fail, then apply the spray sparingly to one component at a time. If or when the problem disappears, replace the last component sprayed.

The *improper* use of freeze spray is to hose down large areas of the board, hoping to luck out and cause the unit to start working. After it heats and fails again, you try one section of the last section of board sprayed. After a few attempts, the component may be identified. However, the sudden change in temperature puts physical stress on the component, like concrete cracking under repeated freezing and thawing, and can cause failure of a previously good part, compounding your problem. The best approach is to use electronic troubleshooting techniques to narrow your search to just a couple of components before spraying.

Using voltage and resistance values and your meter may help identify a problem area. Table 2.3 gives typical voltages and resistances at various points on the main power supply printed circuit board and at the output plugs.

**FIGURE 2.8    Power supply circuit board**

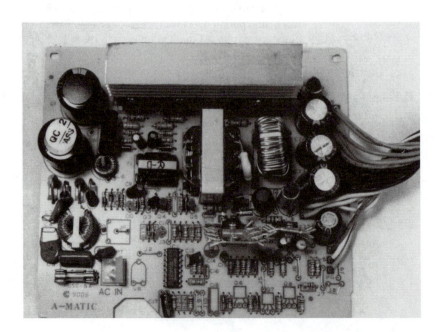

TABLE 2.3   Typical voltages
and resistances, taken with a
multimeter

| Test Point | Voltage | Resistance (forward bias) | Resistance (reverse bias) |
|---|---|---|---|
| Main diodes or bridge | 325 V DC | 100 Ω | infinite |
| Main filter capacitors | 325 V DC | n.a. | n.a. |
| Primary (main transformer) | 370 V AC | 0.3 Ω | n.a. |
| Secondary (main 5 V) | 25 V AC | 0.05 Ω | n.a. |
| Secondary (main 12 V) | 50 V AC | 0.1 Ω | n.a. |
| +5 V | +5 V DC | 2 Ω | 50 Ω |
| −5 V | −5 V DC | 50 Ω | 10 Ω |
| +12 V | +12 V DC | 10 Ω | 50 Ω |
| −12 V | −12 V DC | 50 Ω | 12 Ω |

## 2.5   ASSEMBLING POWER SUPPLIES

You can usually purchase the case with the power supply included. This relieves a real headache of matching form factors. The **form factor** is the *shape* of the supply that physically fits the case. Form factors for cases include:

- PC/XT slide
- PC/XT flip-top
- AT, full
- AT, mini
- AT, baby
- Tower, full (upright)
- Tower, medium
- Tower, mini
- ATX case

The power supply form factor must match the case form factor. The power supply line power plug, the display power plug, the fan, the mounting holes, and the switch must perfectly match the case. You can see how buying the supply in the case is easiest. If replacing a supply, be very careful to confirm a matching form factor before ordering the supply.

The ATX supply connector is strongly keyed and cannot be plugged in incorrectly. (Figure 2.9).

**FIGURE 2.9   The strongly keyed ATX power supply connector**

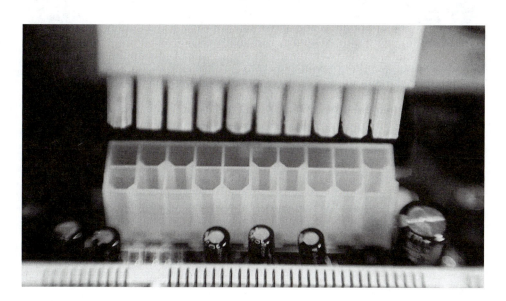

There is no external power switch and no power switch cable to a front panel switch. It is turned on and off from the motherboard. All other supplies have two power supply 6-pin plugs that should have all the black wires together. Remember, "black to black, back to back." If there is no mounted on/off switch, there will be a cable to a power switch on the front panel. **Caution!** Although these wires are insulated, the terminals have 110 V AC and can give a severe shock and burn, burn out the switch, and blow a house circuit breaker. Correct wiring is critical.

Supplies in desktop cases are almost universally fastened by four screws through the rear panel of the case. Standard case supplies additionally have two slots in the bottom that catch tabs on the bottom of the case when sliding it toward the rear panel. These secure the front of the supply. Tower cases have a bracket to hold the supply. Instead of the tabs and slots, they have two screws through a flange in the front bottom of the supply into the bracket. Typically, tower supplies are smaller than standard case supplies. Most tower supplies have a cable to the case front panel power switch rather than a switch on the supply itself.

Figure 2.10 shows the form factor of the XT supply, an early AT "L" supply, a newer style AT supply, and a typical tower supply. Figure 2.11 shows final mounting of a typical XT supply and a typical minitower supply.

ATX power supply

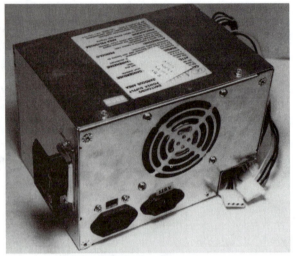

Early AT "L" power supply

Newer AT power supply

Typical tower power supply

**FIGURE 2.10   Shapes or form factors of standard supplies**

XT case

Tower case

**FIGURE 2.11  Mounted supplies**

## 2.6  BASIC TROUBLESHOOTING

Troubleshooting power supplies begins by noting any POST audio or video error codes. Then you can proceed to noting any visual clues and taking appropriate measurements. Details of these steps are as follows:

1. A POST audio error code that is one long continuous beep, repeating short beeps, or no display and no beep may indicate a faulty power supply (Appendix A).
2. A POST video error code of 02$x$ (where $x$ is any numeral) indicates a power supply problem.
3. Measurements.

**Caution:** 110 V AC and 330 V DC are present inside the power supply case.

1. Is the fan working? If so, you have +12 V or –12 V and the comparator is not shutting down the system, so the supply is probably OK. If the fan is not turning, either *it* is dead, or there is no +/– 12 V. If the fan is dead, the supply will also soon be dead from overheating. Continue this sequence until the supply measures good as in Figure 2.1.
2. Is the monitor power light on? If not, check the wall outlet for 110 V AC. If that checks out, check for 110 V AC at the input to the power supply circuit board (or mono monitor outlet on rear of supply). If the wall has 110 and the board does not, remove the power cord and test continuity.
3. Use a multimeter to check +/–5 and +/–12 V and the power good signal on plug 8, pin 1. Use Figure 2.1 for pinouts. If you have these, the supply is good. If not, continue.
4. Remove all power supply connections, insert load resistor, pull pin 14 of ATX supply low, and power-up. If voltages appear, a board or drive was loading down supply. Isolate the culprit by reinserting the system board, adapter cards, and drives one at a time, powering up and down each time. Repair or replace that adapter card. If voltages do *not* appear, users should replace the supply—do not proceed beyond this point.

**Caution:** Only technicians should proceed with the remaining steps.

If voltages are not as in Figure 2.1, remove supply from CPU case and open. Remove case from supply to access board.

5. Is 165 V DC across each of the two large electrolytics, for 330 V DC total in series? If not, check cord, switch, fuse, bridge, and caps. **Warning:** These caps pack quite a charge. Watch your fingers.
6. With a meter, does the primary have 100–200 V AC? If so, the oscillator is working. If not, welcome to the catch-22. Is there no voltage because the oscillator is not working, or is the oscillator not working because there is no voltage?

7. With a scope or a meter, does the secondary have 20–40 V peak to peak? If so, the oscillator is working and the comparator is passing its signal through the optoisolators.
8. With a meter, check the continuity of all outputs to their respective rectifiers.
   a. The +5 V is from plug 9, pins 4, 5, or 6 to the cathodes of a small pair of diodes.
   b. The +12 V uses the two largest diodes (with cathodes connected, mounted on a large 1.5" × 1.5" heat sink). Check continuity from plug 8, pin 3 to the cathodes of a pair of diodes.
   c. Check –5 V anodes to plug 9, pin 3.
   d. Check –12 V anodes to plug 8, pin 4.
9. Use Table 2.3 for general voltage and resistance checks.
10. When appropriate, just replace the whole power supply quickly.

## 2.7 CHAPTER REVIEW/QUIZ QUESTIONS

1. To thoroughly test the power supply you need
   a. Specialized electronic test equipment.
   b. A diagnostic disk.
   c. A multimeter and load resistor.
   d. Only your eyes, ears, and nose.

2. Before adding an additional adapter card or drive, check
   a. The interrupts.
   b. The motherboard jumpers.
   c. For compatibility.
   d. Whether the power supply can handle the additional load.

3. The floppy and hard drive power connectors have
   a. +5, –5, +12, and –12 volts.
   b. +5 and +12 volts.
   c. +5, +12, and 120 volts.
   d. Six connectors.

4. AT power supply connectors P8 and P9 connect to
   a. The hard drives.
   b. The floppy drives.
   c. The motherboard.
   d. The front panel switches.

5. On a totally dead system, verify 110 V AC wall line by a quick check of the
   a. Power supply fan.
   b. Display monitor pilot light.
   c. Floppy drive pilot light.
   d. Audio error "beep" code.

6. The AT power supply delivers what DC voltages?
   a. +5, –5, +12, –12
   b. +5, +12, +24, +48
   c. +12, –12, +24, –24
   d. It delivers only AC voltages.

7. What are the advantages of a switching power supply over a linear power supply?
   a. Heavier, larger, and more powerful.
   b. Faster, but weaker.
   c. Smaller and lighter.
   d. Only economic.

8. To match a case, the power supply must have a compatible
   a. BIOS.
   b. Voltage output.
   c. Input voltage.
   d. Form factor.

9. A first check of the power supply is to feel whether
    a. It is hot.
    b. It is cold.
    c. The fan is working.
    d. It is vibrating.

10. The two most important considerations in replacing power supplies are the
    a. Volts and the amps.
    b. Form factor and the wattage.
    c. Size and the shape.
    d. Brand and the output voltage.

11. The ATX power supply connector is compatible with the older AT power supplies.
    a. True.
    b. False.

12. Which kind of power supply switches on to power a system when the wall line power fails?
    a. UPS.
    b. SPS.
    c. DPS.
    d. OOPS.

13. Which kind of power supply constantly powers a system whether the wall line power fails or not?
    a. UPS.
    b. SPS.
    c. DPS.
    d. OOPS.

14. The ATX power supply delivers what DC voltages?
    a. +5, −5, +12, −12.
    b. +3.3, +5, −5, +12, −12.
    c. +3.3, −3.3, +5, −5, +12, −12.
    d. +1.0, +2.0, +3.0, +4.0, +5.0, +12, −12.

15. The ATX power supply will appear nonfunctional unless it is
    a. In perfect condition.
    b. Loaded.
    c. Loaded and pin 14 is pulled low.
    d. Plugged into a motherboard.

16. Who should open a computer power supply to work on it?
    a. Only power users.
    b. Only the beginning computer technician
    c. Only a trained electronic/computer technician.
    d. Anybody with courage and a bad power supply.

# 3

# Motherboards

**OBJECTIVES**

After completing this chapter you should be able to:

- Describe the major components on the motherboard
- Describe the function of each component
- List the microprocessors used in the PC, the XT, and the AT
- List the data and address bus widths of the 80X86 microprocessors
- Identify the CPU and bus type of a motherboard
- Choose and install a math coprocessor
- Install a motherboard
- Troubleshoot and repair problems

## 3.1 GENERAL DISCUSSION

The motherboard, or system board, is usually the largest printed circuit board (PCB) in the computer system. A PCB is made of a type of fiberglass, often covered with a green or tan coating. Copper traces or stripes on the surface and inside the PCB act as wire between the components that are soldered on it. Many of the components are small plastic rectangles, or **chips,** that hold electronic circuitry. Many different circuit functions are integrated into one large circuit so they are called **integrated circuits** or ICs.

There are two major physical types of chips in the computer system. The first type is dual inline package (DIP) chips. These are small plastic rectangles with a row of leads on two opposing sides. This gives two rows in parallel lines, or dual inline, leads. A DIP chip's leads go through holes in the printed circuit board and are soldered in place. Sometimes a socket is soldered in instead, so the DIP chip can be removed and replaced easily. The DIP must be plugged into the socket in the correct direction. A semicircular indentation on one end or a circular depression identifies pin 1. A similar semicircular notch or corner diagonally cut off identifies pin 1 of the socket. Sometimes pin 1 is labeled on the PCB in white painted wording. The electronics inside a chip is tiny; most of the package size is there so humans can handle the chip.

You need to be careful not to bend the leads of the chip while inserting it in the socket. Because upgrading may require multiple removals and insertions, a zero insertion force (ZIF) socket (Figure 3.1) is sometimes used. The chip leads just drop in between contacts, and a lever arm gently closes the contacts on the pin. Figure 3.2 shows a DIP chip and its socket.

**FIGURE 3.1    ZIF socket with chip out and arm up**

The second physical type of chip is a thin, flat rectangle with leads that stick out the sides in the same plane as the chip, making it almost flat, as shown in Figure 3.3. The leads do not go through the motherboard, as the DIP leads do, but they are soldered on the surface of the board to the copper stripes. These are called **surface mount chips** because they are created with surface mount technology (SMT). They do not require holes drilled through the PCB, and they are usually machine placed and soldered. This allows the packages to be much smaller than DIP chips, so the board constructed with SMT is cheaper and more reliable. One drawback is that the smaller chips have smaller leads that are closer together, making them very difficult or impossible to replace without specialized equipment.

As engineers developed the technical capability to manufacture the electronic circuits inside a chip smaller and smaller and as the complexity of the circuitry in those chips or ICs rose, the scale of integration rose also. Small-scale integration (SSI) uses ten or fewer

**FIGURE 3.2    DIP chip and socket**

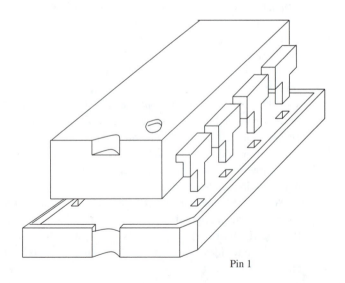

Pin 1

**FIGURE 3.3   Surface mount and pin grid array chips**

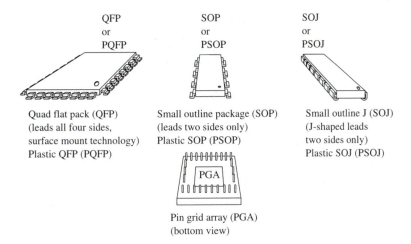

Quad flat pack (QFP)
(leads all four sides,
surface mount technology)
Plastic QFP (PQFP)

Small outline package (SOP)
(leads two sides only)
Plastic SOP (PSOP)

Small outline J (SOJ)
(J-shaped leads
two sides only)
Plastic SOJ (PSOJ)

Pin grid array (PGA)
(bottom view)

transistors, medium scale integration (MSI) has more than ten but fewer than 100 transistors, large scale integration (LSI) has more than one hundred but fewer than 1000 transistors, and very large scale integration (VLSI) has more than 1000 transistors. In fact, we will discuss the 80386, which has more than 1.2 million transistors; the 80586, which has more than 3.1 million; and the Pentium II, which has 7.5 million.

The motherboard holds a number of the major components of the system. We will briefly review these here, then cover some more thoroughly later. Some of the major components of the motherboard are shown in Figure 3.4.

## MOTHERBOARD COMPONENTS

### Microprocessor

The microprocessor is the electrical component that does the "thinking" or calculating. Called the central processing unit (CPU), it is the brain of the motherboard. To think, you must be logical; therefore, the circuitry in the computer system is logic circuitry. Most of the computer circuits use a device called a **transistor.** With many transistors connected in one circuit, many of the circuits are called transistor–transistor logic, or TTL. All other components are present to support the operation of the CPU. Many CPUs are covered later in this chapter.

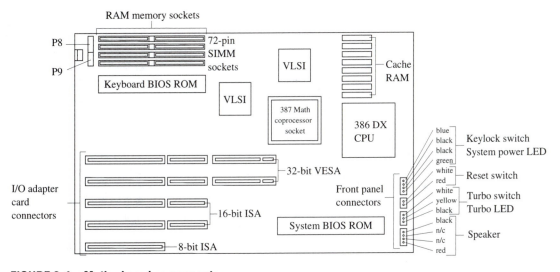

**FIGURE 3.4   Motherboard components**

## Coprocessor

The microprocessor is a general-purpose processor, and like many of us, it is not too good at math. Therefore on some older systems, a math coprocessor or numeric processing unit (NPU) is needed to help the CPU with higher math. Some old boards have an empty chip socket for you to plug in an NPU if desired. In modern processors, the coprocessor is no longer an enhancement, it's a necessity and is built in the CPU.

## Chipset

The CPU does not do all the work. A wide array of circuitry is required to support the CPU, control memory, control peripherals, and much more. In the beginning these circuits were realized in different medium and large scale integrated circuits. With the higher speeds, distances and resistance of copper have become a bottleneck to speed. Therefore, VLSI chips are being created in sets to aid the microprocessor. These are called *chipsets* and must be matched to the CPU being used. This is done by the motherboard manufacturer and cannot be changed in the field by the technician. Still, it is interesting to keep an eye in what sets are being used.

## ROM

Just as your thinking organ, the brain, has a memory, the CPU needs memory also. A number of different kinds of memory are used in the computer. The **read-only memory** (ROM) is a chip that holds the instructions for the CPU to follow to do its job. Such instructions are called a "program." Since this program tells the CPU how to bring information or data in from components and how to send information out to components, it is called the **basic input-output system** (BIOS). The program is written into the BIOS ROM at the factory, and it cannot be changed. It is called "read-only memory" because it can be "read" but not "written" by the user. ROM does not lose its program when the power is turned off. There is also a keyboard ROM on the motherboard that tells the CPU what character was sent from the keyboard.

## RAM

The **random access memory** (RAM) is a chip like the ROM, but the information or data in it can be erased and rewritten by the CPU. Since it can be "read" and "written," access to it is "random." This is the memory the CPU uses, just like your memory, in doing its work. Unlike ROM, RAM needs power to hold its information, and it loses all information when the power is turned off. This dynamic RAM (DRAM) is the memory people are talking about when they ask how much memory your system has.

## Cache

The faster the CPU, the faster the system. But CPUs became too fast for the normal RAM to keep up. So a different, much faster type of RAM was needed. Because the new static RAM (SRAM) takes more power, the bulk of the memory is still low-power RAM. However, to keep the CPU from slowing down, a cache, or small storage area, is made of the faster SRAM. Because it is RAM, it also loses the information it holds when the power is turned off.

## CMOS

As systems became more complex, it became necessary for them to hold information about how the particular system was configured, such as how many and what kind of floppy drives, what kind of hard drive, and what kind of display. The user had to be able to change this information, so it could not be stored in ROM. However, the system had to hold this information even when the power was turned off, so it could not be stored in normal RAM. A different kind of memory was needed. The solution was the complementary metal oxide semiconductor (CMOS), which holds information with very, very little power. It can be powered by a small battery, and it holds information when the system is turned off. There are only 64 bytes of CMOS memory. While nicad (nickel–cadmium rechargeable) batteries can be used, the most common by far are lithium batteries. They can last for three or more years in a computer system and can be replaced.

### I/O Connectors (Slots)

The motherboard cannot do everything by itself. It requires other devices to be a complete, functioning computer system. The keyboard is used to input information and is connected directly to the motherboard. All other devices must have a printed circuit board or adapter card or controller card to interface with the motherboard. These adapter cards for devices used to input and output information to the motherboard are connected via input and output (I/O) connectors, or slots. There are a number of different kinds, as explained throughout this chapter.

### Crystals

These components look like small steel cylinders or slabs. A **crystal** is an electronic device that vibrates or oscillates at a precise frequency. Crystals are used to time all computer operations. The frequency or speed of the crystal determines the speed at which the CPU runs, which determines the speed of the computer system.

### Shunt Jumpers or Switches

The CPU needs to be informed of how you want the system configured. The CMOS tells the CPU some things, but you need to set some motherboard switches to tell the CPU more. Two types of switches are used: the dual inline package switch, or DIP switch, and the shunt jumper switch, or Berg connector.

The shunt jumper switch is much cheaper and, therefore, more common. More will be given on setting these in this chapter. As shown in Figure 3.5, a small plastic "box" shorts together the two pins over which it is installed. In a two-pin switch, the switch is either shorted (on) or not shorted (off). The jumper can be installed on just one pin, hanging off the other side, to save it for possible future use. In a three-pin switch, the shunt can go from pin 1 to 2, or 2 to 3 to make a choice. If you ever change a switch setting, write it down to document it for later use. Tape the paper in or under your computer case so you will always know just where it is.

**FIGURE 3.5   DIP switches, shunt jumpers, and headers**

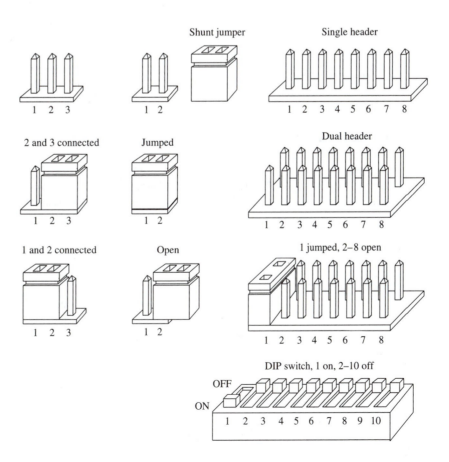

The DIP switch, as seen in Figure 3.5, usually has four, eight, or ten positions, which are each an individual switch. One side of the switch is usually marked "on" or 1, and the other side "off" or 0. You slide the tab to the side wanted. A ballpoint pen is easier to use than a finger, but never use a pencil—pencil lead may get in the switch and ruin it.

### Connectors

Finally, the keyboard, a speaker, and a few lights (light-emitting diodes, or LEDs) and switches on the front of the computer case need to be connected to the motherboard. These connectors are straight pins that look like little nails sticking up out of the motherboard. If they are in one straight line, it is called a **single inline header.** If the pins are in two rows, it is called a **dual inline header.** These are shown in Figure 3.5. Connectors for these will be wired to the front panel of the computer case. Switches and lights there may include:

- Keylock switch, which locks out the keyboard to stop unauthorized use.
- Power LED to show that the system is turned on.
- Turbo switch and LED to show that the system is running at its highest speed.
- Reset switch, which tells the system to drop everything and start again.
- Hard drive LED, which lights when the hard drive is accessed.

## MICROPROCESSORS

The microprocessor, or CPU, is the brain of the motherboard. The motherboard, the main circuit board or card, also holds the memory (RAM and ROM), the I/O ports, and the expansion connectors (printed circuit board edge card connectors to accept adapter cards that add functions to the system). While the power supply may be the foundation, the motherboard is the heart of the computer system. The lines that transfer related information from one point to another on the computer are called a bus. A typical computer system has at least three buses, the data, address, and control buses. The "size" or classification of the CPU is determined by its data bus width. An 8-bit, 16-bit, 32-bit, or 64-bit microprocessor has that sized data bus. The amount of memory the CPU can address is dependent on the width of its address bus. Table 3.1 gives a simple comparison of the current popular microprocessors. See Appendix F for more on all microprocessors covered.

Direct memory addressed is actual RAM that may be installed. Virtual memory is the amount of memory the computer *seems* to have, although it is actually using the hard drive for storing information. Large blocks of information are switched back and forth from the RAM to the hard drive as needed. This is done transparently (the user never notices) and gives the illusion of vast amounts of RAM.

**TABLE 3.1   Microprocessor comparisons**

| CPU | Data Bus Bits | Address Bus Lines | Possible Memory Direct | Virtual | Internal Registers | Math Coprocessor | Internal Cache(s) |
|-----|-----|-----|-----|-----|-----|-----|-----|
| 8088 | 8 | 20 | 1 MB | | 16 | 8087 | |
| 8086 | 16 | 20 | 1 MB | | 16 | 8087 | |
| 80286 | 16 | 24 | 16 MB | | 16 | 80287 | |
| 80386SX | 16 | 24 | 16 MB | | 32 | 80387SX | |
| 80386 | 32 | 32 | 4 GB | 64 TB | 32 | 80387 | |
| 80486SX | 32 | 32 | 4 GB | 64 TB | 32 | 80487SX | 8K instruction and data |
| 80486 | 32 | 32 | 4 GB | 64 TB | 32 | in CPU | 8K instruction and data |
| 80586 | 64 | 32 | 4 GB | 64 TB | 256 | in CPU | 8K instruction and 8K data |
| 80686 | 64 | 32 | 4 GB | | | in CPU | 8K instruction, 8K data, and 256K general |
| Pentium II | 64 | 32 | 4 GB | | | in CPU | 16K instruction and 16K data |
| AMD K6 | 64 | 32 | 4 GB | | | in CPU | 32K instruction and 32K data |
| Cyrix 686MX | 64 | 32 | 4 GB | | | in CPU | 64K unified |

The earliest motherboards had two setup switches, which were pared down to one on the XT and to none on ATs. If your motherboard has switches, they must be set correctly. If it is an AT or 286 or better, then it does not have switches. In that case, the battery-backed-up CMOS memory setup must be set correctly, and perhaps several jumpers must be set for options, such as size of memory chips.

On older systems motherboard switch number 2 (SW2) settings referred to the amount of memory installed in the system. The very first had only 64K on the motherboard, later increased to 256K, although you could add on to both. This added memory was on an expansion card.

If you are repairing an old system and it becomes confusing, write down SW1 and SW2 settings, pull the expansion card, and set SW2 for no additional memory above

**TABLE 3.2   PC SW1 settings**

| SW1 | | | | |
|---|---|---|---|---|
| **Disk Drives Installed** | **Switches:** | **1** | **7** | **8** |
| 0 drives | | ON | ON | ON |
| 1 drive | | OFF | ON | ON |
| 2 drives | | OFF | OFF | ON |

| **Display Installed** | **Switches:** | **5** | **6** |
|---|---|---|---|
| None or EGA or VGA | | ON | ON |
| CGA x 40 characters | | OFF | ON |
| CGA x 80 characters | | ON | OFF |
| Mono | | OFF | OFF |

| **Math Coprocessor** | **Switch:** | **2** |
|---|---|---|
| None | | ON |
| Installed | | OFF |

| **Memory 64K System** | **Switches:** | **3** | **4** |
|---|---|---|---|
| 16K | | ON | ON |
| 32K | | OFF | ON |
| 48K | | ON | OFF |
| 64K | | OFF | OFF |

| **Memory 256K System** | **Switches:** | **3** | **4** |
|---|---|---|---|
| 64K | | ON | ON |
| 128K | | OFF | ON |
| 192K | | ON | OFF |
| 256K | | OFF | OFF |

**SW2 (memory on expansion cards, 6, 7, 8 always off)**

**Memory 64K System**

| On Card | Total | Switches: | 1 | 2 | 3 | 4 | 5 |
|---|---|---|---|---|---|---|---|
| 0 | 64 | | ON | ON | ON | ON | ON |
| 32 | 96 | | OFF | ON | ON | ON | ON |
| 64 | 128 | | ON | OFF | ON | ON | ON |
| 96 | 160 | | OFF | OFF | ON | ON | ON |
| 128 | 192 | | ON | ON | OFF | ON | ON |
| 160 | 224 | | OFF | ON | OFF | ON | ON |
| 192 | 256 | | ON | OFF | OFF | ON | ON |
| 224 | 288 | | OFF | OFF | OFF | ON | ON |
| 256 | 320 | | ON | ON | ON | OFF | ON |
| 288 | 352 | | OFF | ON | ON | OFF | ON |
| 320 | 384 | | ON | OFF | ON | OFF | ON |

TABLE 3.2 *(continued)*

| SW2 (memory on expansion cards, 6, 7, 8 always off) Memory 64K System | | | | | | | |
|---|---|---|---|---|---|---|---|
| **On Card** | **Total** | **Switches:** **1** | **2** | **3** | **4** | **5** | |
| 352 | 416 | OFF | OFF | ON | OFF | ON | |
| 384 | 448 | ON | ON | OFF | OFF | ON | |
| 416 | 480 | OFF | ON | OFF | OFF | ON | |
| 448 | 512 | ON | OFF | OFF | OFF | ON | |
| 480 | 544 | OFF | OFF | OFF | OFF | ON | |
| 512 | 576 | ON | ON | ON | ON | OFF | |
| 544 | 608 | OFF | ON | ON | ON | OFF | |
| 576 | 640 | ON | OFF | ON | ON | OFF | |

*Note:* 576 and the 64 on the motherboard add to 640K

| SW2 (memory on expansion cards, 6, 7, 8 always off) Memory 256K System | | | | | | | |
|---|---|---|---|---|---|---|---|
| **On Card** | **Total** | **Switches:** **1** | **2** | **3** | **4** | **5** | |
| 0 | 256 | ON | OFF | OFF | ON | ON | |
| 32 | 288 | OFF | OFF | OFF | ON | ON | |
| 64 | 320 | ON | ON | ON | OFF | ON | |
| 96 | 332 | OFF | ON | ON | OFF | ON | |
| 128 | 389 | ON | OFF | ON | OFF | ON | |
| 160 | 416 | OFF | OFF | ON | OFF | ON | |
| 192 | 448 | ON | ON | OFF | OFF | ON | |
| 224 | 480 | OFF | ON | OFF | OFF | ON | |
| 256 | 512 | ON | OFF | OFF | OFF | ON | |
| 288 | 544 | OFF | OFF | OFF | OFF | ON | |
| 320 | 576 | ON | ON | ON | ON | OFF | |
| 352 | 608 | OFF | ON | ON | ON | OFF | |
| 384 | 640 | ON | OFF | ON | ON | OFF | |

*Note:* 384 and 256 on the motherboard add to 640K.

what is on the motherboard (Table 3.2). Most newer systems automatically recognize all memory available. Some still include SW1, so the setting guide follows in Table 3.3.

## SYSTEM IDENTIFICATION

The type of motherboard or system board you have can be identified visually by a number of pieces of hardware. The details of this quick explanation are described in the rest of the chapter.

First and most important is the microprocessor. The PC and the XT use the 8088. The AT uses the 80286 or higher. Systems may have a 386DX, 386SX, 486DX, 486SX, 486DX-2, 486DX-4, 586, or one of the overdrive chips. Modern systems will have a 686 or Pentium II. The speed of the CPU is written on the chip, so your first identification factor is the type and speed of CPU. The actual bottom line on speed is the system clock crystal. Dividing this speed by 2 will give you what the system is really running. This is *not* the 14.31818-MHz crystal found on some older motherboards; that is for video.

The second identification factor is the expansion slots. On older systems these may be 8-bit ISA, 16-bit ISA, 32-bit EISA, or two or three 32-bit VESA slots with the rest 16-bit ISA. Mordern motherboards will have PCI slots with an AGP slot and a couple 16-bit ISA slots.

**TABLE 3.3   XT SW1 settings**

| SW1 | | | |
|---|---|---|---|
| **Post** | Switch: | 1 | |
| Enabled | | OFF | |
| Disabled | | ON | |
| **Disk Drives Installed** | Switches: | 7 | 8 |
| 1 drive | | ON | ON |
| 2 drives | | OFF | ON |
| 3 drives | | ON | OFF |
| 4 drives | | OFF | OFF |
| **Display Installed** | Switches: | 5 | 6 |
| None or EGA | | ON | ON |
| CGA x 40 char | | OFF | ON |
| CGA x 80 char | | ON | OFF |
| Mono | | OFF | OFF |
| **Math Coprocessor** | Switch: | 2 | |
| None | | ON | |
| Installed | | OFF | |
| **Memory** | Switches: | 3 | 4 |
| 64K | | ON | ON |
| 128K | | OFF | ON |
| 192K | | ON | OFF |
| 256K | | OFF | OFF |
| *Note:* On some clones, the 256K setting means 256–640K. | | | |

The third identification factor is the installed equipment. This includes:

- Amount and speed of memory installed (DRAM and SRAM cache).
- Type of keyboard and plug (5-pin DIM or PS/2).
- Size and density of floppy drive(s).
- Type and size of hard drive(s).
- Video adapter card and display monitor installed.
- I/O (input/output) adapter card possibilities
  - Serial RS-232 ports
  - Parallel printer ports
  - Game port(s)
- Bus mouse adapter card or PS/2 mouse port on motherboard.
- Modem and/or fax adapter card.
- Optical scanner adapter card.
- Sound and/or music adapter card.

Although there are many more options, these are the major peripherals. Check relevant chapters for further details.

## MATH COPROCESSORS

Although modern CPUs have the math coprocessor built-in, many older CPUs do not. When many complex mathematical operations are performed by the processor, it slows down the overall speed of the program running. Math coprocessors of the 8087 family were designed to take over the higher math functions. Table 3.1 gave the type of coprocessor to use with each processor. The 8087 math coprocessor can use integer numbers up to eighteen digits and floating-point numbers up to 80 bits. While it has the functions of addition, subtraction, multiplication, and division, these may not be noticeably faster than with the CPU alone. Where the math coprocessor really makes a difference is with its sine, cosine, tangent, arc-tangent, square root, and log functions. With these functions, the recalculation of a large spreadsheet (using these higher functions) can be performed in one-tenth the time, or ten times faster! A spreadsheet using the lower functions may not be recalculated any faster at all.

Drawing or computer-aided design (CAD) programs use vectors (rays) to store the information in the drawing, instead of pixels (dots) as a paint program does. The vectors require considerable calculation with the higher math functions. A math coprocessor may speed these computations up by more than 10 to 1, giving a much faster redrawing on the display.

The conclusion drawn from the above observations is that you must carefully consider the kinds of programs you use often to decide if a math coprocessor is worth the cost. If you do not use higher functions in your spreadsheets and do not use a drawing program (as opposed to paint programs), you may see no observable increase in speed on your system.

To install a math coprocessor, choose the one compatible in type and speed, as shown in Table 3.4. On the 8088 system, the 8087 runs the same speed as the 8088, so they are synchronous. On an 80286 system, the 80286 internally divides the system clock by 2; that is, it runs at half the speed of the system clock. The 80287 internally divides the system clock by 3; it runs at one-third the speed of the system clock. It therefore has two-thirds of the microprocessor speed, making them run asynchronously. The 80386 also divides the system clock in half internally, running at half the system clock as labeled on the crystal on the motherboard. The 80387 does exactly the same, so they run at the same speed. The 486SX is a 486DX with the internal math coprocessor disabled. The 487SX is like a 486DX with the internal math coprocessor operating. Therefore, they run at the same speed.

Locate the coprocessor socket on the motherboard. It is usually very close to the CPU and usually the only empty socket. Locate pin 1 for the proper orientation.

**Caution:** Never install or remove a chip or card with the power on! Always take precautions against electrostatic discharge (ESD) by wearing an antistatic wristband or by touching the shiny bare metal of the power supply to ground yourself. This discharges any static from your body. Don't move around while handling the chip. As with printed circuit boards, don't touch the contacts with your fingers, because of grease and salt.

Carefully plug the chip in. On a PC or XT remember to change the motherboard switch on switch bank 1 for the coprocessor. On an AT system, the presence of the processor is automatically recognized. Your diagnostic program may have a test for the math coprocessor (Appendix F).

**TABLE 3.4 Compatible CPUs and math coprocessors**

| CPU | | Math Coprocessor |
|-----|-----|-----|
| 8088 | 8087 | Same speed as CPU or faster |
| 80286 | 89287 | Two-thirds the speed of CPU |
| 386SX | 387SX | Same speed as CPU |
| 386SL | 387SL | Same speed as CPU |
| 386DX | 387DX | Same speed as CPU |
| 486SX | 487SX | Same speed as CPU |
| Internal floating-point unit math coprocessor: | | |
| 486SL, 486DX, 486DX-2, 486DX-4, and Pentium processor | | |

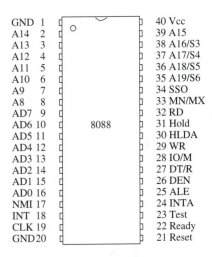

**FIGURE 3.6** **8088 CPU pinout**

```
GND   1              40 Vcc
A14   2              39 A15
A13   3              38 A16/S3
A12   4              37 A17/S4
A11   5              36 A18/S5
A10   6              35 A19/S6
A9    7              34 SSO
A8    8              33 MN/MX
AD7   9              32 RD
AD6  10    8088      31 Hold
AD5  11              30 HLDA
AD4  12              29 WR
AD3  13              28 IO/M
AD2  14              27 DT/R
AD1  15              26 DEN
AD0  16              25 ALE
NMI  17              24 INTA
INT  18              23 Test
CLK  19              22 Ready
GND  20              21 Reset
```

## 3.2 GENERIC SYSTEM OPERATION

### 8088 MOTHERBOARDS

If the microprocessor is the brain of the motherboard, the motherboard is the heart of the computer system. The XT motherboard uses the 8088 for a CPU, as does the PC. Figure 3.6 shows the pinout of the 8088. The 8088 is an 8-bit microprocessor. It has an 8-bit data bus interface but a 16-bit internal bus. Note the eight data lines (AD0–AD7) and the twenty address lines (AD0–AD7 and A8–A15 and A16/S3–A19/S6). These address lines (16–19) double as status lines (3–6).

Remember that $2_{20}$ is 1,048,576, or 1 MB. One megabyte is really that many locations, numbered from 0 to 1,048,575. This is why the twenty address lines give the ability to address 1 MB of total memory.

### The PC

The PC can be identified by the 8088 processor and five 8-bit 62-connector expansion slots. The layout of the PC motherboard is shown in Figures 3.7 and 3.8. Note the two DIN connectors (Deutsche Industrie Norm) J6 and J7 at the rear right. The leftmost (J6) is the cassette tape deck jack for recording and retrieving programs if there is no disk drive. This is now archaic. Note there are only five expansion slots and only 256K memory on the motherboard. The 8088 needed a memory expansion board in one of the slots if you wanted more memory, and, of course, you needed to set SW1 and SW2 to the correct amount. There is a keyboard jack and a jack for the internal speaker. This speaker has no volume control, and so has limited usefulness.

P3 is the speaker jack; its pinout is shown in Table 3.5. Note that a low on pin 1 energizes the voice coil. A series of active low pulses are needed for sound. A normal flashlight battery across the speaker jack (disconnected from the motherboard P3) should produce a static sound. The speaker is 8 Ω.

The PC (and the XT) use the 8088 for a CPU. The 8088 has 16-bit internal registers, an 8-bit external data bus, and a 20-bit external address bus for a maximum of 1 MB of memory (ROM and RAM).

The original IBM PC ran at 4.77 MHz, while clones quickly jumped to 7.14 and 8.00. Some systems had a turbo mode that would switch from the 4.77 MHz to a higher speed via a switch or some keystroke combination. A common keystroke combination was ALT, CTRL, – (the minus key).

The XT can be identified by the 8088 processor and eight 8-bit 62-connector expansion slots. The XT layout in Figure 3.9 shows the eight slots and no cassette port; the keyboard port is now labeled J9. Note that expansion slot number 8 (closest to the supply) is different and needs special cards to give a signal. Unless you are sure a card is designed for slot 8, put it elsewhere.

**FIGURE 3.7    8088 motherboard with setup switches**

**FIGURE 3.8    PC motherboard layout**

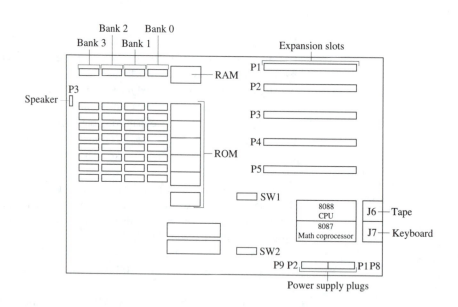

**TABLE 3.5    Speaker jack signals**

| Pin | Signal |
|-----|--------|
| 1   | Audio  |
| 2   |        |
| 3   | Ground |
| 4   | +5 V DC |

**FIGURE 3.9   XT motherboard layout**

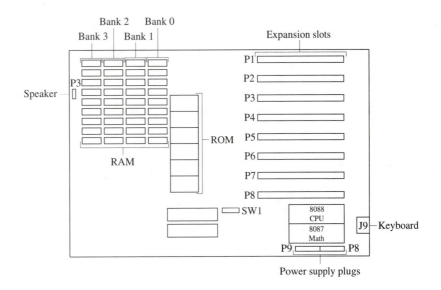

If you install the motherboard, watch for metal standoffs touching the traces on the bottom of the PC board and grounding them to the case. A stiff paper or plastic washer can prevent any problems. One of the standoffs is usually meant to ground the PC board to the case, but you can see that the trace around it is a ground plane.

## Generic 8088 System Block Diagram

(**Note:** This section is recommended for technicians only.)

Figure 3.10 shows a block diagram of a generic 8088 system board. The 8088 is the CPU, the heart of the system. The 8284 clock generator produces the basic system timing

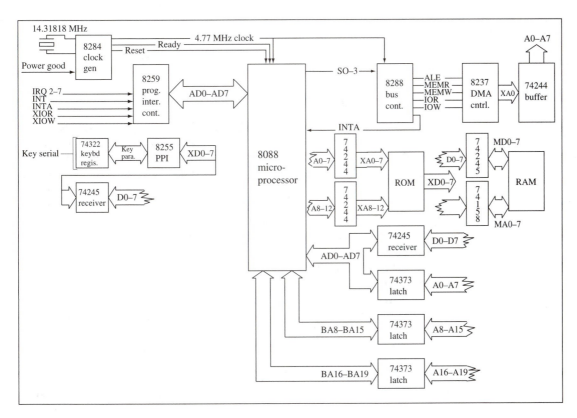

**FIGURE 3.10   Generic system block diagram**

TABLE 3.6    8088 registers

| | |
|---|---|
| AX (16 bits) or AH<br>and AL (8 bits each) | Accumulator |
| BX (16 bits) or BH<br>and BL (8 bits each) | Base |
| CX (16 bits) or CH<br>and CL (8 bits each) | Count |
| DX (16 bits) or DH<br>and DL (8 bits each) | Data |
| SP (16 bits) | Stack pointer |
| BP (16 bits) | Base pointer |
| SI (16 bits) | Source index |
| DI (16 bits) | Destination index |
| CS (16 bits) | Code segment |
| DS (16 bits) | Data segment |
| SS (16 bits) | Stack segment |
| ES (16 bits) | Extra segment |
| IS (16 bits) | Instruction pointer |
| IS (16 bits) | Status flags |

pulses. The 8259 programmable interrupt controller catches the interrupt requests (IRQ0–7) and sends the address to jump to, to service the interrupt to the CPU. The 74322 is an 8-bit shift register used as a serial-in/parallel-out register for the keyboard. It feeds an 8255 programmable peripheral interface that sends keyboard data to the data bus. The 8288 bus controller does just that, with the input of address latch enable (ALE), memory read and write (MEMR and MEMW), and input/output read and write (IOR and IOW). Besides the address, it also sends out interrupt acknowledge (INTA) to show the CPU it received the request. The 74158, 74244, 74245, and 74322 are latches, buffers, or transceivers to protect and isolate system components from the bus. Pinouts of the chips are given later in this chapter.

The 8088 has fourteen internal registers to store information. They are 16-bit registers, although four can be addressed by either the high or low byte. Table 3.6 shows the 8088 registers.

## 8-Bit 62-Connector Expansion Slot

Figure 3.11 shows the expansion slot to hold cards, the printed circuit boards that are used to expand the 8088 system. They are called adapter cards, controller cards, or I/O (input/output) cards. The connector is connected to the I/O bus. Reflecting the architecture of the microprocessor, note the eight data lines D0–D7 (external, remember) and the twenty-line address bus A0–A19. Note also the four different voltages from the power supply are present to power the adapter cards. This slot (with no extensions) identifies either a PC or an XT. Older cards, or cards that are inserted and removed often, sometimes build up oxidation on the contacts and lose electrical contact with the connector. These can be removed and the contacts cleaned with a pencil eraser. Be careful not to touch the contacts with your hands or to scratch them. Brush off the eraser dust before reinstalling.

Signals to inform the microprocessor that a peripheral device needs servicing are called **interrupts,** because they ask the CPU to "interrupt" what it is doing. The lowest number interrupt has the highest priority. Note IRQ2–7 on the expansion connector. These are the lines for interrupt requests. Also note DRQ1–3. These are request lines for devices to request direct memory access. Both IRQs and DRQs are extremely important in setting up peripheral adapter cards in your system so they don't conflict. Interrupt conflicts can take real work to overcome. Table 3.7 gives interrupts and the direct memory access request channels for an 8088 system.

Figure 3.12 shows the pinouts for the 8237 programmable DMA controller used in 8088 systems. On pins 19–16, DREQ0–3 show it has only four channels, and only one is usually available.

**FIGURE 3.11   Expansion slot signals**

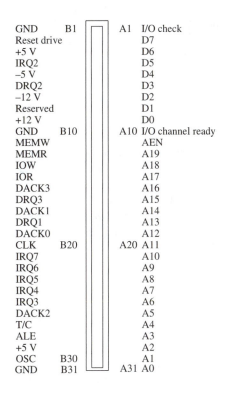

| | | | |
|---|---|---|---|
| GND | B1 | A1 | I/O check |
| Reset drive | | | D7 |
| +5 V | | | D6 |
| IRQ2 | | | D5 |
| –5 V | | | D4 |
| DRQ2 | | | D3 |
| –12 V | | | D2 |
| Reserved | | | D1 |
| +12 V | | | D0 |
| GND | B10 | A10 | I/O channel ready |
| MEMW | | | AEN |
| MEMR | | | A19 |
| IOW | | | A18 |
| IOR | | | A17 |
| DACK3 | | | A16 |
| DRQ3 | | | A15 |
| DACK1 | | | A14 |
| DRQ1 | | | A13 |
| DACK0 | | | A12 |
| CLK | B20 | A20 | A11 |
| IRQ7 | | | A10 |
| IRQ6 | | | A9 |
| IRQ5 | | | A8 |
| IRQ4 | | | A7 |
| IRQ3 | | | A6 |
| DACK2 | | | A5 |
| T/C | | | A4 |
| ALE | | | A3 |
| +5 V | | | A2 |
| OSC | B30 | | A1 |
| GND | B31 | A31 | A0 |

**TABLE 3.7   System interrupts and DMA channels for the 8088 system**

| IRQ | Use | DMA | Use |
|---|---|---|---|
| 0 | Timer | 0 | Refresh dynamic RAM |
| 1 | Keyboard | 1 | Available |
| 2 | Available | 2 | Floppy drive controller |
| 3 | COM2:   COM4: | 3 | Hard drive controller |
| 4 | COM1:   COM3: | | |
| 5 | Hard drive controller | | |
| 6 | Floppy drive controller | | |
| 7 | LPT1: | | |

**FIGURE 3.12   8237 programmable DMA controller**

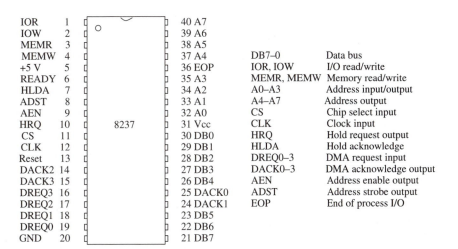

| | | | |
|---|---|---|---|
| IOR | 1 | 40 | A7 |
| IOW | 2 | 39 | A6 |
| MEMR | 3 | 38 | A5 |
| MEMW | 4 | 37 | A4 |
| +5 V | 5 | 36 | EOP |
| READY | 6 | 35 | A3 |
| HLDA | 7 | 34 | A2 |
| ADST | 8 | 33 | A1 |
| AEN | 9 | 32 | A0 |
| HRQ | 10 | 31 | Vcc |
| CS | 11 | 30 | DB0 |
| CLK | 12 | 29 | DB1 |
| Reset | 13 | 28 | DB2 |
| DACK2 | 14 | 27 | DB3 |
| DACK3 | 15 | 26 | DB4 |
| DREQ3 | 16 | 25 | DACK0 |
| DREQ2 | 17 | 24 | DACK1 |
| DREQ1 | 18 | 23 | DB5 |
| DREQ0 | 19 | 22 | DB6 |
| GND | 20 | 21 | DB7 |

| | |
|---|---|
| DB7–0 | Data bus |
| IOR, IOW | I/O read/write |
| MEMR, MEMW | Memory read/write |
| A0–A3 | Address input/output |
| A4–A7 | Address output |
| CS | Chip select input |
| CLK | Clock input |
| HRQ | Hold request output |
| HLDA | Hold acknowledge |
| DREQ0–3 | DMA request input |
| DACK0–3 | DMA acknowledge output |
| AEN | Address enable output |
| ADST | Address strobe output |
| EOP | End of process I/O |

The eight interrupts are handled by an 8259. Figure 3.13 shows the pinouts for the 8259 programmable interrupt controller. Pins 18–25 give eight interrupt request lines. For 80286 and higher systems, two 8259s are needed for sixteen interrupts. D0–7 are bidirectional data lines. Pin 17 sends the interrupt to the CPU, and pin 26 receives the interrupt acknowledge from the CPU.

When information needs to be sent to or received from the outside world, the CPU uses I/O ports. The bit patterns for commands for particular I/O devices are placed in memory locations shown in Table 3.8.

(**Note:** The rest of this section is recommended for technicians.)

A look at the signals available on the I/O connector (Figure 3.11) gives much information about how the system is operating. You can use an oscilloscope or a digital logic probe. Use an empty slot, and place a 3-by-5 card cut to fit in the slot to keep yourself from shorting out across the connector and destroying the motherboard. This card can be marked 1–31 on both sides to facilitate finding the right pin. Check out an operating system first, so you know what's "normal."

**FIGURE 3.13   8259 programmable interrupt controller**

| Pin | Name | | Pin | Name |
|---|---|---|---|---|
| CS | 1 | | 28 | Vcc |
| WR | 2 | | 27 | A0 |
| RD | 3 | | 26 | INTA |
| D7 | 4 | | 25 | IR7 |
| D6 | 5 | | 24 | IR6 |
| D5 | 6 | | 23 | IR5 |
| D4 | 7 | 8259 | 22 | IR4 |
| D3 | 8 | | 21 | IR3 |
| D2 | 9 | | 20 | IR2 |
| D1 | 10 | | 19 | IR1 |
| D0 | 11 | | 18 | IR0 |
| CAS0 | 12 | | 17 | INT |
| CAS1 | 13 | | 16 | SP/EN |
| GND | 14 | | 15 | CAS2 |

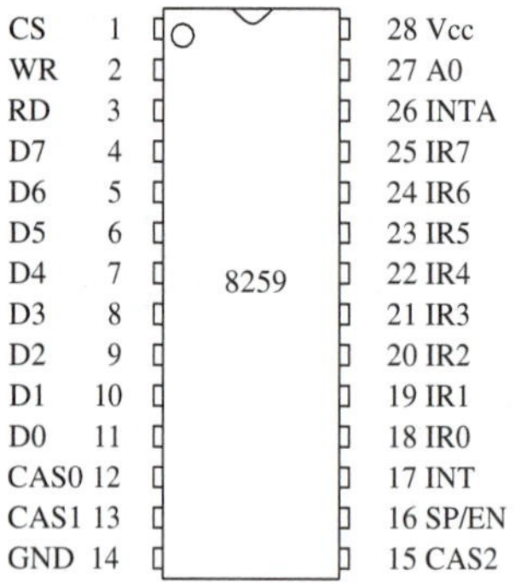

| | |
|---|---|
| D7–0 | Data bus (in/out) |
| RD | Read input |
| WR | Write input |
| A0 | Command select address |
| CS | Chip select |
| CAS2–0 | Cascade lines |
| SP/EN | Slave program/enable buffer |
| INT | Interrupt output |
| INTA | Interrupt acknowledge input |
| IR7–0 | Interrupt request inputs |

**TABLE 3.8   I/O port addresses for an 8088 system**

| Address | Device |
|---|---|
| 000H to 00FH | DMA controller |
| 020H to 021H | Interrupt controller |
| 040H to 043H | Timer |
| 060H to 063H | Programmable peripheral interface |
| 080H to 083H | DMA page register |
| 0A0H to 0BFH | NMI mask register enable |
| 0C0H to 0EFH | Reserved |
| 0F8H to 0FFH | Reserved |
| 200H to 20FH | Game I/O adapter |
| 210H to 217H | Expansion unit |
| 220H to 2F7H | Reserved |
| 2F8H to 2FFH | Serial port 2 |
| 300H to 31FH | Prototype card |
| 320H to 32FH | Hard disk |
| 378H to 37FH | Parallel printer port 1 |
| 380H to 389H | Bisync 2 |
| 3A0H to 3A9H | Bisync 1 |
| 3B0H to 3BFH | Monochrome display adapter |
| 3C0H to 3CFH | Enhanced graphic adapter (EGA) |
| 3D0H to 3DFH | Color graphic adapter (CGA) |
| 3F0H to 3F7H | Floppy disk controller |
| 3F8H to 3FFH | Serial port 1 |

A1 is I/O Check; it is OK if high and signals parity errors by going low. A2–A9 are the eight data bits. A10 is I/O Channel Ready if high, busy if low. A11 is Address Enable; it is OK if low and indicates DMA (Direct Memory Access) is being done if high. A12–A31 are the twenty address lines from MSB (Most Significant Bit) A19 to LSB (Least Significant Bit) A0. These are active when high. So the right side of the expansion slot (viewed from above, the component side, with A1 at the right rear) is mostly data and addresses.

B1, B10, and B31 are grounds; B3, B5, B7, and B9 are +5, –5, –12, and +12 V DC for the system. B4, B25, B24, B23, B22, and B21 are Interrupt Requests IRQ2 (highest priority) to IRQ7 (lowest priority). Usually low, they are pulled high to get the processor's attention. IRQ0 and IRQ1 are conspicuous by their absence. They are not on the I/O bus but are used on the motherboard only. IRQ0 is used for the timer-counter for the time-of-day clock. IRQ1 is used by the keyboard to interrupt when a scan code is sent. B18, B6, and B16 are Direct Memory Access Requests DRQ1 (highest priority) to DRQ3. Normally low, if you want DMA, pull one of these high and hold until you get a DACK. B19, B17, B26, and B15 are DMA Request Acknowledges DACK0–DACK3. DACK0 is used for refreshing DRAM and DACK1–DACK3 to acknowledge DRQ1–DRQ3. B11 and B12, Memory Write and Memory Read, are usually high and go low to write data from the data bus to memory or to read data from memory to the data bus. B13 and B14, I/O Write and I/O Read, are usually high and go low to write data from the data bus to the I/O device or to read data from the I/O device to the data bus. B20 is the System Clock and is important. This is the CPU clock speed. It is 4.77 MHz for the PC and XT, and 12, 16, or higher on 286s and 386s. B27 is Terminal Count for the DMA. B28 is Address Latch Enable, which latches addresses from the CPU. B30 is the 14.31818 MHz oscillator used for CGA video.

In Figure 3.14, note the pinouts for the 74322 serial-to-parallel keyboard buffer. Pin 12 Q/H goes out to IRQ1, the keyboard interrupt. On the PC, an 8048 sends a serial signal from the keyboard that comes through the keyboard jack J7, pin 2 and enters the 74322 on pin 17, data in 1. The 8-bit parallel data exits the 74322 on pins QA–QH.

Figure 3.15 shows the pinouts of the 8284 clock generator used in the 8088 systems. Power good comes from the power supply plug 8, pin 1 to the 8284 pin 11. This holds the clock in a constant reset if the power fails. The system crystal is connected across pins 16 and 17. Pin 10, reset out, goes to the 8088 pin 21. Pin 8, CLK, goes to the 8088 pin 19, and to I/O slot pin B20, the system clock. Pin 12, oscillator, goes to I/O slot pin B30.

Figure 3.16 shows the pinout of the 8288 bus controller used in 8088 systems. Lines S0–2 receive the status of the CPU. Lines 7, 8, 9, 11, 12, 13, and 14 are control lines out, making up part of the control bus commands. Pins 4, 5, 16, and 17 are control signals also.

**FIGURE 3.14  74322 serial-to-parallel keyboard buffer**

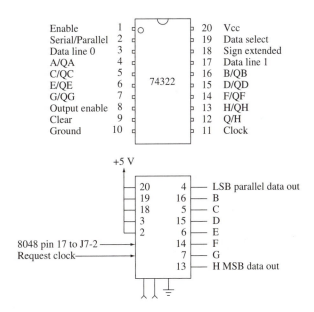

**FIGURE 3.15    8284 clock generator**

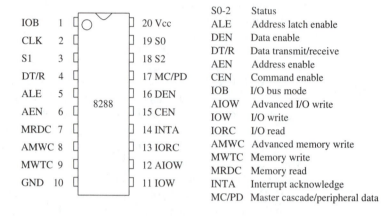

| | | |
|---|---|---|
| CYSNC | 1 | 18 Vcc |
| PCLK | 2 | 17 X1 |
| AEN1 | 3 | 16 X2 |
| RDY1 | 4 | 15 ASYNC |
| Ready | 5 | 14 EFI |
| RDY2 | 6 | 13 F/C |
| AEN2 | 7 | 12 OSC |
| CLK | 8 | 11 RES |
| GND | 9 | 10 Reset |

| | |
|---|---|
| X1-2 | Crystal |
| F/C | Clock source select |
| EFI | External clock input |
| CSYNC | Clock sync input |
| ASYNC | Ready sync select |
| RDY1-2 | Ready signal |
| AEN1-2 | Address enabled/RDY1, 2 |
| RES | Reset input |
| Reset | Synchronized reset output |
| OSC | Oscillator output |
| CLK | MOS clock for processor |
| PCLK | TTL clock for peripherals |
| Ready | Synchronized ready output |

**FIGURE 3.16    8288 bus controller**

| | | |
|---|---|---|
| IOB | 1 | 20 Vcc |
| CLK | 2 | 19 S0 |
| S1 | 3 | 18 S2 |
| DT/R | 4 | 17 MC/PD |
| ALE | 5 | 16 DEN |
| AEN | 6 | 15 CEN |
| MRDC | 7 | 14 INTA |
| AMWC | 8 | 13 IORC |
| MWTC | 9 | 12 AIOW |
| GND | 10 | 11 IOW |

| | |
|---|---|
| S0-2 | Status |
| ALE | Address latch enable |
| DEN | Data enable |
| DT/R | Data transmit/receive |
| AEN | Address enable |
| CEN | Command enable |
| IOB | I/O bus mode |
| AIOW | Advanced I/O write |
| IOW | I/O write |
| IORC | I/O read |
| AMWC | Advanced memory write |
| MWTC | Memory write |
| MRDC | Memory read |
| INTA | Interrupt acknowledge |
| MC/PD | Master cascade/peripheral data |

## CMOS AND BATTERIES

The 8088 system motherboards have setup switches to set for monitors, memory, and drives. The 286 systems have shunt jumper switches for choosing a color or mono display, the size of the BIOS ROM, the amount of RAM on each RAM chip, internal or external CMOS battery, and its voltage. The 286 systems also have a battery-backed-up CMOS memory that holds some information on system configuration. The CMOS battery is often a 3.6 V lithium battery soldered into the board. A 4-pin single inline header is usually found in the same area of the motherboard as the battery. It is for connecting an external battery if the soldered in one fails. Six volts is a standard, although 4.5 and 3.6 V are also used. Some boards have shunt jumpers in the same location as the header to choose internal or external battery and voltage.

All batteries die sooner or later, and replacing the CMOS battery will have to be done someday. If you can plug in an external battery, do so. If not, desolder and replace the battery only if you are a very experienced solderer, and take great care. Sloppiness or roughness here can destroy the motherboard.

When the system is first assembled, when the battery has failed, or when the system components are changed, the CMOS system configuration settings must be correctly set. Once they are, a SHIFT-PRINTSCRN will dump them to a printer for future reference. The POST on the newer systems may show a CMOS memory error, meaning the physical system and the CMOS settings are different. This may only mean a backup battery failure, and after replacement the SETUP program must be run. On a typical system, while booting, just after the memory check, you are instructed on the display to press DELETE to run SETUP. This must be done in a short time after the message is displayed, or the system must be rebooted to try again. The earliest 286 system boards had a SETUP program on disk, but the new ROM-based ones are certainly more handy.

## 80286 MOTHERBOARDS

The 80286 is a 16-bit microprocessor. Figure 3.17 shows a 286 motherboard. These boards vary immensely depending on manufacturer; however, there are some similarities. For one

**FIGURE 3.17    80286 motherboard**

thing, they use the 80286 for a CPU. The 80286 still has 16-bit internal registers, but it has a 16-bit external data bus compared to 8 on the 8088. The external address bus is expanded to 24 bits. Because the address bus is 4 bits wider, and there are sixteen unique combinations in 4 bits of binary (0 through 15), the system can address 16 MB of total memory. Running at half the speed of the motherboard crystal, a 20-MHz 286 can run at 3.5 million instructions per second (MIPS).

When a 286 system resets, the system starts executing at address 0FFFF0Hex, which is 1,048,560 decimal. This is just 16 bytes from the end of the 1 MB conventional RAM and ROM area. These bytes hold a jump instruction to another part of the BIOS.

In case of keyboard failure on 286 and higher systems, check the keyboard lock and cables from it to the lock connector pins on the system board.

All 286/386/486 boards have a high degree of integration and make heavy use of surface mount technology components. For serious component level motherboard repair, pinouts are essential. Chips and Technologies Inc. is a major manufacturer of these chip sets. (Their address and phone number are given in Appendix B under Schematics and Documentation.)

## SETUP PROGRAM

The SETUP program is entered in different ways on different manufacturers' systems. On many, when there is a setup problem the system will prompt you, asking you to press the F1 function key to enter SETUP directly or giving a menu asking what level of setup (light or advanced) you want to perform. ALT-CTRL-ESC held together enters SETUP on some systems. On others try ALT-CRTL-S or ALT-CTRL-INSERT or pressing DELETE on powering up or holding down INSERT while powering up. If all else fails, try ALT-CTRL-INSERT after ALT-CTRL-DELETE.

Some CMOS setups allow you to set a password to keep unauthorized users from possibly destroying your data. Three options given are "always," "setup," or "disable." Choosing "always" will allow no one to operate your system without knowing the password. Choosing "setup" will allow others to use your system but only those who know the password to change the CMOS setup. Choosing "disable" stops all requests for the password, and anyone can do anything on your system. If you forget your password, about the only option is to disconnect the CMOS battery and wait before reconnecting, so the CMOS forgets everything. Then you will either be able to reenter the CMOS setup or, on an American Megatrends Incorporated (AMI) system, enter the default password, AMI.

The information you must provide for a standard CMOS setup is as follows.

1. DATE/TIME: If it is a new system, if the battery has failed, or if the date and time are wrong for any reason, enter the proper date and time.
2. DISPLAY: The monitor type should be checked. Most systems automatically sense the proper display, but check it and set it if necessary.
3. RAM: The RAM memory setup is usually automatic. If it is in error, suspect bad or improperly installed RAM.
4. FLOPPY DRIVES: The floppy setup is just a matter of telling the system what capacity drives A: and B: are.
5. HARD DRIVES: The hard drive is next, and you need to know the parameters of your drive(s). See Chapter 9 or Appendix E.
6. KEYBOARD: Why would you not install the keyboard in a network server or a dedicated system that controls equipment and gets input from that equipment?
7. OTHER: The system may ask you about "shadow RAM" and "buffering." If so, either accept the defaults or see Chapter 5. In general, these will speed up your system.

On some systems there is an advanced setup. This is for highly optimizing your system and can make the system appear to fail if it is set up incorrectly. If this occurs, you can check documentation on clearing the CMOS or choose the option default settings if you are able. Following are typical settings for a major ROM BIOS manufacturer.

1. TYPEMATIC: If you hold down a key on the keyboard for a certain length of time, it will start repeating the character automatically. You can enable or disable this feature, set the length of time delay before it starts repeating, and set the speed at which it repeats.
2. MEMORY TEST: To speed up booting, some people disable the POST memory test of RAM above 1 MB. I always enable this test just for safety.
3. NUMLOCK: You may be able to set whether NUMLOCK is ON or OFF on boot-up. NUMLOCK ON tells the system to read the numeric keypad keys as numbers. This is standard on the 101 keyboards because they have another set of cursor keys. If the keyboard does not have separate cursor keys, OFF may be a better choice.
4. NUMERIC PROCESSOR: If you have installed a math coprocessor, choose INSTALLED. You may get another option asking if it is a WEITEK coprocessor. If so, make that option INSTALLED, instead.
5. FLOPPY BOOT: Attempting to boot the floppy each time you boot is a waste of time. However, if the hard drive fails to operate some dark morning, booting from the floppy may help troubleshoot the system. Enable this choice.
6. BOOT SEQUENCE: This is when you cause the floppy boot above to not waste time. Select the hard drive, then the floppy, as in C:, A:.
7. CACHE RAM: This is the faster static RAM that is used to buffer the slower dynamic RAM to speed up the system. 386s and 486s include this RAM internally in the microprocessor. Most modern motherboards also have external cache RAM on the system board. Select both, if offered, to speed up the system.
8. SHADOW RAM: ROM is slower than RAM. Making a copy (shadow) of the system ROM BIOS and video ROM BIOS in RAM speeds up the system. Shadowing adapter ROM from hex location C8000 up to EFFFF depends on whether you have an adapter card that supports shadowing. ROM from F0000 up to FFFFF is the system BIOS ROM and should be shadowed for higher system speed. In general, the only bad thing about shadowing is that it consumes RAM. This explains why two identical systems with different amounts of RAM run different speeds. The one with more RAM can shadow ROMs and run faster. For experienced technicians or programmers, Chapter 4 gives an introduction to programming the CMOS yourself. Practice on a test system first, not your main machine.

On modern systems SETUP has been expanded to include the following choices:

Standard CMOS Setup
BIOS Feature Setup
Chipset Feature Setup
Power Management Setup

PCI Configuration Setup
Load Setup Defaults
Password Setting
IDE HDD Auto Detection
Save and Exit Setup
Exit Without Saving

Most choices are self-explanatory. Chipset, Power Management, and PCI have the following functions:

Chipset Feature Setup: DRAM timing, PCI Control, System and Video Caching, IDE primary and Secondary Master/Slave, On-Board Serial and Parallel Port Setup.

Because of the EPA's Energy Star green standards, new CMOS SETUP programs have a menu choice for Power Management Setup. A typical example screen follows:

|  | Min Saving | Max Saving | User Def | Disable |
|---|---|---|---|---|
| Doze mode | 1 hr | 1 min | 1 min to 1 hour | Y/N |
| Standby mode | 1 hr | 1 min | 1 min to 1 hour | Y/N |
| Suspend mode | 1 hr | 1 min | 1 min to 1 hour | Y/N |
| HDD power-down | 15 min | 1 min | 1 min to 1 hour | Y/N |

The modern high-speed system takes advantage of the speed of the PCI bus as follows:

Slot (1–3) – IRQ# – A,B,C,D,Auto
PCI IRQ Activation (level/____)
PCI IDE IRQ Map to :(ISA)

While there were many kinds of 80286 motherboards, many similarities exist. Some are the CMOS battery, the presence of an 80287 math coprocessor socket, the absence of system setup switches, and the 98-pin edge card connector (the original 62 pins of the 8088 system plus the extra 36 pins needed for the increased data and address lines) for the 16-bit expansion slots. These have the original 62-pin slot, and below it (closer to the front of the system motherboard), the 36-pin extension expansion slot. Figure 3.11 shows the expansion slots and signals to be expected for the PC and XT. The 286 and higher boards have the extension also. Figure 3.18 shows only the extension expansion slot for 16-bit systems. The expansion cards for the ISA systems are shown in Figure 3.19. Note the 8-bit 8088 version has only the 31-contact connector (per side), while the 16-bit for the 286 and higher systems has both the 31-contact and the 18-contact extension connectors.

**FIGURE 3.18   16-bit extension slot signals**

16-bit extension slot

| | | | |
|---|---|---|---|
| MEMCS16 | D1 | C1 | SBHE |
| I/OCS16 | | | LA23 |
| IRQ10 | | | LA22 |
| IRQ11 | | | LA21 |
| IRQ12 | | | LA20 |
| IRQ15 | | | LA19 |
| IRQ14 | | | LA18 |
| DACK0 | | | LA17 |
| DRQ0 | | | MEMR |
| DACK5 | D10 | C10 | MEMW |
| DRQ5 | | | SD08 |
| DACK6 | | | SD09 |
| DRQ6 | | | SD10 |
| DACK7 | | | SD11 |
| DRQ7 | | | SD12 |
| +5 V | | | SD13 |
| Master | | | SD14 |
| Ground | D18 | C18 | SD15 |

FIGURE 3.19   ISA expansion
cards

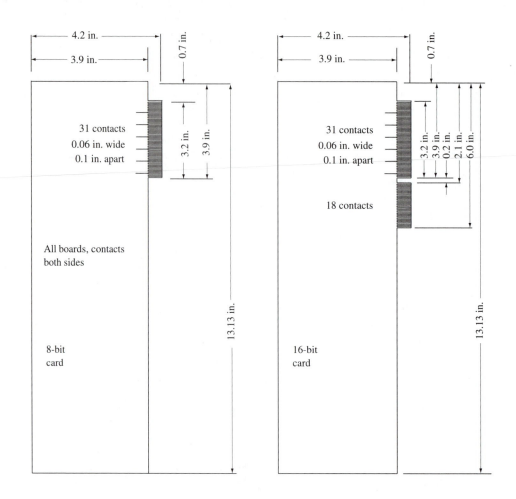

Noting the 80286 pinout in Figure 3.20, on the top, D0–D15 give the sixteen data lines for a 16-bit data bus. On the right, note A0, A1, and A2, then A3 through A13. On the bottom are A14 through A21, then A22 and A23 for a 24-bit address bus. N.C. means no connection. The other pins are for power and ground, and the control bus.

Note in Figure 3.21 that the card on the right is an 8-bit card and that it drops down after the 8-bit edge connector, protruding into the area where the 16-bit extension socket is. This physical incompatibility is about the only reason an 8-bit card will not work in a 16-bit socket. At the left and the top are 16-bit ISA cards.

## 16-Bit 36-Connector Extension Slot

The 16-bit extension slot shown in Figure 3.18 has the additional signals needed for the 286 and higher CPUs. It is designed so that ISA 8-bit cards can still plug in and work (if they don't have a "skirt" hanging where the 16-bit extension socket is). This makes the 16-bit ISA I/O connector backward compatible. A 16-bit ISA will not fit in an 8-bit ISA slot, but the older 8-bit ISA will fit in the newer 16-bit ISA slot.

Note the additional interrupt requests from 10 through 12 and 14 and 15. The NMI (Non-Maskable Interrupt) is not on the expansion connector, nor are IRQ0 (used for the system counter-timer) and IRQ1 (used for the keyboard). Since more interrupts were needed for the AT class of computers, a second 8259 was added, and its output, pin 17 (int), was sent to IRQ2, pin 20, of the first 8259. This way IRQ2 is the access to IRQ8–15. When IRQ2 is given, it is redirected to IRQ9, so IRQ9 appears as IRQ2. This means IRQ2 is available.

Arranging the 8259s this way is called "cascading." More Direct Memory Access (DMA) channels are made available also. DMA requests and acknowledgments 0 and 5–7 are on pins D8 and D15. DMA 4 is reserved and so is unavailable. Channels 0–3 are the low DMA channels, and channels 5–7 are the high DMA channels. Having a 16-bit

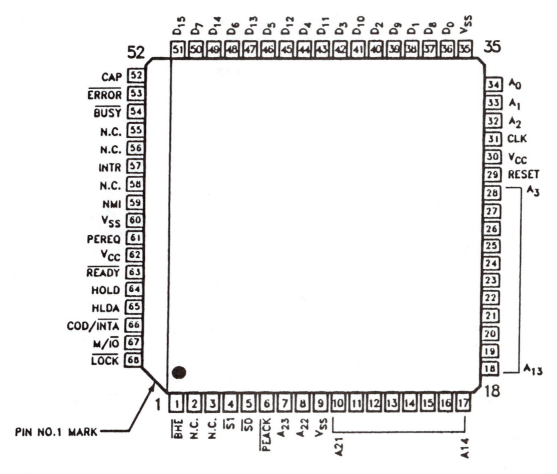

FIGURE 3.20    80286 pinout from top, reprinted by permission of Intel Corp. Copyright © 1987 by Intel Corp.

FIGURE 3.21    8-bit and 16-bit
ISA cards

data bus on the 80286 and higher boards (compared to eight data bits on the 8088), we have the additional eight data lines from SD08 through SD15. The additional latchable address lines are LA17 through LA23.

286 system interrupts and DMA requests are shown in Table 3.9, I/O port addresses in Table 3.10, and BIOS interrupt vectors are in Table 3.11. The system interrupts, I/O port addresses, and direct memory access requests (DRQs) are important for you to know for any adapter board installation. With a parallel printer port, a serial port, a mouse, a game port, a fax/modem, an optical scanner, a sound card, a CD-ROM drive, and perhaps a plotter, conflicts must be avoided to keep from system crashes. Keep all documentation that comes with peripherals that you add to your system, because with future additions it may be necessary to change the system interrupts, DMA channels, and I/O addresses to make all the equipment work together.

(**Note:** The remainder of this section is recommended for technicians.)

The I/O addresses are also necessary if you program for input or output to devices. The BIOS interrupt vectors are the locations to "call" to cause the BIOS to perform some function, such as read or write a floppy drive. They are usually only necessary when writing an assembly level program.

In Figure 3.22, note the pinouts for the 146818 real-time clock. In newer systems this may be incorporated in the VLSI chip set. Pins 4–11 are used as an 8-bit path for both the data and the address. Pin 19 gives the real-time clock interrupt to the second 8259 pin 19, IRQ8.

In Figure 3.23, note the 82284 clock generator used on many 286/386/486 systems. See Appendix F for more details.

Figure 3.24 shows the pinouts for the 82288 bus controller used in 286/386 systems.

**TABLE 3.9 DMA channels for the 80286 system. These hardware interrupts for IRQs are actual lines or copper traces.**

| IRQ | Use |
|-----|-----|
| NMI | System board, parity |
| 0 | System counter-timer |
| 1 | Keyboard |
| 2 | Cascade from second IRQ controller |
| 3 | COM2: COM4: |
| 4 | COM1: COM3: |
| 5 | LPT2: (was hard drive controller on 8088 systems) |
| 6 | Floppy drive controller |
| 7 | LPT1: |
| 8 | Real-time clock |
| 9 | Available to 8-bit card (IRQ2 redirected) |
| 10 | Available to 16-bit card |
| 11 | Available to 16-bit card |
| 12 | Available to 16-bit card |
| 13 | 80X87 math coprocessor |
| 14 | Hard drive controller 80X86 system |
| 15 | Available to 16-bit card |

| DMA | Use |
|-----|-----|
| 0 | Available to 16-bit card (not 8-bit; see slot pins) |
| 1 | Available to 8-bit card |
| 2 | Floppy drive controller |
| 3 | Available to 8-bit card |
| 4 | Hold request from first 8237 DMA controller |
| 5 | Available to 16-bit card |
| 6 | Available to 16-bit card |
| 7 | Available to 16-bit card |

**TABLE 3.10 Port addresses for 286 system**

| Address | Device |
|---|---|
| 000H to 01FH | DMA controller |
| 020H to 03FH | Interrupt controller |
| 040H to 05FH | Timer |
| 060H to 06FH | Keyboard controller |
| 070H to 07FH | Real-time clock |
| 080H to 09FH | DMA page register |
| 0A0H to 0BFH | Interrupt controller |
| 0C0H to 0DFH | DMA controller 2 |
| 0F0H | Clear 80287 NPX busy |
| 0F1H | Reset 80287 NPX |
| 0F8H to 0FFH | 80287 NPX |
| 1F0H to 1F8H | Hard disk controller |
| 200H to 207H | Game I/O adapter |
| 278H to 27FH | Parallel printer port 2 |
| 2F8H to 2FFH | Serial port 2 |
| 300H to 31FH | Prototype card |
| 360H to 36FH | Reserved |
| 378H to 37FH | Parallel printer port 1 |
| 380H to 38FH | Bisync 2 |
| 3A0H to 3AFH | Bisync 1 |
| 3B0H to 3BFH | Monochrome display adapter |
| 3C0H to 3CFH | Enhanced graphic adapter (EGA) |
| 3D0H to 3DFH | Color graphic adapter (CGA) |
| 3F0H to 3F7H | Floppy disk controller |
| 3F8H to 3FFH | Serial port 1 |

**TABLE 3.11 BIOS interrupt vectors**

| Interrupt Number | Function | Address |
|---|---|---|
| 0 | Divide by zero | |
| 1 | Single-step | |
| 2 | Nonmaskable | F000:E2C3 |
| 3 | Breakpoint | |
| 4 | Overflow | |
| 5 | Print screen | F000:FF54 |
| 6 | Unused | |
| 7 | Unused | |
| 8 | System timer | F000:FEA5 |
| 9 | Keyboard | F000:E987 |
| A | Unused | |
| B | Serial com | |
| C | Serial com | |
| D | Fixed disk | |
| E | Floppy disk | F000:EF57 |

TABLE 3.11 *(continued)*

| Interrupt Number | Function | Address |
|---|---|---|
| F | Printer prn: | |
| 10 | Video I/O | F000:F065 |
| 11 | Equipment check | F000:F84D |
| 12 | Memory size | F000:F841 |
| 13 | Floppy disk I/O | F000:EC59 |
| 14 | Communication I/O rs232 | F000:E739 |
| 15 | Cassette I/O | F000:F859 |
| 16 | Keyboard I/O | F000:E82E |
| 17 | Printer I/O | F000:EFD2 |
| 18 | Cassette basic | F600:0000 |
| 19 | Power on reset, boot-strap loader | F000:E6F2 |
| 1A | Time of day | F000:FE6E |
| 1B | Control keyboard break | |
| 1C | Timer control | F000:FF53 |
| 1D | Video initialization table pointer | F000:F0A4 |
| 1E | Floppy parameter table pointer | F000:EFC7 |
| 1F | Video graphics character generator pointer | |
| 20 | DOS program terminate | 00B1:0011 |
| 21 | DOS function call | 00B1:0015 |
| 22 | DOS terminate address | 02F7:01FF |
| 23 | DOS control break exit address | 02F7:0204 |
| 24 | DOS critical error handler | 02B1:019B |
| 25 | DOS absolute disk read | 0060:0015 |
| 26 | DOS absolute disk write | 0060:0018 |
| 27 | DOS terminate and stay resident (TSR) | 02B1:0187 |

**FIGURE 3.22    146818 real-time clock**

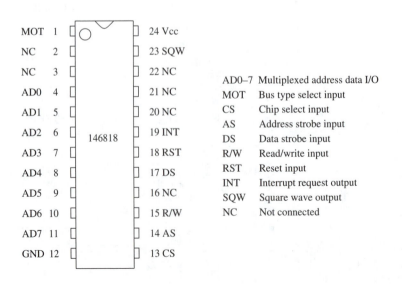

| | | | |
|---|---|---|---|
| AD0–7 | Multiplexed address data I/O |
| MOT | Bus type select input |
| CS | Chip select input |
| AS | Address strobe input |
| DS | Data strobe input |
| R/W | Read/write input |
| RST | Reset input |
| INT | Interrupt request output |
| SQW | Square wave output |
| NC | Not connected |

**FIGURE 3.23   82284 clock generator**

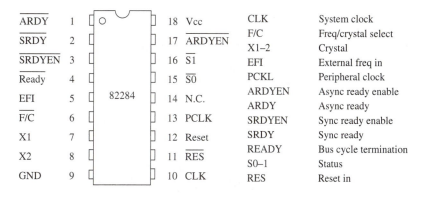

| | | | |
|---|---|---|---|
| $\overline{\text{ARDY}}$ | 1 | 18 | Vcc |
| $\overline{\text{SRDY}}$ | 2 | 17 | $\overline{\text{ARDYEN}}$ |
| $\overline{\text{SRDYEN}}$ | 3 | 16 | $\overline{\text{S1}}$ |
| $\overline{\text{Ready}}$ | 4 | 15 | $\overline{\text{S0}}$ |
| EFI | 5 | 14 | N.C. |
| $\overline{\text{F/C}}$ | 6 | 13 | PCLK |
| X1 | 7 | 12 | Reset |
| X2 | 8 | 11 | $\overline{\text{RES}}$ |
| GND | 9 | 10 | CLK |

| | |
|---|---|
| CLK | System clock |
| F/C | Freq/crystal select |
| X1–2 | Crystal |
| EFI | External freq in |
| PCKL | Peripheral clock |
| ARDYEN | Async ready enable |
| ARDY | Async ready |
| SRDYEN | Sync ready enable |
| SRDY | Sync ready |
| READY | Bus cycle termination |
| S0–1 | Status |
| RES | Reset in |

**FIGURE 3.24   82288 bus controller**

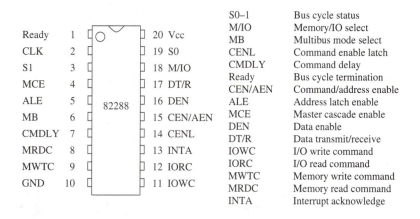

| | | | |
|---|---|---|---|
| Ready | 1 | 20 | Vcc |
| CLK | 2 | 19 | S0 |
| S1 | 3 | 18 | M/IO |
| MCE | 4 | 17 | DT/R |
| ALE | 5 | 16 | DEN |
| MB | 6 | 15 | CEN/AEN |
| CMDLY | 7 | 14 | CENL |
| MRDC | 8 | 13 | INTA |
| MWTC | 9 | 12 | IORC |
| GND | 10 | 11 | IOWC |

| | |
|---|---|
| S0–1 | Bus cycle status |
| M/IO | Memory/IO select |
| MB | Multibus mode select |
| CENL | Command enable latch |
| CMDLY | Command delay |
| Ready | Bus cycle termination |
| CEN/AEN | Command/address enable |
| ALE | Address latch enable |
| MCE | Master cascade enable |
| DEN | Data enable |
| DT/R | Data transmit/receive |
| IOWC | I/O write command |
| IORC | I/O read command |
| MWTC | Memory write command |
| MRDC | Memory read command |
| INTA | Interrupt acknowledge |

## 80386 MOTHERBOARDS

The 80386 is a 32-bit microprocessor. Figure 3.25 shows the 80386 and its math coprocessor socket. Note the top right corner of the socket is cut off, identifying pin 1. With the original

**FIGURE 3.25   80386 and coprocessor socket**

286 ATs running at 2.7 MIPS, this bus expansion from 16 to 32 bits greatly increased processor speed. Like the 80286, the 80386 runs at half the crystal speed. Running at 33 MHz gives 12 MIPS. The 40-MHz 386 became the most popular 386.

The 386 CPU uses more power, and therefore runs hotter than previous CPUs. Keeping the chip cool will extend its life, so little cooling fans (Figure 3.26) were designed to sit on top of the CPU and blow away excess heat. A cooling fan may be purchased for the 486 and 586 also.

Figure 3.27 shows the pinout of the 80386DX. Note lines D0–D31 for the 32-bit external data bus and lines A0–A31 for a 32-bit address bus. The 32-bit address bus is an 8-bit increase

**FIGURE 3.26　CPU cooling fan**

**FIGURE 3.27　80386DX pinout from pin side, reprinted by permission of Intel Corp. Copyright © 1988 by Intel Corp.**

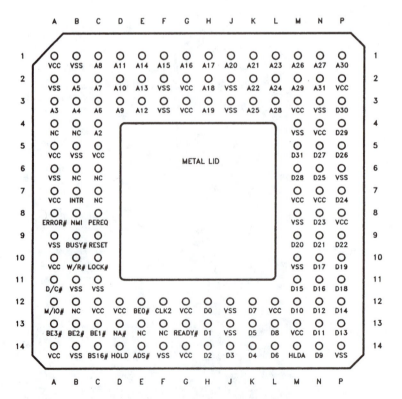

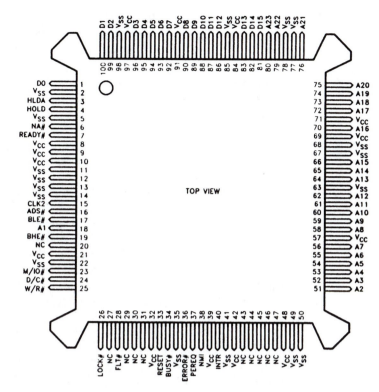

**FIGURE 3.28    80386SX pinout from top side, reprinted by permission of Intel Corp. Copyright © 1994 by Intel Corp.**

**TABLE 3.12    The 386 family**

| Chip | Function |
|------|----------|
| 386SX | 16-bit data, 24-bit address, inexpensive version |
| 386DX | 32-bit data, 32-bit address |
| 386SL | 16-bit data, 24-bit address, low-power version |
| 387SX | Math coprocessor for SX |
| 387DX | Math coprocessor for DX |
| 387SL | Math coprocessor for SL |

from the 24 of the 80286. These 8 bits give 256 more possibilities, which, multiplied by the 16 MB we had, now gives 4096 MB, or 4 GB, of total possible memory. At about $40 a megabyte, this is more than most could afford to install.

The 80386SX is a 16-bit external bus version of the 80386. This compromise severely limits the system speed and gives away much of what you wanted the 386 for in the first place. Note the pinout of the 386SX in Figure 3.28. Data lines D0–D15 show it has a 16-bit external data bus. Address lines A0–A23 give a 24-bit address bus. This shows that on the outside the 386SX is just like the 286.

The 386SL is a low-power version of the 386. The processor puts itself into a "low-power mode" when demand on it is low. When the CPU is not in use at all, it goes into a "sleep" mode. For notebooks and laptops, a 3.3 V version uses half the power of the 5 V version, and extends length of use between battery charges.

Table 3.12 summarizes the functions of the 386 family of CPUs.

## 80486 MOTHERBOARDS

The 80486DX has 1.2 million transistors and runs at the speed of the crystal. It is similar to a 386 with a 387 math coprocessor built in. It has an 8-K internal level 1 cache of fast SRAM for data and instructions. Being internal, this memory can be accessed much faster than external memory. Both the data bus and the address bus are 32 bits. Since 32 bits is 2 bits more than 30, and $2^2$ is 4, a 32-bit bus ($2^{32}$) can handle numbers four times larger than a 30-bit bus, or numbers up to 4 GB. The 50 MHz version uses 0.8-micron structures to reduce internal capacitance so it responds to signal changes more quickly.

The 80486DX has the math coprocessor built in, which gives a speed boost for programs doing a lot of computation. Complex instruction set computing (CISC) needs many clock cycles to perform most instructions. This architecture is what most of the microprocessors you hear about are based on. Reduced instruction set computing (RISC) architecture runs many instructions in one clock cycle. The 486 incorporates some RISC techniques, and this helps the 486 to give twice the performance of a 386 running at the same clock speed, because the 386 needs multiple clocks for many instructions.

With higher and higher speeds, microprocessors have come to the point where the 33-MHz bus from the CPU to the memory is a bottleneck. The microprocessors can run faster, but the motherboard memory bus is running at top speed. To address this problem, one of the major manufacturers designed a microprocessor that runs twice as fast internally as it communicates with the world via the memory bus. The 50 and 66 MHz 80486DX-2s have an internal clock doubler, which allows them to run at twice the speed of the motherboard clock (25 and 33 MHz). The 75 and 100 MHz 486DX-4 has an internal clock tripler, for three times the speed. Also, in a system motherboard that can support a 486DX-33, you can plug a ODPR486-33 (popularly called the overdrive), which runs at 66 MHz internally but 33 MHz externally over the memory bus (Figure 3.29). In fact, there is a whole family of chips, as shown in Table 3.13. Advanced Micro Devices makes 486s up to the AM486DX4-120, and Texas Instruments makes them up to the TI486DX4-100 (Figure 3.30).

**FIGURE 3.29  Intel i80486DX-2 and upgrade socket**

**TABLE 3.13  The 486 family of chips**

| Chip | Function |
|---|---|
| 486SX | 486 with no math coprocessor included. |
| 486DX | Full 486, 8K internal cache, 25 and 33 MHz. |
| 486DX-2 | Full 486 with internal speed doubled. This is to build into motherboards originally. |
| 486DX-4 | Full 486 with internal speed tripled. On-board cache doubled to 16K. Speeds are 75 and 100 MHz. |
| 486SL | Low-power version, still 32-bit buses. |
| ODP486SX-33 | Plugs into the 487 math coprocessor socket on a 486SX system. |
| -25 | This adds the coprocessor and doubles the CPU speed to 66 MHz. |
| -20 | MHz for the 33, 50 for the SX25, and 40 for the SX20. |
| ODP486DX-33 | Plugs into the overdrive socket for 486DX systems |
| -25 | that were built with that socket. |
| ODPR486DX-33 | For DX motherboards without the overdrive socket. |
| -25 | These boards have no processor sockets, so the Overdrive Processor Replacement replaces the socketed 486DX chip you remove. |

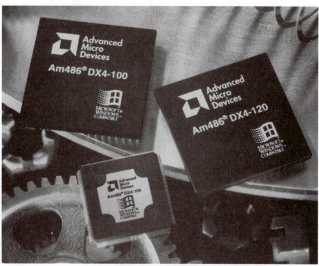

AMD 486 chips

TI 486DX4 chip

A glance at the pins of the 80486DX-33 (and -25) in Figure 3.31 shows D0–D31 for a full 32-bit data bus. The 50 MHz chip changes A3 to TCK (Test Clock), A14 to TDI (Test Data Input), B14 to TMS (Test Mode Select), and B16 to TDO (Test Data Output). These are not connected on the 25 and 33 MHz versions. A2–A31 give only thirty address lines, but BE0–3 (Byte Enable) are used with these for a full 32-bit address bus.

Looking at the pinout of the 80486SX in Figure 3.32, we see D0–D31 for a full 32-bit data bus. A2–A31 give the full 32-bit address bus. Note the 486SX does not have a smaller bus width than the 486DX (as the 386SX does compared to the 386DX). It is only missing the math coprocessor.

## 80586 MOTHERBOARDS

In April 1993, Intel announced the new microprocessor to succeed the 80486. They named it the Pentium processor, although many have reasonably named it the 80586. Figure 3.33 shows an 80586 motherboard, with PCI and 16-bit ISA I/O slots, a cooling fan, four 72-pin, 36-bit SIMM sockets in the bottom right corner and a cache of eight chips in the bottom left corner. It has 3.1 million transistors integrated into a package smaller than a quarter coin. It is about five times faster than a 486, depending on system clock speed. The address bus is still at 32 bits, like the 386 and 486. With RAM at $40 a megabyte, that is $160,000, to fill the 4 GB that 32 bits will address. There seems no need to expand the address bus width for the foreseeable future.

**FIGURE 3.31    80486DX-33** pinout from pin side, with pin grid array, reprinted by permission of Intel Corp. Copyright © 1988 by Intel Corp.

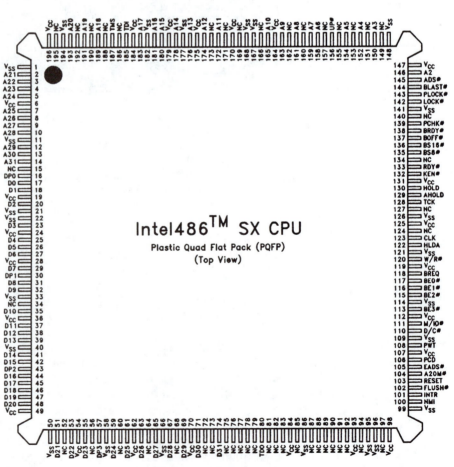

**FIGURE 3.32    80486SX** pinout from top side, reprinted by permission of Intel Corp. Copyright © 1994 by Intel Corp.

**FIGURE 3.33  Pentium 735\90 processor and 80586 motherboard, reprinted by permission of Intel Corp. Copyright © 1994 by Intel Corp.**

Pentium 735/90

80586 motherboard

The external data bus has been expanded to 64 bits. Since 64 bits is 4 bits more than 60 bits, and $2^4$ is 16, a 64-bit data bus can handle a number sixteen times larger than a 60-bit bus, or 16 Exabytes, or 16 quintillion. These numbers are not the main benefit of the data bus's being wider: speed is the main benefit. This bus can transfer eight 8-bit bytes of data at the same time. It has two 8K internal caches and two instruction pipelines, allowing it to process two instructions in one clock cycle.

From Figure 3.34, note the address pins on the top right of the chip, ending with A31 at location AJ33. Pins 0 to 31 give thirty-two lines. On the bottom from right to left, the data pins start at D0, at location K34, and end with D63, at location N3.

The price of 486DX motherboards are about $60. The 80586 (Pentium processor) motherboards cost about $150 to $850.

Motherboards for the higher-end Pentium processors may have any combination of I/O slots. Combinations found on boards now include four PCI, six 16-bit ISA, and two VESA slots, or three PCI and five 16-bit ISA slots, or four PCI and four EISA slots. These have four 72-pin SIMM sockets.

**FIGURE 3.34   Pentium processor pinout from pin side**

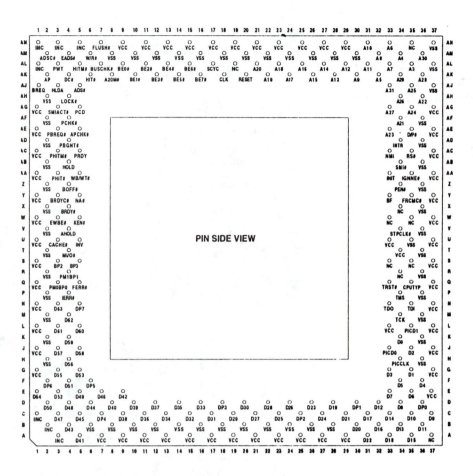

A number of companies ship their 486 boards with a 238-pin "Pentium-based over-drive" socket to upgrade when the owner decides to move up. The P54C is a newer and smaller Pentium processor that runs at 90 or 100 MHz.

Cyrix Corporation makes the Cyrix 5x86, as shown in Figure 3.35. It has 64-bit internal buses and an 80-bit floating-point unit. It has a 16K internal cache and runs on 3.3 V to reduce heat. It is pin compatible with the Pentium processor.

Advanced Micro Devices, well known for their AM486s, makes the AMD K5 as their 80586 processor. It is pin compatible with the Pentium processor.

**FIGURE 3.35   Cyrix 5x86, reprinted by permission of Cyrix Corporation**

NexGen makes the Nx586. As with all 586s, it has 64-bit internal buses but has doubled the size of the two internal caches to 16K each. The Nx586 also has a controller on the chip for the second-level (external) cache. The bus to the cache is 64 bits wide. Extensive testing has proved this chip to be fully software compatible with the Intel Pentium processor. To keep cost low for the user, it has an external math coprocessor. Chip sets for Nx586 motherboards are being made by VLSI Technology Inc. Regretfully, it is not *pin* compatible with the Pentium processor. A photo of the Nx586 is shown in Figure 3.36.

Table 3.14 summarizes the 80586 family of chips.

## System Speed Standards

Since speed has become so important, some standards to measure speed of computer systems exist. One standard is how many million instructions per second (MIPS) the system can perform. Remember, mega means million, giga means billion, and tera means trillion. What kind of instructions being run for the test becomes important for consistency. A **Dhrystone** is a program to "benchmark" a system's overall processing speed. It runs a mixture of instructions and outputs the number of times the program can be run in a second. A **Whetstone** is the same type of benchmark program that tests only floating point operations. Floating (decimal)-point operations are called *flops*. Whetstone measures a system's math speed. Winstone 95 is a speed test that marks performance time running thirteen popular business programs in Windows. Landmark Speed is another performance rating test program.

To put this in perspective, Table 3.15 shows the speed of some personal computers and supercomputers, with ratings under various performance test programs for comparison. The Intel

**FIGURE 3.36    NexGen Nx586, reprinted by permission of NexGen**

**TABLE 3.14    The 586 family**

| Manufacturer | Chip |
|--------------|------|
| Intel | 60/66 MHz versions of the Pentium processor |
| Intel | 90/100 MHz version, SL enhanced |
| Intel | 120/133 MHz versions |
| Intel | P54C newer and smaller, 90 and 100 MHz |
| Cyrix | 5x86 |
| NexGen | Nx586 |

**TABLE 3.15  System speeds**

| System CPU | CPU (MHz) | Dhrystone MIPS | Whetstones | Landmark | Winstone 95 |
|---|---|---|---|---|---|
| 286 PC AT | 16 | 2.7 | | | |
| 286 PC AT | 20 | 3.5 | | | |
| 386 | 33 | 12 | | | |
| 386 | 40 | 10.7 | 0.213 | | |
| 486DX | 25 | 20 | 5.5 | | |
| 486SX+487 | 33 | 27 | 5.5 | | |
| 486DX | 50 | 41 | 12 | | |
| 486DX-2 | 66 | 54 | 25 | | 45 |
| 586 | 66 | 112 | 67 | | |
| 586 | 100 | 174 | 95 | 570 | 130 |
| Cyrix 5x86 | 100 | | | 460 | |
| NexGen Nx586 | 100 | | | 570 | |
| Intel P6 | 133 | | | 190 | |
| Cyrix 6x86 | 100 | | | | 207 |

| Floating-Point Performance Comparison | |
|---|---|
| The Macintosh II | 0.23 Mf (megaflops) |
| The Sun Sparc 2 | 3.8 Mf |
| The IBM System 6000 (RISC) | 50 Mf |
| The Intel 860XR | 80 Mf or 65 Dhrystone MIPS |
| The DEC VAX 9000 | 125 Mf |
| The IBM 3090 | 414 Mf |
| Intel's Touchstone Delta | 13.9 Gf (gigaflops) |
| The Cray Research Y-MPC90 | 16 Gf |

Corporation's 860XR is a 64-bit RISC microprocessor. Its integer performance is 114 kilo-Dhrystones or 65 Dhrystone MIPS. Its floating-point peak performance is 80 megaflops (Mf). Cray produces the fastest, most powerful computers in the world. The Cray Research Y-MPC90, with sixteen parallel processors, has a peak performance of 16 gigaflops. The parallel processor idea is to break large, complex jobs down into smaller, simpler jobs and have numerous processors working on their tasks simultaneously. Intel's Touchstone Delta has a peak performance of 13.9 gigaflops. It has an amazing 528 coprocessors.

## 80686 MOTHERBOARDS

The Intel P6 or Pentium Pro has internal 8K instruction and 8K data caches, and a 256K internal cache. This requires another die (chip of silicon), so it has dual cavities in the ceramic package. The CPU has 5.5 million transistors crammed into a square of silicon 0.7" on each side. It has a separate piece of silicon (die) for a cache made of 15.5 million transistors in a rectangle about 0.7" by 0.5". The internal connections are only one-half of a millionth of an inch wide. It has 387 pins to connect to the motherboard, with an external bus of 64 bits for data and 32 bits for address, the same as the Pentium processor or 586. The system (motherboard) bus can run up to 66 MHz, and the CPU will run at two, three, or four times the system bus speed. First versions will run at 133 MHz. Up to four P6s may be paralleled on a single external bus for higher speeds using parallel processing. It runs on 2.9 V to reduce heat, and because of more efficient operation, it should run about twice as fast as 586s at the same clock speed. The dual cavity design with the separate cache is shown in Figure 3.37.

FIGURE 3.37 P6 processor
with two die, reprinted by
permission of Intel Corp.
Copyright © 1995 by Intel Corp.

FIGURE 3.38 Cyrix 6x86,
reprinted by permission of Cyrix
Corporation, 1996

Cyrix has produced the Cyrix 6x86, in direct competition with the P6. According to Cyrix, the 6x86's Winstone performance is about 207 with a 100-MHz clock, compared to 175 for a 150-MHz P6 and 190 for a Pentium processor at 133 MHz. The Cyrix 6x86 is not using RISC architecture, but staying with what Cyrix calls "native x86" processing. This may give it a speed advantage in running existing software. It is not pin P6 compatible, and it does not have the level 2256K cache. It is shown in Figure 3.38.

## MMX

MMX is a set of 57 multimedia instructions that perform parallel operations on eight data items at the same time. The purpose is to speed up the special kind of processing needed for multimedia intensive applications. Since most applications are centered around multimedia effects, processors with MMX added are faster than the basic processor.

### Cyrix 686MX

The Cyrix 686MX Processor PR233 has the 57 MMX instructions for faster multimedia processing. It has an internal cache of 64 K and the standard 64-bit external data bus, and 32-bit address bus. It fits the Socket 7 321-pin socket and runs at 2.9 V with 3.3 V I/O. Winstone 97 for Windows 95 rates the Cyrix 686MX233 at 52.3 compared to 51.8 for the Pentium II Processor 233 MHz.

### Cyrix MII 300

Cyrix makes the MII, which runs at 300 MHz as direct competition for the Pentium II. It incorporates all the features of its earlier processors, and still is compatible with Socket 7 motherboards.

### AMD-K6 MMX

AMD has incorporated MMX technology in the AMD-K6 MMX Enhanced Processor. It has a 64 K L1 (level 1) cache, with 32K for instructions and 32K for data. It comes in a 321-pin CPGA (Ceramic Pin Grid Array) which is Socket 7 compatible. The K6 comes in 233, 266, and 300 MHz speeds. It has 8.8 million transistors and is 0.25 micron technology.

### Pentium MMX

Intel added the MMX instruction set to the Pentium for the Pentium MMX. It comes in a Socket 7 package (293-pin PGA) at speeds up to 233 MHz. The Pentium MMX runs a multimedia application up to 60% faster than a standard Pentium of the same clock speed. It has an L1 cache of 32K, 16K for data, and 16K for instructions.

### Pentium II

In the quest for ever-increasing speed, Intel created the Pentium II processor. It has the 57 MMX instruction enhancements to speed up multimedia applications. Built with 7.5 million transistors, it has internal connectors of only 0.25 micron width. It comes in speeds of 233, 266, 300, 333, 350, and 400 MHz. It has dual independent buses (DIB) that the processor can access simultaneously. It can access the L2 cache bus and at the same time the motherboard system bus. Intel also designed a 100 MHz Front Side Bus, raising motherboards from 66 MHz to 100 MHz. The 64-bit data and 32-bit address bus remain the same as previous processors. These boards need PC100–compliant memory to keep up with the system. It is packaged in a Single Edge Contact (SEC) cartridge and plugs into a "Slot 1" connector on the motherboard as shown in Figure 3.39. These motherboards have the ATX form factor. Although the CPU voltage has dropped to 3.3 V, the power may reach 35 watts. Relative speeds using Intel's own "iCOMP Index 2.0" place the Pentium 200 MHz at 142, the Pentium MMX 200 MHz at 182, the Pentium Pro 200 MHz at 220, and the Pentium II 233 MHz at 267. In some Pentium IIs, the cache chips or modules have an amazing 5 nanosecond access time.

### INTEL Celeron™

For less expensive systems, Intel removed the L2 cache from the Pentium II and the SEC cartridge, and has the separate processor bare board plug into the Slot 1. Called the Celeron, it comes in a Single Edge Processor Package (SEPP) and saves money. It does have a 32K level 1 cache and the MMX instruction set. Of course, without the L2 cache the CPU is slower than those with a cache, but it is an inexpensive way to get close to Pentium II performance. Celeron supports a 66 MHz bus, not the 100 MHz bus. The Celeron is shown in Figure 3.40.

**Figure 3.39    Intel's Pentium II in a SEC cartridge plugged into a Slot 1 connector on an ATX motherboard**

**Figure 3.40  The Intel Celeron™ front and back side views.**

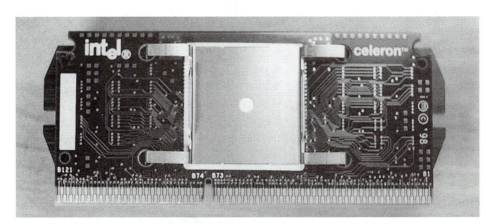

## CHIPSETS

Chipsets may support a mixture of the following features, all in a couple compatible chips, reducing the hundreds of individual logic chips needed in older boards to support the system. They may include memory controller, PCI bridge, EIDE controller, real time clock (RTC), DMA controller, keyboard controller, PS/2 mouse controller, secondary cache controller, and CMOS SRAM.

The Intel 440EX AGP set works well with the Celeron for a 66 MHz bus system. It supports ACPI (Advanced Configuration and Power Interface), Ultra DMA/33 hard drives and USB, and 256 MB of SDRAM or EDO (Extended Data Output) RAM.

A popular chipset for Pentium II motherboards is the Intel 82440BX, nicknamed the BX Chipset. This set supports the Accelerated Graphics Port (AGP), Ultra DMA/33 for 33MB/sec transfer time for Ultra DMA hard disks, the PCI bus, the Universal Serial Bus (USB), and a 100 MHz system bus. The CPU may run at multiples of the system bus up to six times as fast (600 MHz).

The Intel 430HX runs single or dual processors, and supports four banks of RAM for up to 512MB of memory. It supports a 512K cache, PCI slots, EIDE, USB, and EDO RAM.

Another popular chipset is the VIA MVP3. This set supports Socket 7 CPUs such as the Pentium MMX 266, the Cyrix/IBM 6X86MZ PR 233, and the AMD-K6-2/300 with PCI, 100 MHz bus, AGP, ACPI (Advanced Control and Power Management Interface), Ultra DMA-33 hard drives, and more.

SiS has a 5597 chipset for Socket 7 boards. It includes on board video controllers so the video port can be built into the motherboard.

OPTi makes a 750 chipset that supports up to 512 MB of RAM, video support, ECC, and four SDRAM slots.

## ATX MOTHERBOARDS

A new type of motherboard has appeared for the new generations of chips. The ATX motherboard appears to be rotated 90 degrees, making it longer from side to side than front to back. The full ATX board is 12 inches wide and 9.6 inches front to back. The mini-ATX is 11.2 inches wide and 8.2 inches front to back. On an ATX motherboard up to seven I/O expansion slots, ISA and PCI, are in the same place as on older motherboards, but the new taller microprocessors are to the right rear, by the power supply, allowing full-length cards. This also allows changing the microprocessor without removing overhanging expansion cards. Figure 3.41 shows the layout of an ATX motherboard.

The power supply is rotated to blow in and over the microprocessor for increased cooling efficiency. RAM memory sockets are DIMM, and may be oriented front to rear between the seventh expansion slot (from the left) and the microprocessor or oriented side to side between the microprocessor and the front of the board. Hard and floppy disk connectors are suggested

**Figure 3.41 ATX motherboard layout**

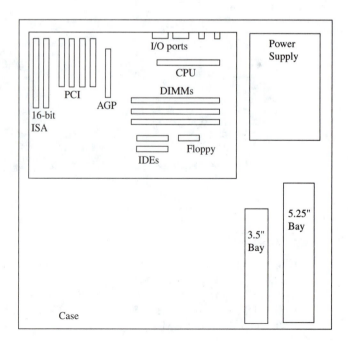

**Figure 3.42   The rear panel I/O connectors of the ATX motherboard.**

to be to the right of the expansion slots, and toward the front of the board, but to the left of the drive bays. Front panel I/O connectors (power LED, disk LED, etc.) are along the front of the card. The rear panel I/O connectors for serial (USB), parallel, video, keyboard (PS/2), mouse (PS/2), audio, and game port are supposed to be in an area about 6 inches wide by 2 inches high at the rear right side of the board, behind the microprocessor, and to the right of the expansion slots (Figure 3.42).

From the left in Figure 3.42, note the keyboard and mouse PS/2 ports, two rectangular USB ports, and a parallel over two RS-232 serial ports. This area is tall enough for connectors to be stacked two high. Most of this I/O circuitry is on the motherboard, eliminating a number of expansion cards. This of course also may limit upgrading unless there are jumpers to deactivate a card, such as a motherboard-bound video circuitry, and allow an add-on card to function for video upgrade. These jumpers, if they exist, would also allow repair by replacing a failed motherboard video with an expansion video card. On the positive side, this eliminates a number of expansion cards and should minimize cost of the board. There is a new single power connector which includes 3.3 V (Chapter 2, Power Supplies). The new AGP has become standard on these motherboards.

## BUSES

The lines that transfer related information from one point to another on the computer are called a **bus.** The eight data bits D0–D7 running from one chip to another make up an 8-bit-wide data bus. On 8088 machines, lines A0–A19 make up the 20-bit-wide address bus. All of the various lines that are used to coordinate and control the system are called the **control bus.** An explanation of the various buses in current use in personal computer systems follows.

### The 8-Bit ISA Bus

The 8-bit ISA bus was used on the 8088 systems that had only 8-bit data buses. Many newer motherboards still have one or two 8-bit I/O expansion connectors, but only to save money and be compatible with an 8-bit card with a skirt. The bus was designed for a theoretical maximum data transfer rate of 8.33 MB/sec. The data transfer rate is also called the throughput, and it is the limit of how fast the system can communicate through the I/O bus to the cards in the I/O connectors.

A glance at the signals in Figure 3.11 will show the data and address lines. Because it was the original type of card, newer buses have been designed as backward compatible, so the 8-bit ISA adapter card can be used in 16-bit ISA, EISA, or VESA slots or connectors.

### The 16-Bit ISA Bus

The 16-bit ISA bus came into being with the first AT class computer, using an 80286. This microprocessor had a 16-bit data path, and therefore needed a 16-bit bus. The

16-bit I/O expansion connector in Figure 3.18, discussed earlier, shows the added data, address, and control signals. With sixteen lines for data, this bus can transfer 2 bytes of information at a time, or twice the amount of the 8-bit bus. This gives 16.7 MB/sec throughput.

The 16-bit ISA adapter card can be used in EISA or VESA slots also, but usually not in 8-bit slots. I say "usually" because some 16-bit ISA video cards are designed to work in either 8- or 16-bit ISA slots. Yes, it is unusual to have half a card dangling in space, but they do work.

### The 32-Bit EISA Standard

With the microprocessors now having a 32-bit external bus, the 16-bit expansion slot no longer works efficiently. In 1988 a group of major computer manufacturers announced a new standard. The extended industry standard architecture bus is a 32-bit EISA bus, backward compatible with ISA expansion cards. This means 8- and 16-bit ISA cards will fit in the EISA slot, but the EISA card will not fit in any slot but its own. Because it has thirty-two data bits, or four times the number of the original, it can transfer four 8-bit bytes at a time. Therefore, it will handle data transfer rates up to 33.3 MB/sec.

Figures 3.43 and 3.44 show the 32-bit EISA expansion slot and connector, respectively. Note the addition of LA2–LA16 and LA24–LA31 latchable address lines. LA17–23 are on the C side of the front (18-pin) 16-bit extension connector. BE0–BE3 are byte enable lines, letting the processor know whether this is a 16- or 32-bit word. Note also the additional data lines SD16–SD31, giving a full 32-bit (0–31) data bus. MSBURST is the master burst signal and SLBURST is the slave burst signal. The LOCK line locks other requesters out of the memory while the bus master is using it. EX16 and EX32 run by the slave to show the bus master that the slave supports 16- or 32-bit data transfers. M I/O shows the bus master whether an IN or OUT cycle is occurring. MREQx and MAKx are used by a bus master to request access to bus and by the system to acknowledge (and grant) the request.

Intel created an EISA chip set using VLSI to implement an entire system board in five major chips. To support the 80486 processor, it added an 82352 bus buffer, an 82355 bus master interface controller, an 82357 integrated system peripheral, and an 82358 bus controller. Add a handful of minor support chips, such as a 8742 keyboard controller, and RAM and ROM, and you can build an EISA system. (Just remember, nothing is as easy as it sounds.)

Note that the EISA expansion card is slightly larger overall than the ISA card. The EISA edge connectors are the same length, but they are deeper to provide for the second row of contacts. Figure 3.45 shows the EISA 32-bit card. Note the EISA access keys, which allow the EISA card to slip into the lower slot but keep the ISA card up from the bottom of the connector. This allows the 8/16-bit ISA card to be used in the 32-bit connector but only to contact the upper row of the system connector contacts. The EISA card mates with the EISA access keys and fully seats, so the upper and lower rows of card contacts all contacts. The EISA card cannot fit into the old 16- or 8-bit ISA slots, but the new EISA slots are backward compatible, and the 8- and 16-bit ISA cards can fit and run.

**FIGURE 3.43    EISA expansion connector**

**FIGURE 3.44　EISA connector**

Rear 31-contact section:

| F1 (Bottom) | B label (ISA top) | B pin | A pin | A label (ISA top) | E1 (Bottom) |
|---|---|---|---|---|---|
| GND | GND | B1 | A1 | I/O Check | CMD |
| +5 V | Reset drive | 2 | 2 | D7 | Start |
| +5 V | +5 V | 3 | 3 | D6 | EXRDY |
| RSVD | IRQ2 | 4 | 4 | D5 | EX32 |
| RSVD | –5 V | 5 | 5 | D4 | GND |
|  | DRQ2 | 6 |  | D3 |  |
|  | –12 V | 7 | 7 | D2 |  |
| RSVD | Reserved | 8 | 8 | D1 | EX16 |
| RSVD | +12 V | 9 | 9 | D0 | SLBURST |
| MIO | GND | B10 | A10 | I/O CH Ready | MSBURST |
| LOCK | MEMW | 11 | 11 | AEN | W/R |
| RSVD | MEMR | 12 | 12 | A19 | GND |
| GND | IOW | 13 | 13 | A18 | RSVD |
| RSVD | IOR | 14 | 14 | A17 | RSVD |
| BE3 | DACK3 | 15 | 15 | A16 | RSVD |
|  | DRQ3 | 16 |  | A15 | GND |
| BE2 | DACK1 | 17 | 17 | A14 |  |
| BE0 | DRQ1 | 18 | 18 | A13 | BE1 |
| GND | DACK0 | 19 | 19 | A12 | LA31 |
| +5 V | CLK | B20 | A20 | A11 | GND |
| LA29 | IRQ7 | 21 | 21 | A10 | LA30 |
| GND | IRQ6 | 22 | 22 | A9 | LA28 |
| LA26 | IRQ5 | 23 | 23 | A8 | LA27 |
| LA24 | IRQ4 | 24 | 24 | A7 | LA25 |
|  | IRQ3 | 25 |  | A6 | GND |
| LA16 | DACK2 | 26 | 26 | A5 |  |
| LA14 | T/C | 27 | 27 | A4 | LA15 |
| +5 V | ALE | 28 | 28 | A3 | LA13 |
| +5 V | +5 V | 29 | 29 | A2 | LA12 |
| GND | OSC | B30 | | A1 | LA11 |
| LA10 | GND | B31 | A31 | A0 | GND |
| LA8 |  |  |  |  | LA9 |

Front 18-contact section:

| H1 (Bottom) | D label (ISA top) | D pin | C pin | C label (ISA top) | G1 (Bottom) |
|---|---|---|---|---|---|
| LA6 | MEMCS16 | D1 | C1 | SBHE | LA7 |
| LA5 | I/OCS16 | 2 | 2 | LA23 | GND |
| +5 V | IRQ10 | 3 |  | LA22 | LA4 |
| LA2 | IRQ11 | 4 | 4 | LA21 | LA3 |
|  | IRQ12 | 5 | 5 | LA20 | GND |
| SD16 | IRQ15 | 6 |  | LA19 | SD17 |
| SD18 | IRQ14 | 7 | 7 | LA18 | SD19 |
| GND | DACK0 | 8 | 8 | LA17 | SD20 |
| SD21 | DRQ0 | 9 | 9 | MEMR | SD22 |
| SD23 | DACK5 | D10 | 10 | MEMW | GND |
| SD24 | DRQ5 | 11 | 11 | SD08 | SD25 |
| GND | DACK6 | 12 | C10 | SD09 | SD26 |
| SD27 | DRQ6 | 13 | 13 | SD10 | SD28 |
|  | DACK7 | 14 | 14 | SD11 | GND |
| SD29 | DRQ7 | 15 | 16 | SD12 | SD30 |
| +5 V | +5 V | 16 | 17 | SD13 | SD31 |
| +5 V | Master | 17 | 18 | SD14 | MREQx |
| MACK | Ground | D18 | 19 C18 | SD15 |  |

*Notes:*

1. Labels closest to connector are ISA labels.

2. Labels farthest from connector are EISA labels.

3. Connector pin numbers on left are ISA pin numbers.

4. Connector pin numbers on right are EISA pin numbers.

5. Connector side designators A, B, C, and D are top contacts.

6. Connector side designators E, F, G, and H are bottom contacts.

7. Bars across slot at E6, E16, E25, G6, and G15 are EISA access keys. These restrict ISA cards from fully seating in lower slot. Thus ISA cards only contact upper ISA contacts.

8. Bar below A31 is divider between rear 31 contact and front 18 contact slots.

　　Note the EISA connector in Figure 3.43. It is taller than the standard 16-bit ISA connector, but the same length.

**FIGURE 3.45　EISA and VESA local bus cards**

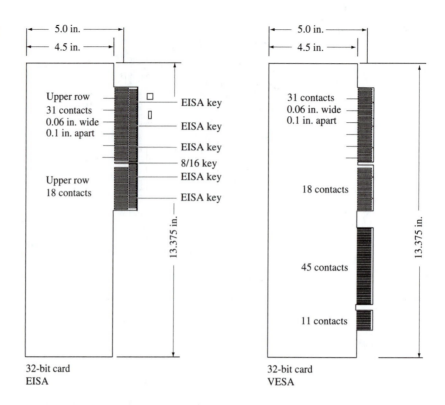

| | |
|---|---|
| 32-bit card | 32-bit card |
| EISA | VESA |

## The 32-Bit VESA Standard (Local Bus)

Because of the increased speed, wider, faster buses are desirable. The 32-bit EISA standard mentioned before is one choice. The 32-bit VESA (Video Electronics Standards Association) standard is another. Instead of deeper or taller connectors, the 32-bit VESA bus standard leaves the 16-bit ISA connector as is and adds another connector closer to the front of the motherboard, more or less making a longer I/O slot. This means the VESA I/O connector also is backward compatible, and the 8- and 16-bit ISA cards fit right in the old slot. Of course a VESA adapter card cannot run in the 8- or 16-bit ISA slot; even though one end fits in those sockets, the other would dangle in space. EISA and VESA cards are definitely not compatible with each other's connectors or the 8- or 16-bit ISA connector.

Figure 3.45 also shows the VESA adapter card. Figure 3.46 compares the ISA bus to the VESA bus. Note that anything connected to the VESA bus slots, for example the video and hard drive, may now run faster. Figure 3.47 shows a VESA video card.

Typical motherboards with processors older than the 586 include two VESA slots, because only two cards can be supported. The remaining slots are ISA 16-bit. While the ISA 16-bit bus was restricted to 8.33 MHz, the VESA expansion slots may run at 33.3 MHz. Thus, being 32 bits, four times wider than the original bus, and running four times faster, it is sixteen times as fast as the 8-bit ISA bus. This gives it a throughput of 133.3 MB/sec. Although the video and the hard drive may realize a great improvement in speed, the floppies, restricted in speed by the mechanics, would not be much improved. If you must work from floppies, I suggest creating a RAM disk, copying the floppy into the electronic RAM disk, working, and saving the RAM disk to a floppy when terminating.

A photo of the VESA expansion connector (slot) is shown in Figure 3.48. Note the additional connector of the VESA bus and that the pins are half the size or twice as close together.

## The PCI Bus

The 32-bit peripheral component interconnect (PCI) bus is faster and smarter than any of the previous buses. BIOSs for PCI boards have Plug-and-Play (PnP) campatibility, automatically recognized PnP cards. Non-PnP cards are called Legacy cards. Like the VESA bus, the PCI bus also may run at 33 MHz, giving it a throughput of 132 MB/sec. It is supposed to eliminate DMA channel and IRQ interrupt conflicts and require no jumpers to set it up. The physical

**FIGURE 3.46  VESA local bus versus ISA bus**

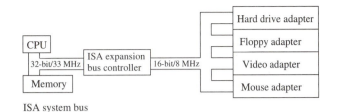

ISA system bus

VESA local system bus

**FIGURE 3.47  VESA video card**

**FIGURE 3.48  VESA expansion connector**

PCI I/O connector is similar to the 8-bit ISA, but has 120 pins, 60 on each side, with a 2-pin-wide keying slot 11 pins from the end. Counting the keying slot as 2 pins per side, that brings the total to 124 pin locations.

The cards are similar to the 8-bit ISA cards, but the components are on the opposite side of the card, so the PCI bus is easily identifiable by having the chips on the "wrong" side.

A 64-bit version is available, with an additional 64-pin connector, 32 per side. This PCI version 2 bus runs at 100 MHz. PCI slots and cards, shown in Figures 3.49 and 3.50, are totally incompatible with all other slots and cards. One motherboard has 216-bit ISA, two 32-bit PCI, and two 64-bit PCI slots.

**FIGURE 3.49   PCI bus connector**

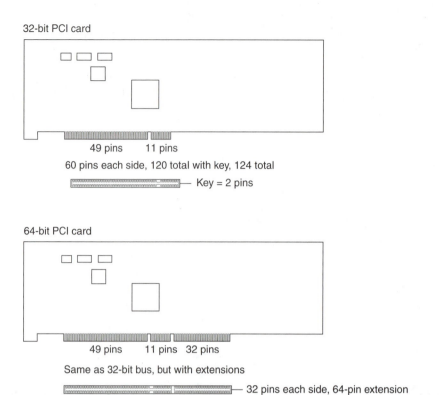

32-bit PCI card

49 pins      11 pins

60 pins each side, 120 total with key, 124 total

Key = 2 pins

64-bit PCI card

49 pins      11 pins   32 pins

Same as 32-bit bus, but with extensions

32 pins each side, 64-pin extension

**FIGURE 3.50   32-bit PCI card and connectors**

## The PCMCIA Bus

The Personal Computer Memory Card International Association (PCMCIA) has developed a standard for a physically small I/O connector. The devices to be plugged in are almost as small as credit cards. These are obviously very desirable for portables, laptops, or notebook computers because of their small size. Because most of their parts are custom-made, repairing these computers is not addressed specifically in this book.

PCMCIA devices currently available include memory, fax/modem, network, and hard drive cards. The PCMCIA bus uses a 68-pin dual header as a connector. These credit-card-sized peripherals are undoubtedly the wave of the future and already have had many revisions, giving PCMCIA types I, II, III, and IV. The cards attempt to be backward compatible to previous types, but that is not guaranteed. Of course, they are totally incompatible with all other buses.

There is a device to mount in a 3.5" bay of your desktop that accepts a PCMCIA card. The cost is about $300.

The PCMCIA connector is shown in Figure 3.51, and the pinout for some memory cards is given in Table 3.16. Note that D0–D15 show a 16-bit bus, and in this case A0–A21 give a 22-bit address bus that can address 4 MB of memory.

To help fit all these bus connectors into an overall picture, the 16-bit ISA, 32-bit EISA, 32-bit VESA, and 32-bit PCI connectors are shown in Figure 3.52.

**FIGURE 3.51  PCMCIA connector**

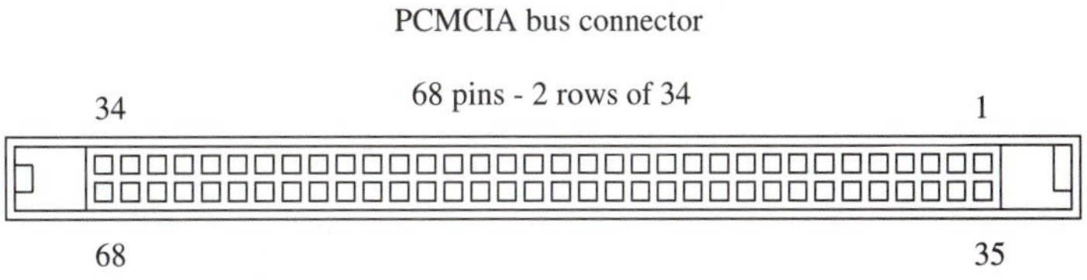

PCMCIA bus connector

68 pins - 2 rows of 34

**TABLE 3.16  PCMCIA pinout**

| Pin | Use | Pin | Use | Pin | Use | Pin | Use |
|-----|-----|-----|-----|-----|-----|-----|-----|
| 1 | GND | 18 | VPP1 | 35 | GND | 52 | VPP2 |
| 2 | D3 | 19 | A16 | 36 | CD1 | 53 | NC |
| 3 | D4 | 20 | A15 | 37 | D11 | 54 | NC |
| 4 | D5 | 21 | A12 | 38 | D12 | 55 | NC |
| 5 | D6 | 22 | A7 | 39 | D13 | 56 | NC |
| 6 | D7 | 23 | A6 | 40 | D14 | 57 | NC |
| 7 | CSL | 24 | A5 | 41 | D15 | 58 | NC |
| 8 | A10 | 25 | A4 | 42 | CSH | 59 | NC |
| 9 | OE | 26 | A3 | 43 | NC | 60 | NC |
| 10 | A11 | 27 | A2 | 44 | NC | 61 | NC |
| 11 | A9 | 28 | A1 | 45 | NC | 62 | ALM1 |
| 12 | A8 | 29 | A0 | 46 | A17 | 63 | ALM2 |
| 13 | A13 | 30 | D0 | 47 | A18 | 64 | D8 |
| 14 | A14 | 31 | D1 | 48 | A19 | 65 | D9 |
| 15 | PGM | 32 | D2 | 49 | A20 | 66 | D10 |
| 16 | NC | 33 | WP | 50 | A21 | 67 | CD2 |
| 17 | Vcc | 34 | GND | 51 | Vcc | 68 | GND |

**KEY:**
VPP1 = programming voltage for lower
      bytes of EPROM
VPP2 = programming voltage for upper
      bytes of EPROM
PGM = program enable
OE = output enable
CSL = chip select for low bytes
CSH = chip select for high bytes
CD1 = card detect
CD2 = card detect
ALM1 = low-battery alarm
ALM2 = low-battery alarm
WP = write protect

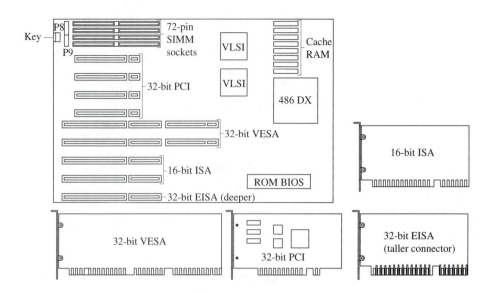

**FIGURE 3.52  Top view of all slots and cards**

## OVERALL OPERATION

(**Note:** This section is recommended for technicians.)

If you suspect motherboard failure, remove all cards and the keyboard so that you have nothing but the power supply and the speaker connected to the motherboard. (On a 286 or higher, leave the front panel switches, the CMOS battery, and the LEDs hooked up.)

**Caution:** Be very careful not to short from side A to B when probing the I/O connector. A 3-by-5 card cut to fit the slot and numbered for pins will help both in finding the right pin quickly and in not shorting them out.

Recheck the power supply outputs at B1, B3, B5, B7, B9 and the system switch settings. Remember to check the power good signal on power supply plug P8, pin 1. This signal usually goes to an 8284 or an 82284 clock generator and will shut down the system clock if the power supply is not good.

If the motherboard is not operating now, the problem is obviously something on that board.

To get a "feeling" for what is happening on the board, let's look at the expansion slot bus under different circumstances. First, on the motherboard with only the power supply connected—no CPU (8088), no ROM, and no RAM (which you removed with the power off)—what should we see with the oscilloscope? (A digital logic probe with pulse detector can also work.) On the expansion slot bus, B3, B5, B7, and B9 show 5, −5, −12, and 12 V DC, respectively, and B1 and B31 are ground. This never changes.

In the following discussion, all pins are low unless otherwise stated. A low will only be stated as a change. "Active" in this discussion will mean "with activity" or being pulsed. Now we see that B4, B8, B11–15, B17, B21–26, and B29 are high. B20 has a clock of the speed of the CPU as in 4.77 MHz for the original IBMs. B30 has the video oscillator clock of 14.31818 MHz. On the right side of the slot, A1, A10, and A12–31 are high. A12–A31 are the twenty address lines, and all are inactive because the processor (CPU) is not there to address anything. A2–A9 being low shows that no data is present.

If your bare board does not have these signals, find out why now. The power good signal could be stopping the clock generator circuit if absent.

Now power-down, add the CPU, and power-up. What changes do we see? B11 and B12 (which were high) are now active, showing that the CPU is operating and trying to access memory with reads and writes. A2–A9 (which were low) are now high, showing that data is loaded and ready. A12–A15 (which were high) now are active (with pulses about every 110 milliseconds, so look closely with the scope or it may look low). A16–A31 are also active, showing all address lines being actively manipulated by the CPU. If your bare board plus CPU shows these signals, things are optimistic. If not, try a different CPU (meaning a *good* 8088 or 80286, not a different type).

Now power-down, add the ROM, and power-up. What changes do we see? B11 (which was active) is now high, while B12 stayed active, showing that the CPU is reading memory (ROM). B13 (which was low) is now active, showing an attempted I/O write. B19 (which was high) is now active, showing by Data Acknowledge Zero that the CPU is trying to refresh the RAM, which in all these systems is dynamic and will lose information if not refreshed periodically. Of course the RAM is not installed yet, but the CPU (8088/80286/80386SX) keeps trying. On the right side of the slot bus, A2–A9 (which were high) are low again. A11 (which was high) is now active, showing Address Enable for DMA (Direct Memory Access) control of the address and data bus changing. A14 (which was active) is much more active, and A15 (which was active) is now low, showing that the addresses are changing from what they were without the ROM. If you see these signals, at least the ROM is not in total failure, although it may still be bad, because just one bad memory location can cause a fatal error.

Now power-down, add the RAM, and power-up. What changes do we see? Well, the system is fully operational, with no outside I/O hooked up. B11 (which was high) is now active again, showing RAM is being written now. B28 (which was low) is now active. This is the Address Latch Enable, latching addresses from the microprocessor. On the right side of the slot, A2–A9 are finally active, showing data is being sent back and forth. A12–15 are very active now, making the entire address bus active addressing ROM and RAM. If you see all these signals in this progression, there is a good possibility that you have a working motherboard.

Whatever the microprocessor on your board, some chips that are very likely to be there are the bus buffers and transceivers. If a data or address line is not active when it is supposed to be, a line of the buffer or transceiver may be out. Some pinouts of the most popular ones, the 74244 and the 74245, are shown in Figures 3.53 and 3.54, respectively.

The speaker on a system is more important than you might first imagine, because it often gives the first sign of trouble with the audio error codes or beeps. Beyond that, many programs beep on errors to warn the operator to enter the command or data again. Finally, music or sound effects from games and other programs (some serious) add to the enjoyment of the system. Program Listing 3.1, which follows, is a program to test the speaker. If it fails to produce sound, check the wires from the speaker to the motherboard. With the system off, a multimeter set on resistance scale touched across the speaker will make a scratching noise. If this is absent, replace the speaker. If it works, check the output transistor.

**FIGURE 3.53    74244 octal line driver/buffer**

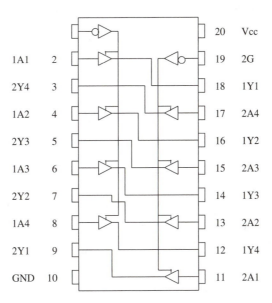

**FIGURE 3.54**  74245 octal transceiver

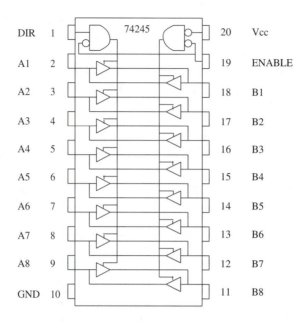

**PROGRAM LISTING 3.1**
**Speaker test**

```
10 REM                 This demonstrates the SOUND command
20 REM SOUND a,b       where a = 37-32767 is freq in hertz and
30 REM                 b=0-65535 is duration in clock ticks
40 REM                 very roughly in 10ths of a second.
50 SOUND 100,10
60 BEEP
70 SOUND 200,20
80 BEEP
90 SOUND 400,30
100 BEEP
110 FOR I = 100 TO 10000 STEP 50
120 SOUND I,2
130 NEXT I
```

## 3.3 ASSEMBLING SYSTEM BOARDS

System board or motherboard form factors include not only the physical size of the board but also the mounting holes. Expansion slot mating with the case could be included as a consideration, but that is hardly ever a problem. System board form factors include:

- PC/XT full size
- AT full size
- AT half size
- AT baby size
- ATX

Follow these steps to install the motherboard.

1. Insert the RAM and cache in the motherboard before mounting the board in the case. The motherboard is usually much easier to reach physically when out of the case. See Chapter 4 on memory for more details on installing memory if needed. Remember to seat DIMMs firmly, and to pull up and in on side clips to secure memory. Be careful of DIMM speed. Slower DIMMs that work in PC66 boards will *not* work in PC100-compliant boards.

**Installing Socket 7 CPUs**  The CPU will need to be installed in the motherboard if purchased separately. The two types of CPUs currently in use fit Socket 7 and Slot 1 connectors.

A Socket 7 zero insertion force (ZIF) is shown in Figure 3.55. Note the raised arm that opens the contacts so the chip can drop into place. Using an antistatic wristband or other ESD protection, identify pin 1 on the socket and the chip. Carefully insert the CPU in the socket as shown in Figure 3.56. Note the white dot above pin 1. Finally, lower the arm. There will be a snap or clip to lock it in place and secure the CPU. Figure 3.57a shows the

**Figure 3.55   The empty Socket 7 with arm raising ready to receive a CPU.**

**Figure 3.56   The CPU in the socket with the arm still raised.**

**Figure 3.57   (a) Socket 7 CPU fully installed;**

*(continued on next page)*

chip fully installed. A heat sink and fan are usually installed now to remove heat and lengthen the life of the chip. The fan and heat sink will have some means of clipping on to the socket. You may use silicone heat sink compound between the sink and the CPU to increase thermal conduction and make the sink work better. Figure 3.57b shows a heat sink and fan. Note there must be a power connection to the fan, and it is either to a motherboard power header, or with a drive power supply tap as shown. Figure 3.57c shows heat sink installed, and the power jumper connected between a drive and the power supply.

**Figure 3.57**  *(continued)*
**(b) A CPU heat sink and fan with a power tap jumper; (c) The installed Socket 7 CPU heat sink and fan.**

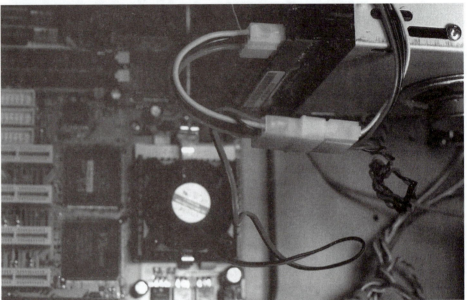

**Installing Slot 1 CPUs**   The other type of CPU socket is the Pentium II, which fits in a Slot 1. Slot 1 has a tab on one end to polarize the plastic bracket that holds the CPU. Figure 3.58a shows the tab on one end of the slot. Note the screw coming from the motherboard to secure the bracket. There is a screw at each corner. Figure 3.58b shows the notch in the plastic bracket and the holes to fit over the screws on one end of the slot. There should be four nuts with the motherboard bracket kit to secure the bracket to the board.

Figure 3.59 shows the bracket installed and ready to receive the CPU. Note the slot and CPU edge connector are keyed for polarity and will not go in the wrong way. Carefully slide the CPU down the bracket into the connector and press down firmly. There will be some kind of retention clip to hold the installed CPU to the bracket. Finally, there should be a cable for power to the CPU fan. Connect this to the CPU fan jack and to the motherboard header for the CPU fan as shown in Figure 3.60a and b.

The completely installed Slot 1 CPU is shown in Figure 3.39.

**Jumpers/Jumperless**   Some motherboards are jumperless, depending on the chipset, manufacturer, and version. Jumperless motherboards usually use the CMOS BIOS SETUP program to setup the chipset, CPU and bus speed, and other features. For those that do not, you should set or check the jumpers now. Documentation may be silk-screened on the printed circuit board itself or included in a booklet, or downloadabale from the Web. A typical Socket 7 motherboard with the VX-PRO chipset requires some jumper settings. Relate the settings below to the photos in the figures.

**Figure 3.58    (a) The Slot 1 tab for the CPU retaining bracket. (b) The CPU bracket showing the notch to match the tab on the slot.**

**Figure 3.59    The CPU bracket in position.**

Jumper JP1 discharges the CMOS, in case a password is forgotten or you just want to totally clear the CMOS. The default is of course 1-2, so the battery maintains the CMOS. You would set it 2-3 for just a moment to clear the CMOS, then return it to 1-2 for normal running. See Figure 3.61a pin connection

1  Battery
2  CMOS
3  Ground

Figure 3.60 (a) The CPU fan
power cable connected to the
CPU fan. (b) The CPU fan power
cable connected to the
motherboard header.

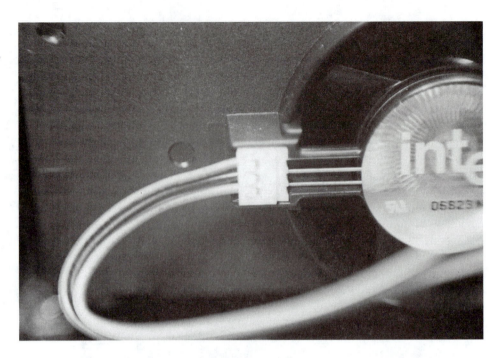

Jumper JP2 (not shown) allows selecting SDRAM DIM, or EDO DRAM DIMM.
Jumpers JP3A,B,C set the CPU clock speed according to the following selections.
Remember the CPUs have internal clock multipliers; for instance, a 66-MHz system
clock may be doubled in the CPU to 133.

| JP3 | A | B | C |
|---|---|---|---|
| 50 MHz | 2–3 | 2–3 | 2–3 |
| 55 | 1–2 | 2–3 | 2–3 |
| 60 | 2–3 | 2–3 | 1–2 |
| 66 | 2–3 | 1–2 | 2–3 |
| 75 | 1–2 | 2–3 | 1–2 |
| 83 | 1–2 | 1–2 | 2–3 |

**Figure 3.61** (a) JP1, CMOS
maintain or clear jumper;
(b) JP3 CPU clock settings;

Jumper JP3-D sets the PCI clock at either 1/2 the CPU clock or at 33 MHz.

| JP3-D | PCCLOCK |
|-------|-------------|
| 1-2 | CPU clock/2 |
| 2-3 | 33 MHz |

See the silk-screen printing and jumpers for JP3 in Figures 3.61b and c.
Jumper JP5 sets the CPU internal clock speed multiplier for different CPUs, as shown
in Figure 3.61d.

| JP5 | 1.5X | 2.0X | 2.5X | 3.0X |
|-----|------|------|------|------|
| A | 1–2 | 2–3 | 2–3 | 1–2 |
| B | 1–2 | 1–2 | 2–3 | 2–3 |

**Figure 3.61** *(continued)* **(c)**
**JP3 jumpers; (d) JP5 to set CPU**
**speed multiplier;**

Jumper JP6 sets the CPU core voltage. This must be learned from the CPU supplier or manufacturer. See Figure 3.61e for settings.

| JP6 | ON |
|-----|------|
| A | 3.5 V |
| B | 3.2 V |
| C | 2.9 V |
| D | 2.8 V |
| OFF | 2.5 V |

When you have checked or set the jumpers on the motherboard, you are ready to install the board in the case.

2. Mount the board in the case. Almost all motherboards are mounted by plastic or metal standoffs that snap into holes on the motherboard itself and slide into slots on the case. On most cases, there is at least one metal standoff to screw to the motherboard and case to secure it physically and electrically. In Figure 3.62 note the plastic standoff mounted in the motherboard. It is one of approximately six to nine that slide into the slots in the case, also shown. In Figure 3.63a note the one metal standoff and how it is used to secure

**Figure 3.61** *(continued)*
**(e) JP6 setting the CPU core voltage.**

**FIGURE 3.62    Plastic standoff and slots**

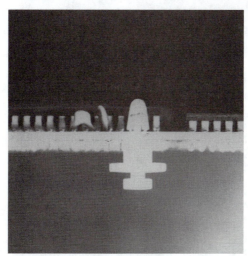

Plastic standoff

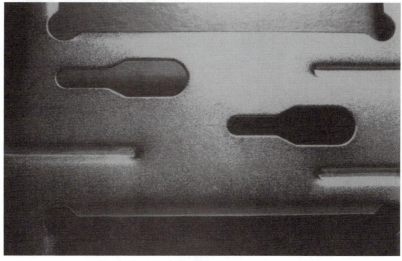

Slots

**Figure 3.63 (a) Metal hex-shaped standoff for electrical grounding of motherboard to case; (b) screw for metal standoff with fiber washer for the top of the motherboard;**

the motherboard. Figure 3.63b shows the screw with the fiber washer to protect the top of the motherboard. Figure 3.63c shows the final assembly secured to the case.

3. On older systems connect power supply plugs 8 and 9. Two connectors from the supply attach to a male Molex™ connector. Plug 8, with the three red wires, goes toward the front of the case; plug 9 goes toward the rear. The black wires on each should

**Figure 3.63** *(continued)*
**(c) screw with fiber washer securing motherboard to case.**

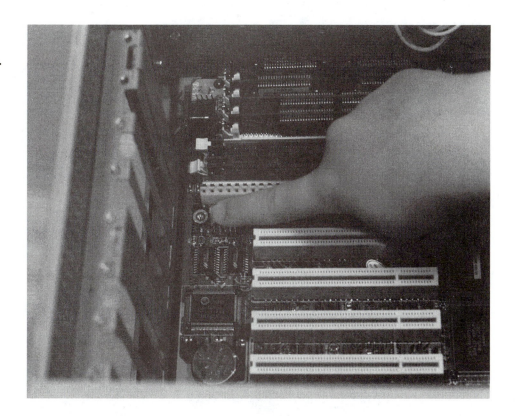

**Figure 3.64 Speaker plug and header.**

be together, giving from front to rear, three reds, white, two blacks, two blacks, blue, yellow, red, and orange. On ATX systems, connect the single power supply plug as shown in Figure 2.9 in Chapter 2.

4. Connect the speaker. It is almost always a 4-pin header. Check silk-screening on the motherboard for the proper plug. The speaker is nonpolarized, meaning it can go on either direction. Figure 3.64 shows the speaker plug and its 4-pin header. Do not think of the speaker as just a sound source. It is a troubleshooting device.

5. At this point, check the orientation of the power supply connectors and power-up the system. If the motherboard is operational, it should run the POST and give audio error codes or beeps for numerous problems, such as no keyboard, and no boot device. These beeps are good news, because they show that the board is operational. If there is no sound, power-down and test the speaker with a meter on resistance function. When the probes are rubbed on the speaker contacts, it should make a

**FIGURE 3.65   Motherboard mounting holes and a mounted 386**

Mounting holes

Mounted 386

noise like static on a radio. If it does, it is fine, and the motherboard has a problem. (See Basic Troubleshooting in this chapter.)

6. Connect front panel switches. There will be connectors for the power LED, keylock switch, turbo switch, turbo LED, and possibly more. These are usually identifiable by labels silk-screened on the board. If not, try logical choices. This is often finished up by trial and error. Test them and change until they are operational. ATX boards with ACPI may not turn on at all unless these are connected correctly.

In Figure 3.65, note the mounting holes on the motherboard for the plastic standoffs. Below is a mounted 80386 motherboard.

In Figure 3.66, note the CMOS battery with external connections if needed. Below are the reset, turbo LED, turbo switch, speaker, keylock, and color/mono jumpers.

CMOS battery

Front-panel connectors

## 3.4 BASIC TROUBLESHOOTING

Troubleshooting motherboards begins by noting any POST audio or video error codes. Then you can proceed by noting any visual clues and taking appropriate measurements. Details of these steps are as follows.

a. POST audio error code of one long and one short beep.
b. POST video error code of 1xx.
c. Measurements.

1. DEAD BOARD, GOOD SUPPLY. Check switch and jumper settings. Check RAM and cache for type and orientation.
2. Remove all adapter cards, keyboard, leave only speaker and power supply. Check B3, B5, B7, B9, or power supply plugs P8 and P9 for proper voltages to see that the motherboard is not dragging down the supply.
3. **Technicians only:** Check B30 for video clock and B20 for CPU clock. Remember that the power good signal can stop the clock if absent. Use a POST diagnostic card in I/O connector (watch orientation). These are well worth the money.
4. **Technicians only:** Check CPU for power, ground, and clock. Check expansion bus for proper activity described in discussion.
5. Add video adapter card and display monitor. Are BIOS header and POST RAM test seen? If bus clocks and signals are present but system seems dead, replace first bank

of RAM. Some systems won't give audio error codes or any video if RAM is bad. If a card has dirty oxidized contacts, clean them with a pencil eraser and reseat.

6. Try holding down INSERT when powering-up or ALT-CTRL-INSERT. If the CMOS battery is dead, some systems will appear dead. See the section in this chapter on SETUP.

7. Use a POST diagnostic card. These display the last operation of the POST or the error code it gives. If the error message is not useful (which happens), this is the time to replace the CPU if you have another.

8. Add keyboard. If system has bus activity but still seems dead, change BIOS ROM. ROM must be for same processor, but different vendors usually work at least well enough to show something. Recheck ROM size setting for 128/256K chips. Adjust with shunt jumpers. Be sure new ROM chip type is compatible with the motherboard.

9. Change keyboard BIOS ROM. BIOS ROMs must be compatible. Changing a few pairs can sometimes jump-start a system enough for you to at least get an error message.

10. Add floppy controller and drive. Run a diagnostic program or Windows boot disk.

11. Swapping power supplies is easier than swapping motherboards just in case the supply is doing something unusual. You can plug P8 and P9 of the new supply in without physically removing the old until you know it is the problem.

12. Remember that random intermittent errors may be power line fluctuation or spikes. Try filtering line.

## 3.5 CHAPTER REVIEW/QUIZ QUESTIONS

1. The "brain" of the motherboard is the
   a. Power supply.
   b. CPU.
   c. RAM.
   d. ROM.

2. The only practical repair of the motherboard is replacing the
   a. VLSI SMT chips.
   b. Whole system board.
   c. The RAM, BIOS ROM, keyboard ROM, CMOS battery, or CPU.
   d. Expansion connectors.

3. You can immediately classify a motherboard as an 8- or 16- or 32-bit system from afar by
   a. Counting the copper stripes in the address bus.
   b. Counting the number of VLSI SMT chips.
   c. Checking the physical size of the microprocessor.
   d. Glancing at the expansion connectors.

4. To determine the speed of the microprocessor you must
   a. Run the diagnostic program system speed check.
   b. Read the marking on the CPU.
   c. Check the documentation.
   d. Ask the distributor or dealer you bought it from.

5. One way an 80486DX is different from the 80386DX is
   a. The address bus width.
   b. The data bus width.
   c. The presence of a math coprocessor.
   d. Amount of RAM it can address.

6. The PC, XT, AT are based, respectively, on the
   a. 8088, 80286, and 80386.
   b. 8088, 80286, and 80286.
   c. 8088, 8088, and 80286.
   d. Z80, 6502, and 68000.

7. 16-bit adapter cards cannot fit in
   a. AT expansion connectors.
   b. EISA expansion connectors.
   c. XT expansion connectors.
   d. Any expansion connectors.

8. 32-bit adapter cards can only fit in
   a. ISA expansion connectors.
   b. EISA expansion connectors.
   c. AT expansion connectors.
   d. XT expansion connectors.

9. From the names of the pins of the CPU you can tell the
   a. Microprocessor speed.
   b. Address and data bus width.
   c. Power requirement.
   d. Number of wait states.

10. Most motherboards have a jumper to set the
    a. Address bus width.
    b. Color/mono display selection.
    c. CMOS battery voltage selection.
    d. Front panel LED system speed display.

11. ISA are initials for
    a. Indicating Signal Alarm.
    b. Industry Standard Architecture.
    c. Integrated System Analysis.
    d. Internal Signal Arching.

12. EISA are initials for
    a. Existing Internal Signal Arching.
    b. Expanded Integrated System Analysis.
    c. Extended Industry Standard Architecture.
    d. External Indicating Signal Alarm.

13. The CPU is the
    a. Calculated Power Used.
    b. Central Processing Unit.
    c. Cooling Power of the Unit.
    d. Cold Pick Up.

14. The 8088 has what width address bus?
    a. 16
    b. 20
    c. 24
    d. 32

15. The 8088 can address how many bytes of total memory?
    a. 640K
    b. 1 MB
    c. 16 MB
    d. 4 GB

16. The 80286 has what width address bus?
    a. 16
    b. 20
    c. 24
    d. 32

17. The 80286 can address how many bytes of total memory?
    a. 640K
    b. 1 MB
    c. 16 MB
    d. 4 GB

18. The 80386 has what width address bus?
    a. 16
    b. 20
    c. 24
    d. 32

19. The 80386 can address how many bytes of total memory?
    a. 640K
    b. 1 MB
    c. 16 MB
    d. 4 GB

20. The 80486 has what width address bus?
    a. 16
    b. 20
    c. 24
    d. 32

21. The 80486 can address how many bytes of total memory?
    a. 640K
    b. 1 MB
    c. 16 MB
    d. 4 GB

22. The 80586 has what width address bus?
    a. 16
    b. 20
    c. 24
    d. 32

23. The 80586 can address how many bytes of total memory?
    a. 640K
    b. 1 MB
    c. 16 MB
    d. 4 GB

24. The 80686 has what width address bus?
    a. 16
    b. 20
    c. 24
    d. 32

25. The 80686 can address how many bytes of total memory?
    a. 640K
    b. 1 MB
    c. 16 MB
    d. 4 GB

26. The 8088 motherboard was set up with
    a. A CMOS setup program.
    b. Switches on the motherboard.
    c. An oscilloscope.
    d. A meter.

27. The 286, 386, 486, or 586 motherboard is set up with
    a. A CMOS setup program.
    b. Switches on the motherboard.
    c. An oscilloscope.
    d. A meter.

28. The main circuit board of the computer system is the
    a. CPU.
    b. Hard drive.
    c. Motherboard.
    d. Video display.

29. The 80286 has what width data bus?
    a. 8
    b. 16
    c. 32
    d. 64

30. The 80386SX has what width data bus?
    a. 8
    b. 16
    c. 32
    d. 64

31. The 80386DX has what width data bus?
    a. 8
    b. 16
    c. 32
    d. 64

32. The 80486SX has what width data bus?
    a. 8
    b. 16
    c. 32
    d. 64

33. The 80486DX has what width data bus?
    a. 8
    b. 16
    c. 32
    d. 64

34. The 80586 has what width data bus?
    a. 8
    b. 16
    c. 32
    d. 64

35. The 80686 has what width data bus?
    a. 8
    b. 16
    c. 32
    d. 64

36. The "chipset" on a motherboard contains the
    a. CPU and main system RAM chips.
    b. Motherboard cache.
    c. EIDE, DMA, keyboard, mouse, and cache controllers, and more.
    d. DIMMs.

37. MMX refers to
    a. Multiple Machine Transfer (X).
    b. Multi-Media eXtension.
    c. Main Math Transistor (X).
    d. Modern Math Crossover.

38. Doze, Standby, and Suspend are three levels of
    a. EPA's Energy Star green standards.
    b. Computing power.
    c. Computing speed comparisons.
    d. Operator mental states.

39. The Pentium II processor has what width address bus?
    a. 24
    b. 32
    c. 64
    d. 128

40. The Pentium II processor has what width data bus?
    a. 24
    b. 32
    c. 64
    d. 128

41. SIMMs and DIMMs are socket compatible.
    a. True
    b. False

42. Socket 7 processors and Slot 1 processors are socket compatible.
    a. True
    b. False

43. PC66 RAM and PC100 RAM are socket compatible.
    a. True
    b. False

44. If an ATX motherboard is jumperless, the chipset, CPU and bus speed, and other settings are set in the
    a. Windows Control Panel SYSTEM window.
    b. CMOS BIOS setup program.
    c. Hard drive autoboot sectors.
    d. Cache.

45. Which type of CPU needs brackets and a retainer clip installed?
    a. Older 40-pin DIP CPUs.
    b. Newer Socket 7 CPUs.
    c. Pentium II processors.
    d. Only Celeron processors.

# 4 Memory

After completing this chapter you should be able to

- Identify different types of memory
- Select the appropriate memory to install or add
- Install memory
- Set up memory for most efficient use
- Troubleshoot and replace memory

## 4.1 GENERAL DISCUSSION

In Chapter 3 we briefly discussed the different kinds of memory on the motherboard. In this chapter we will examine those again, plus one more—video RAM (VRAM). This is used in the display adapter card. RAM on the motherboard and on the video adapter card is socketed to allow for repair (by replacement) or for upgrade and expansion. Today's software is memory hungry, and 16 MB is almost not enough RAM for a fast, efficient system running high-end software. ROM on the motherboard is socketed for replacement or upgrading, but on many adapter cards it is not.

The speed with which the ROM, DRAM, SRAM, and VRAM can be read and, except for ROM, written is called their access time. Access times for typical chips from slowest to fastest are: ROM = 150 ns, DRAM = 60 ns, VRAM = 25 ns, SRAM = 12 ns, and SDRAM = 8 ns.

### ROM

Memory that can have its information read but that does not allow new information to be written into it is called read-only memory (ROM). Memory that can be read and written is called RAM.

In your computer system ROM holds the program that the CPU runs when it is first turned on. This program is the **BIOS** (Basic Input-Output System), and it includes a number of functions. One function is to tell the processor how to interact with the rest of the system, such as getting information from the keyboard and giving information to the display. Another function of the BIOS is the **POST** (Power-On Self-Test), which is a program that tells the microprocessor to examine itself and the rest of the system to verify that it is in good operating order. The POST reports errors through beeps on the speaker and error messages on the display. A third function is the bootstrap loader, which accesses the floppy or hard disk to load DOS first and then more programs for you to actually run and accomplish work with. These are called applications.

ROM errors do not happen often, but they do happen. The POST may give you a ROM error warning, but some types of ROM errors will not allow the CPU to run at all. A diagnostic program may be useful, if the error will let you load DOS and the diagnostic program.

One very useful tool for ROM errors is the diagnostic card or POST card. This adapter card fits in an I/O slot and, upon powering up, gives a number on an LED display that indicates the last operation the system accomplished before it failed. Using a manual showing the system's BIOS error codes, you may be able to determine if the ROM is causing the problem. For further signal checking, you can check for power and ground on the chip itself with a meter or logic probe, for activity on the address and data bus, and that the chip is selected by the signal on the chip select pin. ROM is factory programmed, so if you find a bad ROM, all you can do is replace it. Replacement ROMs can be purchased from various vendors (Appendix B).

## PROM

While we call them ROMs, the chips in the personal computer system are a special kind of ROM. A ROM is manufactured by the chip manufacturer with a program etched in the silicon. They must be ordered by the thousands to be cost efficient. Programmable ROM, or PROM, can be purchased with no program installed. It is like a matrix of fuses. PROM is written (blown like a fuse, programmed, burned) at the computer factory or by any tech with a ROM burner. Once a PROM is programmed, the fuses are blown and it cannot be changed.

## EPROM

Another kind of ROM is the erasable programmable read-only memory, or EPROM. EPROMs can be erased by ultraviolet light and rewritten, saving money if a BIOS change is needed and many EPROMs were already programmed and in stock. The EPROM has a glass window in the top to allow the ultraviolet light in. This must have a light-proof label placed over it after programming to protect the program.

## EEPROM

Even more convenient is the electrically erasable PROM (EEPROM), which can be erased by the programming device or burner. This is easier than putting the EPROM in an ultraviolet erasing device and then in an EPROM burner. The EEPROM does not need a window in the top. If you have a known good ROM of that same or a compatible BIOS version and a programmer or ROM burner you can put the good ROM in the programmer and read it, then put a new EPROM or EEPROM in the programmer and write it, thus creating a copy of the good one.

## Flash EPROM

Flash EEPROMs (FEEPROMs) allow erasing to be done in blocks or on the entire chip at once. Access time can be 120 ns. These are considered read/write nonvolatile RAM; 10,000 program/erase cycles are possible. These are replacing ROMs because of their ability to be reprogrammed while in the motherboard. There is an inherent danger that in doing so, if the system crashes, the system ROM (Flash EEPROM) will have a scrambled BIOS. On the other hand, this allows downloading of BIOS upgrades from the Internet. As with other ROMs, the designation is in bits, organized by bytes, so a 2-MB chip is 256K by 8-bit bytes. Flash ROM can be a little faster than ROM, with a read access time of 120 ns. The 256-K size can be written in ten seconds. Because of the updating capability, many motherboards are using these chips for their ROM BIOS.

### ROM Designations

Identifying exactly which ROMS are in your system can be done by matching the designation or number printed on the ROM chip itself or by calculation from the number of pins. Table 4.1 gives some designations of different manufacturers to help you identify what ROMs you have in your system. Thousands could be included; these are just a few.

### ROM Banks

The number of BIOS ROM sockets varies from one to eight, although most modern systems have two at the most. With only two ROMs on a system board, banks may not seem

**TABLE 4.1   EPROM and EEPROM designations**

| Capacity in Bits | Addressable Locations × Bits per Address | EPROM Designation | EEPROM Designation | Flash EPROM Designation |
|---|---|---|---|---|
| 64K | 8K × 8 | TMS2764, MSM2764, µPD27HC65, MBM27C64 | TMS28C64, KM2864, MBM28C64 | |
| 128K | 16K × 8 | TMS27C128, MSM27128, MBM27C128 | | |
| 256K | 32K × 8 | TMS27C256, MSM27256 | µPD28C256 | |
| 512K | 64K × 8 | MSM27512 | | |
| 1 MB | 128K × 8 | µPD27C1001 | µPD23C1000 | 29EE0201, |
| 2 MB | 256K × 8 | µPD27C2001 | | MT28F002CIVG |

critical. However, if there are two ROM BIOS chips, they must be inserted in the correct sockets. The typical designations of the two banks are A and B, high and low, or even and odd. If you are replacing ROMs, note which chip came from which socket.

## ROM Pinouts

The most common ROMs are the 4732, 4764, 47128, and 47256 (Figures 4.1–4.4). As with all chips, orientation is important. Pin 1 is denoted by the circle, or left top with the notch up. On each chip note the power (Vcc) and ground (GND or Vss) pins. Vcc is +5 V, and ground should be 0 V. Let's examine the 4732 as an example of the rest. It is in a 24-pin package. The 32 at the end of its designation means that it is a 32-Kbit chip. Since a byte is 8 bits, dividing 32 Kbits by 8 gives 4K. Four kilobytes is 4096 bytes, or $2^{12}$ bytes. This means we need twelve lines that can go high or low, so we must have twelve address lines for these locations. Those twelve lines are labeled A0–A11.

Since a byte is 8 bits, we must have eight lines for data to go in and out of the chip. These are labeled D0–D7. CS1 and CS2 are chip select lines, for the CPU to choose this particular chip out of two or four chips.

The 4764 has all the same lines except for one. The last two digits of the number show us that it is a 64-Kbit chip, and to fit in a clone, it must be 8 bits per byte; 64 Kbits divided

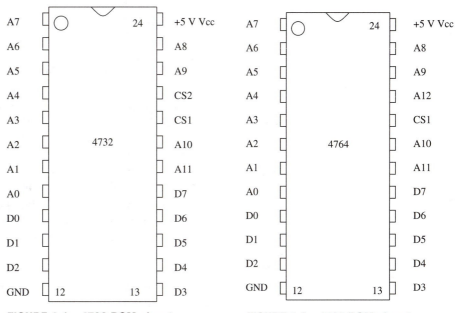

FIGURE 4.1   4732 ROM pinout          FIGURE 4.2   4764 ROM pinout

117

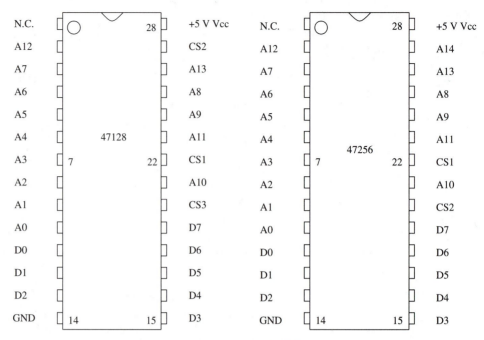

| | | | |
|---|---|---|---|
| N.C. | 28 | +5 V Vcc | |
| A12 | | CS2 | |
| A7 | | A13 | |
| A6 | | A8 | |
| A5 | | A9 | |
| A4 | 47128 | A11 | |
| A3 | 7    22 | CS1 | |
| A2 | | A10 | |
| A1 | | CS3 | |
| A0 | | D7 | |
| D0 | | D6 | |
| D1 | | D5 | |
| D2 | | D4 | |
| GND | 14    15 | D3 | |

**FIGURE 4.3    47128 ROM pinout**

**FIGURE 4.4    47256 ROM pinout**

by 8 bits per byte gives 8K of memory. That is 8192 locations, or $2^{13}$, so we need thirteen address lines. Note that pin 21 has been changed from the chip select it was on the 4732 to another address line, A12. A0 through A12 give the thirteen lines we need to address 8K.

To address more memory, we need more lines to the chip than the 24-pin packages can offer, so the 47128 and 47256 are in 28-pin packages. The 47128 is a 128-Kbit memory arranged as 16K of eight bits each. 16K is $2^{14}$, and we see it has those labeled A0–A13.

The 47256 is a 256-Kbit memory arranged as 32K of 8 bits each. 32K is $2^{15}$, and we see those labeled as A0–A14 on the 47256.

To see what symptoms a totally failed ROM would give a system, remove the ROM and examine the buses with a digital logic probe or oscilloscope. With no ROM installed, all bits of the data bus are high, with no activity at all, showing that the CPU is not being given any instructions and is not addressing any device. Similar symptoms may indicate a failed ROM.

## Addressing ROM

Each CPU has a certain amount of total system memory that it can address depending on the width of its address bus. ROM and RAM share this address space.

The 8088 used in the PC/XT has a 20-bit-wide address bus; $2^{20}$ is 1 MB of total memory address space, or 00000H to FFFFFH (0–1 MB). On the PC/XT, ROM is placed at addresses C0000H to FFFFFH, which is 768K to 1 MB, or the last 256K of the total system memory area. Certain locations in that area are reserved for certain purposes. The ROM BIOS or motherboard ROM uses only part of that space, as shown by the ROM memory map in Table 4.2. Figure 4.5 shows a photo of a ROM BIOS chip.

AT systems use the 286 and newer processors. The 80286 and 80386SX have a 24-bit-wide address bus, and $2^{24}$ is 16 MB for total memory address space. (The second part of Table 4.2 shows the 80286 ROM memory map.) Since hexadecimal is base 16, one digit can replace four binary digits, so for the four extra bits, one more FH is needed. The total memory in hex is 000000H to FFFFFFH (0–16 MB). But for compatibility, the ROM remained at the end of the first 1 MB. This is true in all subsequent x86 processors, as well as in the Pentium processor and the Intel P6.

The 80386DX has a 32-bit address bus, which is $2^{32}$ lines. It can address 4 GB of total memory address space. This stays the same for all processors through the P6. At $40 a megabyte, that is $160,000 of memory. It does not appear to be necessary to increase the address bus for the foreseeable future.

TABLE 4.2    ROM memory maps

| ROM Memory Map for 8088 System | |
|---|---|
| C0000–C7FFF | System expansion |
| C8000–CBFFF | Hard disk control |
| CC000–EFFFF | System expansion |
| F0000–F5FFF | Open |
| F6000–FDFFF | ROM cassette basic (IBM only) |
| FE000–FFFFF | ROM BIOS |
| ROM Memory Map for 80286 System | |
| C0000–DFFFF | Adapter card ROM |
| E0000–EFFFF | Optional ROM reflected at FE0000 |
| F0000–FFFFF | ROM BIOS reflected at FF0000 to FFFFFF |

## DRAM

Memory that can be read (information comes out) and written (information is put in) can be accessed both ways, in and out. It is therefore called random access memory, or RAM. There are four different types of RAM in your computer system: DRAM, SRAM, VRAM, and CMOS.

The DRAM (Dynamic Random Access Memory) forgets its information quickly. It is basically a small RC (Resistor-Capacitor) circuit, which slowly loses its charge and needs to be recharged (refreshed) every so often. This is done by reading the memory and writing back the same information. While this is an inconvenience, DRAM does not consume much power, so the hassle of refreshing is accepted.

Dynamic RAM has had to evolve to keep pace, and some new types of DRAM are found on modern motherboards of higher-end processors.

### Fast Page Mode Dynamic RAM

Fast Page Mode Dynamic RAM (FPMDRAM) was the standard for main memory in the 1980s and early 1990s. In the main memory matrix, a row is activated and then read column by

FIGURE 4.5    ROM BIOS chip

column. While switching columns, the output is turned off. The processor must wait until the next column is addressed and accessed. It the fast page mode it assumes you are going to address the next column so it accesses and prepares for that automatically to read more quickly. The operation of this standard RAM is not synchronized to the system bus, so it is asynchronous memory. The system may have to wait for some clock cycles while the RAM addresses and transfers the data. This obviously slows down the CPU and the system. Typical access speed is 60 or 70 ns. You can see why around 20 MHz system clock speed, they gave up on the main memory keeping up with the CPU. This is why they added the SRAM cache.

## EDRAM

Enhanced Dynamic Random Access Memory (EDRAM) uses a small cache of SRAM in the DRAM to speed operation. If the information the CPU is seeking is in the EDRAM's internal cache (a hit), it may be retrieved in 15 ns. If not (a miss), it comes from the DRAM portion.

## CDRAM

Cached Dynamic RAM (CDRAM) also uses a small SRAM cache in the DRAM to speed up overall memory operation. Similar to EDRAM, CDRAM is supposed to be about 20 percent more efficient in its caching technique, speeding up average access time.

## EDODRAM

Extended Data Output RAM (EDORAM) holds the data on the output longer, giving the CPU more time to access that data. This allows faster access than FPMRAM and comes in speeds of 50 or 60 ns.

## Burst EDORAM

Burst EDORAM (BEDORAM), in addition to holding the data on the output longer, addresses 1 byte and the next 3 are read in one clock cycle each, in a burst. This makes a 4–byte burst, for 32 bits at a time, making it faster than nonburst EDORAM.

## Pipelined RAM

Pipelined memory means the functions are set in stages, and each stage has its own input and output buffer. The stages work independently or in parallel, allowing maximum throughput. The output of one stage is the input of the other, like a number of pipes joined to make a "pipeline." Again, this is to increase the average access speed of the memory.

## SDRAM

Synchronous Dynamic Random Access Memory (SDRAM) is designed to address the need for memory to keep up with the high speed CPUs. SDRAM synchronizes all operations with the processor clock signal to have the accuracy required for nanosecond operation. This eliminates the wait states imposed on the CPU and system by asynchronous memory; therefore, it allows higher system speeds. An internal clock frequency of 400 MHz in a CPU gives it a period of 2.5 ns, or 2.5 thousandths of a millionth of a second. Because the range of frequencies an electronic circuit can perform properly in is the bandwidth, you may hear high-speed memory referred to as having a higher bandwidth. It also has a burst counter to increment the column counter and take 4 bytes sequentially. This gives it a faster transfer speed between the microprocessor and the memory. Most are Low-Voltage TTL (LVTTL) 3.3–V compatible. Commonly available in 10 ns for 100 MHz operation, 8 ns is required for PC100 compliance.

## Buffered DIMMS

The inputs of electronic circuits have a given impedance, so when a signal is applied, the voltage creates a current through the impedance. The output of a circuit is designed to be able to deliver a certain current to a "normal" load. Realizing the outputs may be used to drive more than one other circuit, engineers designed the outputs to be strong enough to drive a number of normal loads. The number the circuit is designed to drive is called "fanout." If a circuit is loaded beyond its fanout, it will overheat and fail. If you need to attach more loads than the fanout allows, you must put between the output and the load another circuit that does not change the signal but can deliver more current, increasing the effective fanout. This circuit is called a buffer. It protects the circuits before and after it from each other.

As memory size has increased to perhaps 384 MB on a motherboard, the load on the system circuitry accessing that memory has increased considerably. Unbuffered DIMMs may load down a system to the point it cannot deliver the proper signal voltage. This is overcome with buffered DIMMs. They have the buffering circuitry to protect the motherboard from overload. However, some motherboards are designed to handle the load and do not need the buffering. Check your system documentation for its requirements for DIMM.

## Gold and Tin Contacts

As odd as it may seem, gold is not always the best contact material. Some companies coat the contacts of SIMM and DIMM connectors with gold, but some coat them with tin. At least two major companies warn against mixing tin-coated DIMMs in gold contact sockets, and vice versa. Apparently tin transfers to the gold surface and oxidizes, interfering with the electrical connection. Gold is more resistant to oxidation, but tin is cheaper. The wisest path is to use tin-plated memory contacts in tin contact sockets and gold in gold. Of course, note the documentation that comes with the motherboard.

## PC100 Memory

With higher CPU and memory speeds, it became necessary to improve the memory bus speed from 66 MHz to 100 MHz. Memory that meets the timing specifications is known as PC100 compliant. You must use PC100-compliant memory in a 100-MHz bus PC. These modules have six-layer printed circuit boards (PCBs) with a power and ground plane. With the FM radio band starting at 88 MHz, you can see why strict specifications are needed in trace length, capacitive loading, component placement, and trace impedance, and why shielding of these radio frequencies is required. Do not install PC66 memory in a system requiring PC100-compliant memory. It must be SDRAM-8 (nanoseconds) or PC-100 and should be unbuffered.

The standards are formalized by the JEDEC (Joint Electron Device Engineering Council), the semiconductor engineering standardization committee of the Electronic Industries Alliance (EIA). This trade association represents all areas of the electronics industry. JEDEC standards are available at www.jedec.org. The files are in the Portable Document Format (.PDF) read by Adobe Acrobat, which is available at www.adobe.com.

## Parity

Parity is an error checking technique used to verify data in memory. In some cases, parity can be chosen as either even or odd. Even parity checking adds up the number of bits that have the value 1 in a byte; if that number is even, add a parity bit of 0 so the number of all high bits is still even. This requires one additional bit per 8-bit byte, therefore, one more chip. If the number of high bits is odd, parity circuitry will add a 1 in the parity bit, forcing the total of high bits to be even. This data is then sent to its destination, which checks the parity. If the total of high bits is even, the system assumes the data is correct. If the total is not even, it detects an error, and usually automatically requests a repeat of that data. You can see that while this has some value, if two bits go bad at the same time, one high, one low, the total is still even, and the faulty byte would go by undetected. Unfortunately, often a parity error will cause a system crash. Repeated parity errors over time may indicate a failing memory module. SIMMs with parity have an extra chip for the parity bit, which of course costs a little more. You may use parity SIMMs in nonparity systems because they just ignore the extra chip. You may use nonparity SIMMs in parity systems if you are allowed to set the system for nonparity. Otherwise this will produce memory errors. For each 8 bits, parity must add one parity bit, so an 8-bit bus would need 9 bits on the memory. A 16-bit bus needs 18 bits. A 32-bit bus needs 36 bits of memory, and a 64-bit bus needs 72 bits for data and parity.

## ECC

Error Checking and Correcting (ECC) checks data for errors and automatically corrects it. Single-bit errors can be caught and corrected. Dual errors, often undetected by parity, are caught. This is done automatically and is a great improvement in reliability. Regretfully, ECC requires 5 bits per 8 bits of data, a very high overhead indeed. But it only requires 6 bits for 16 of data, 7 bits for 32 of data, 8 bits for 64 of data. Note since $8 \times 8 = 64$, 64 bits of data would require 8 bits for parity, or 72 bits total. So the overhead of ECC is no worse than parity at 64 bits of data. Therefore, at 64 bits, parity SIMMs may be used for ECC on some motherboards if so designed. When buses go to 128 bits, ECC will require 9 bits, parity a whopping 16, making ECC the clear choice.

## DRAM Designations

The numbers with which manufacturers sometime label their chips are the chips' designations. The designations of some companies make sense, such as 4164 for a 1-bit-wide 64K chip and 4464 for a 4-bit-wide 64K chip. However, some designations are extremely cryptic, giving no clue to the depth or width of the chip. The width of the particular chip is usually 1, 4, 8, or 16 bits, but how do you know which? Table 4.3 gives DRAM designations and widths to help you identify chips in your system.

Each chip is socketed for easy replacement. Access times for DRAM can be as fast as 60 ns. Be sure all replacement chips are of the same speed or access time.

**TABLE 4.3    DRAM designations**

| DIP Chips (capacity in bits) | Addressable Locations × Bits per Address | Access Time (ns) | Designations | |
|---|---|---|---|---|
| 64K | 64K × 1 | 200 | 4164-20 | |
| 256K | 64K × 4 | 150 | 4464-15 | |
| 256K | 64K × 4 | 100 | 71C464-10 | |
| 256K | 256K × 1 | 120 | 41256-120 | |
| 1 MB | 256K × 4 | 80 | 414256 | 71C4256-80 |
| 1 MB | 1 MB × 1 | 60 | P21010-06 | |

*Note:* SIMMs follow. These are the SMT chips on each simm. The number before the multiplication sign is the number of pieces on each SIMM PCB.

| SMT Chips (capacity in bits) | Addressable Locations × Bits per Address | Access Time (ns) | Designations |
|---|---|---|---|
| 256K × 9 | 256K × 4 | 80 | 2 × AAA1M2045-08H* |
| | 256K × 1 | 80 | + 1 × AAA2801J-08 |
| 256K × 9 | 256K × 4 | 80 | 2 × KM44C256CJ-8 |
| | 256K × 1 | 80 | + 1 × KM41C256J-8 |
| 256K × 9 | 256K × 4 | 70 | 2 × M514256B-70J |
| | 256K × 1 | 70 | + 1 × V53C256 |
| 256K × 9 | 256K × 4 | 100 | 2 × TMS44C256-10N |
| | 256K × 1 | 100 | + 1 × MCM6256BP10 |
| 1 MB × 9 | 1 MB × 4 | 70 | 2 × KM44C1000BJ |
| | 1 MB × 1 | 70 | + 1 × KM41C1000CJ |
| 1 MB × 9 | 1 MB × 4 | 70 | 2 × MT4C4001JDJ |
| | 1 MB × 1 | 70 | + 1 × MT4C1024DJ |
| 1 MB × 9 | 1 MB × 4 | 70 | 2 × HY514400J |
| | 1 MB × 1 | 70 | + 1 × HY531000J |

*Total SIMM

### DRAM Banks

RAM is organized in banks. A bank is a group of bits as wide as the CPU's data bus, which must be filled for the CPU to operate.

### 8088 Banks

The 8088 had an 8-bit data bus, so a bank for DRAM main memory was 8 bits wide, plus one for parity. The 8088 motherboard usually had four banks (rows) of 9 bits (columns, as each DRAM held 1 bit). You have to fill memory by banks, otherwise the system may seem totally dead.

### 286 and 386SX Banks

The 286 and the 386SX have a 16-bit data bus, and they need a 16-bit memory plus one parity bit for each 8 bits, for a total of two parity bits. So, instead of banks 1–4 (or 0–3), their motherboards have banks 0 and 1, each with eighteen 1-bit chips or four 4-bit chips with two 1-bit chips for parity. You have to fill one bank for the CPU to function. If the motherboard has four 30-pin 8-bit SIMM sockets, you must fill two sockets to get 16 bits.

### 386DX, 486SX, and 486DX Banks

The 386DX and all 486s have 32-bit data buses. Adding one parity bit for each 8-bit byte gives a total of 36 bits needed. Since the system will not run without at least one full bank of RAM, a 32-bit CPU with eight 9-bit SIMM sockets or four 36-bit SIMM sockets must have one of the DRAM configurations in Table 4.4. If using 8-bit (9, with parity) SIMMs, 4

**TABLE 4.4   DRAM configurations**

| 32-Bit Boards (80386DX and 80486) | | | | | |
|---|---|---|---|---|---|
| **8 Total 30-Pin SIMM Sockets** 30-Pin 9-Bit SIMMs (1 bit for parity) | | | **2 Total 72-Pin SIMM Sockets** 72-Pin 36-Bit SIMMs (4 bits for parity) | | |
| **Bank 0** 4 SIMMs Capacity Each | **Bank 1** 4 SIMMs Capacity Each | **Total** | **Bank 0** 1 SIMM Capacity Each | **Bank 1** 1 SIMM Capacity Each | **Total** |
| $256K \times 9$ | | 1 MB | $1\ MB \times 36$ | | 4 MB |
| $256K \times 9$ | | 2 MB | $2\ MB \times 36$ | | 8 MB |
| $1\ MB \times 9$ | | 4 MB | $1\ MB \times 36$ | $1\ MB \times 36$ | 8 MB |
| $256K \times 9$ | $1\ MB \times 9$ | 5 MB | $1\ MB \times 36$ | $2\ MB \times 36$ | 12 MB |
| $1\ MB \times 9$ | $1\ MB \times 9$ | 8 MB | $4\ MB \times 36$ | | 16 MB |
| $4\ MB \times 9$ | | 16 MB | $2\ MB \times 36$ | $2\ MB \times 36$ | 16 MB |
| $1\ MB \times 9$ | $4\ MB \times 9$ | 20 MB | $1\ MB \times 36$ | $4\ MB \times 36$ | 20 MB |
| $4\ MB \times 9$ | $4\ MB \times 9$ | 32 MB | $2\ MB \times 36$ | $4\ MB \times 36$ | 24 MB |
| | | | $8\ MB \times 36$ | | 32 MB |
| | | | $4\ MB \times 36$ | $4\ MB \times 36$ | 32 MB |
| | | | $2\ MB \times 36$ | $8\ MB \times 36$ | 40 MB |
| | | | $4\ MB \times 36$ | $8\ MB \times 38$ | 48 MB |
| | | | $8\ MB \times 36$ | $8\ MB \times 36$ | 64 MB |

*Note:* Calculating the total for 32-bit simms is as follows:
8 MB × 32 per chip × 1 chip per bank = 8 MB × 32 bits.
For 8-bit bytes, 32 ÷ 8 = 4, so 8 MB × 32 bits × 4 8-bit bytes ÷ 32 bits = 32 MB × 8 bits, or 32 MB.
One parity bit for each 8 bits makes all the above 32 bits into 36 bits when parity is included.

*(continued on next page)*

**TABLE 4.4** *(continued)*

| 64-Bit Boards (80586 and 80686) 4 Total 72-Pin SIMM Sockets 72-Pin 36-Bit SIMMs (4 bits for parity) | | |
|---|---|---|
| Bank 0 2 SIMMs Capacity Each | Bank 1 2 SIMMs Capacity Each | Total |
| 1 MB × 36 | | 8 MB |
| 2 MB × 36 | | 16 MB |
| 1 MB × 36 | 1 MB × 36 | 16 MB |
| 1 MB × 36 | 2 MB × 36 | 24 MB |
| 4 MB × 36 | | 32 MB |
| 2 MB × 36 | 2 MB × 36 | 32 MB |
| 1 MB × 36 | 4 MB × 36 | 40 MB |
| 2 MB × 36 | 4 MB × 36 | 48 MB |
| 8 MB × 36 | | 64 MB |
| 4 MB × 36 | 4 MB × 36 | 64 MB |
| 2 MB × 36 | 8 MB × 36 | 80 MB |
| 4 MB × 36 | 8 MB × 38 | 96 MB |
| 8 MB × 36 | 8 MB × 36 | 128 MB |

*Note:* Calculating the total for 32-bit simms is as follows:
8 MB × 32 per chip × 2 chips per bank = 16 MB × 32 bits.
For 8-bit bytes, 32 ÷ 8 = 4 so 16 MB × 32 bits × 4 bytes = 64 MB × 8 bits.
There are 2 banks of 2 chips in the example given above, so 2 banks × 64 MB per bank
= 128 MB of 8-bit bytes.

| 64-bit Boards (89585 and 80686) 3 Total 168-pin DIMM sockets 168-Pin 72-Bit DIMMs (8 for ECC or parity) | | | |
|---|---|---|---|
| Bank 0 1 DIMM | Bank 1 1 DIMM | Bank 2 1 DIMM | Total |
| 128 MB | 0 | 0 | 128 MB |
| 128 MB | 128 MB | 0 | 256 MB |
| 128 MB | 128 MB | 128 MB | 384 MB |

are needed for any operation, because $4 \times 8 = 32$ bits. Two 8-bit SIMMs in the system will give the appearance of a totally dead motherboard. If so, socketed, one 72-pin 36-bit SIMM will do.

## 586 and 686 Banks

The 586 and 686 microprocessors and the Intel P6 (Pentium Pro) have 64-bit data buses. CPUs are classified by their data bus width, not their address bus width. These CPUs must therefore have a 64-bit-wide memory. This would require two of the 32-bit (36, with parity) 72-pin SIMMs in a bank or one 168-pin 64-bit (72 with ECC) DIMM. Both SIMMs in a bank must be of the same capacity, 256K, 1 MB, 4 MB, and so on.

For DIPs, 30-pin 9-bit SIMMs, 72-pin 36-bit SIMMs, and 168-pin 64-bit DIMMs, Figure 4.6 shows bank layout and possible configurations.For these DIMMs to have parity or ECC, they need to be 72-bit DIMMs.

By studying Table 4.4 and Figure 4.6, you should get a better understanding of how to calculate the amount of RAM installed.

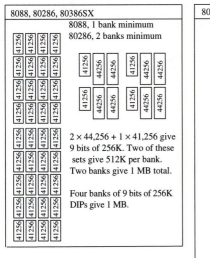

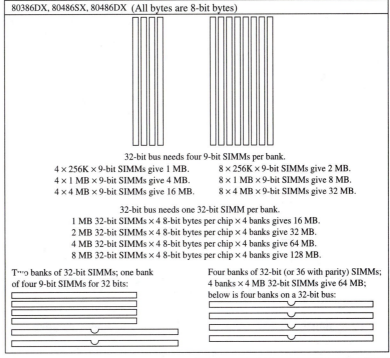

80586, 80686
64-bit bus needs two 32-bit SIMMs per bank; above are two bank (each of two SIMMs) on a 64-bit bus.
64-bit bus needs one 64-bit 168-pin DIMM per bank.

**FIGURE 4.6   DRAM bank layouts**

## DRAM PINOUTS

### DIPs

The Dual Inline Package (DIP) has been the mainstay of main memory to the mid-1990s. Figures 4.7–4.10 show some of the more common DIP DRAM pinouts: 4116, 4164, and 41256. Pin 1 on DIPs is identified by a small circular indent, or left top with notch up.

Looking at Figure 4.7, we notice the 4116 DRAM has a power and ground pin just like ROMs. But, although a 4732 ROM is designated by bits, RAM is designated differently. For a 4116, the "16" means bytes, and the chip is 16K × 1 bit wide. We can see this chip is only 1 bit wide by the fact that it has one data line out and one data line in. The ROM did not need an incoming line for data, but the RAM does because it can be written, and we see a WE (Write Enable) line on pin 3.

Nine chips are needed for an 8088 bank 9 bits wide. The addresses are from A0–A6 or 7 bits wide. Now $2^7$ is 128, which is nowhere near the 16,384 bits in 16K. But note the RAS (Row Address Strobe) and CAS (Column Address Strobe) lines. The memory is in a row × column matrix 7 bits deep and 7 bits wide; $7 + 7 = 14$, $2^{14}$ is 16K, and the chip is 16K × 1 bit wide. The other DRAM pinouts follow the same pattern. In Figure 4.11, note the combination 18-pin and 20-pin socket that allows either chip to be used. Below is a 16-pin DIP, and at the bottom is a 20-pin DIP.

### SIMMs and SIPs

For many years all RAM was in DIPs, but now **SIPs** (Single-In line Packages) and **SIMMs** (Single Inline Memory Modules) also are prevalent. The last two typically consist of nine or more surface mount chips in a single row mounted on a little PC board that either clips in (SIMM) or plugs in (SIP).

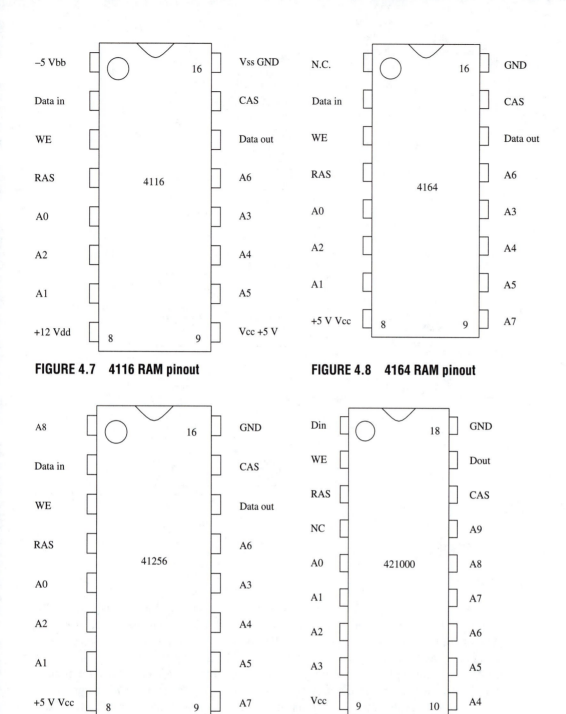

**FIGURE 4.7    4116 RAM pinout**

**FIGURE 4.8    4164 RAM pinout**

**FIGURE 4.9    41256 RAM pinout**

**FIGURE 4.10    421000 1-MB
RAM pinout**

Because of poor documentation on the motherboard, matching pin 1 of the SIP memory with pin 1 on the motherboard can be frustrating. SIMM can only plug in one way. Most SIMM sockets have plastic spring clips on the sides. First, gently bend these left and right away from the SIMM, then rotate the SIMM forward (toward the component side) to remove.

**FIGURE 4.11** 18-pin/20-pin RAM combination socket and 16-pin socket and 20-pin DIP

18-pin/20-pin RAM combination socket

16-pin socket

20-pin DIP

Figure 4.12 shows SIP pinouts and Figures 4.13 and 4.14 show SIMM pinouts.

Examining Figure 4.13, the 4 MB 9-bit SIMM, we see address lines from A0–A10. This gives eleven lines for RAS and eleven for CAS, a total of twenty-two lines. Twenty-two bits gives us 4 MB of possible addressable memory locations.

Studying Figure 4.14, the 256K 36-bit 72-pin pinout, we see A0–A8 for 9 bits of address, doubled with RAS and CAS signals, for $2^{18}$ or 256K of memory. Lines D0–31 give 32 bits of data. The MP0–MP3 lines are for parity of each of four banks, like the RAS and CAS 0–3. The four parity bits add to the thirty-two data bits to give 36-bit designation.

Figure 4.15 is a closeup of a 30-pin SIMM, and at the bottom, a 72-pin SIMM.

**FIGURE 4.12   30-pin 9-bit SIP pinout**

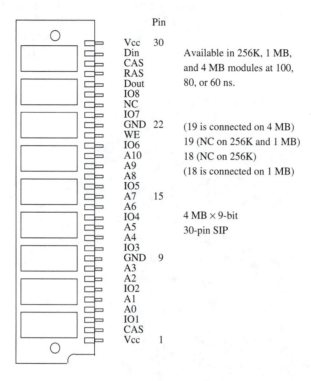

| | Pin | |
|---|---|---|
| Vcc | 30 | |
| Din | | Available in 256K, 1 MB, |
| CAS | | and 4 MB modules at 100, |
| RAS | | 80, or 60 ns. |
| Dout | | |
| IO8 | | |
| NC | | |
| IO7 | | |
| GND | 22 | (19 is connected on 4 MB) |
| WE | | 19 (NC on 256K and 1 MB) |
| IO6 | | 18 (NC on 256K) |
| A10 | | (18 is connected on 1 MB) |
| A9 | | |
| A8 | | |
| IO5 | | |
| A7 | 15 | |
| A6 | | |
| IO4 | | 4 MB × 9-bit |
| A5 | | 30-pin SIP |
| A4 | | |
| IO3 | | |
| GND | 9 | |
| A3 | | |
| A2 | | |
| IO2 | | |
| A1 | | |
| A0 | | |
| IO1 | | |
| CAS | | |
| Vcc | 1 | |

**FIGURE 4.13   30-pin 9-bit SIMM pinout**

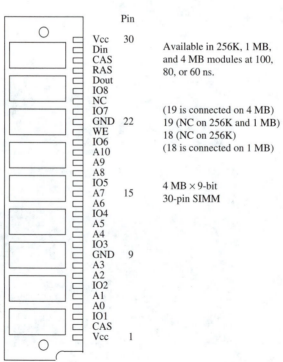

| | Pin | |
|---|---|---|
| Vcc | 30 | |
| Din | | Available in 256K, 1 MB, |
| CAS | | and 4 MB modules at 100, |
| RAS | | 80, or 60 ns. |
| Dout | | |
| IO8 | | |
| NC | | |
| IO7 | | (19 is connected on 4 MB) |
| GND | 22 | 19 (NC on 256K and 1 MB) |
| WE | | 18 (NC on 256K) |
| IO6 | | (18 is connected on 1 MB) |
| A10 | | |
| A9 | | |
| A8 | | |
| IO5 | | 4 MB × 9-bit |
| A7 | 15 | 30-pin SIMM |
| A6 | | |
| IO4 | | |
| A5 | | |
| A4 | | |
| IO3 | | |
| GND | 9 | |
| A3 | | |
| A2 | | |
| IO2 | | |
| A1 | | |
| A0 | | |
| IO1 | | |
| CAS | | |
| Vcc | 1 | |

**FIGURE 4.14   72-pin 36-bit 256K SIMM pinout**

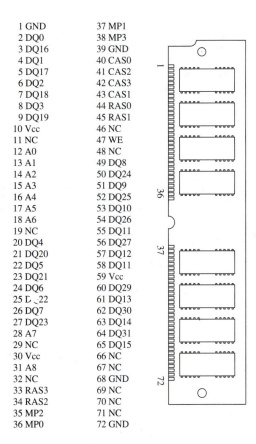

| | |
|---|---|
| 1 GND | 37 MP1 |
| 2 DQ0 | 38 MP3 |
| 3 DQ16 | 39 GND |
| 4 DQ1 | 40 CAS0 |
| 5 DQ17 | 41 CAS2 |
| 6 DQ2 | 42 CAS3 |
| 7 DQ18 | 43 CAS1 |
| 8 DQ3 | 44 RAS0 |
| 9 DQ19 | 45 RAS1 |
| 10 Vcc | 46 NC |
| 11 NC | 47 WE |
| 12 A0 | 48 NC |
| 13 A1 | 49 DQ8 |
| 14 A2 | 50 DQ24 |
| 15 A3 | 51 DQ9 |
| 16 A4 | 52 DQ25 |
| 17 A5 | 53 DQ10 |
| 18 A6 | 54 DQ26 |
| 19 NC | 55 DQ11 |
| 20 DQ4 | 56 DQ27 |
| 21 DQ20 | 57 DQ12 |
| 22 DQ5 | 58 DQ11 |
| 23 DQ21 | 59 Vcc |
| 24 DQ6 | 60 DQ29 |
| 25 Dꞏ22 | 61 DQ13 |
| 26 DQ7 | 62 DQ30 |
| 27 DQ23 | 63 DQ14 |
| 28 A7 | 64 DQ31 |
| 29 NC | 65 DQ15 |
| 30 Vcc | 66 NC |
| 31 A8 | 67 NC |
| 32 NC | 68 GND |
| 33 RAS3 | 69 NC |
| 34 RAS2 | 70 NC |
| 35 MP2 | 71 NC |
| 36 MP0 | 72 GND |

## DIMMs

With the advent of 64-bit CPUs, the need for 64-bit-wide memory arose. This need was met with the 168-pin DIMM. The module is 5.25 inches wide by 1.25 inches high. Two key notches control polarity, with pins 1 to 10, then a notch, pins 11 to 40, then a second notch, and pins 41 to 84 (Figure 4.16). The DIMM sits vertical in its socket and is pushed down while the side retainer clips are pulled up and in. Press down firmly. While the SIMM has seventy-two contacts per side, they are connected front to back, and therefore only seventy-two contacts. The DIMM has eighty-four contacts on the front, and eighty-four separate contacts on the back, giving the total of 168 contacts. While you may wonder why there are so many if only 64 bits are needed for data, don't forget the need for an address bus. In the case of the 586 and 686, this is 32 bits wide, giving 96 pins for those two buses alone.

The data bus consists of contacts D0 to D63 for the 64-bit width and the address bus of A0 to A12 (pins 33–39 and 117–123) for a 14-bit address bus. Remember that the 14-bit address is read on RAS(115) and on CAS(111), effectively giving a 28-bit address bus for a possible 256-MB main memory. BA0 and BA1(122,39) are used for bank address also. The clocks are CK0 to CK3(42,125,79,163) (though CK1 is unused). DQMB0 to DQMB7(28,29,46,47,112,113,130,131) are data mask pins. The numerous VDD pins are for 3.3 V DC power, and VSS pins are for ground. Motherboards are showing up with a variety of RAM sockets. One combination is four 72-pin SIMM sockets along with two 168-pin DIMM sockets. Some motherboards have only two, three, or four DIMM sockets. With 128 MB DIMMs, the three socket board maxes out at $3 \times 128$ or 384 MB of main memory. Just a few years ago, the question was whether you needed 8 or 16 MB of RAM. Now 32 MB is not considered large at all.

## SODIMM

Small Outline DIMM (SODIMM) is smaller and designed to connect to 32-bit data buses. The smaller size is suitable for use in notebook computers.

**FIGURE 4.15    30-pin SIMM and
72-pin SIMM**

30-pin SIMM

72-pin SIMM

## SRAM

Static RAM, based on flip-flops that do not need refreshing, is faster than DRAM, but it consumes more power because it holds data until the power is removed. It is available from 8 ns to 20 ns, while DRAM is hovering at 60 ns.

SRAM was used in some of the original PCs, but it went out of favor as size of memory increased and power consumption became prohibitive. So DRAM took over. With its increased circuitry for refreshing, it was still better because of its decreased power supply size and cost.

Now the need for higher speed is bringing back SRAM in the form of caches. SRAM designers have been pushing the access time down to keep pace with the newer processors. The typical SRAM has transferred data asynchronously. A new pipelined synchronous burst static RAM (SBSRAM) can approach 8 ns cycle times and is showing up in the caches on higher-end motherboards. A typical board may have a 256K cache of SBSRAM, interfacing the processor with a main memory of 16 MB of EDORAM.

The last digits of SRAM designation sometimes show total bits, not bytes, so you must know the organization. How deep and how wide is it? Table 4.5 gives some typical sizes and designations to help you decipher the manufacturer's numbering and find the size of your SRAM cache. If you cannot find the cache size by designation, the section on "Cache RAM," later in this chapter, tells how to calculate it. On 386DXs and higher, the cache is 32 bits wide without parity, or 36 with parity. Caches use DIP SRAM. Table 4.6 shows SRAM configurations for 32-bit boards.

**Figure 4.16 (a) A 168-pin 64 bit DIMM; (b) the DIMM installed in its socket**

Figure 4.17 shows some possible SRAM bank configurations, and Figure 4.18 shows an SRAM cache.

The number of sockets per bank and the number of banks must be combined with the capacity of the chips inserted to determine the amount of cache SRAM. Follow documentation that came with the board. Usually one or two 22-pin or 28-pin chips are present for the ninth parity bit.

1. For example, a 486 system may have two banks of four 28-pin sockets holding four $32K \times 8$-bit chips in one four-socket bank and four empty sockets in bank two. That is $4 \times 32K = 128K$ bytes of 8 bits. It may also have two extra 22-pin sockets. If each of those holds a $16K \times 4$-bit chip, that is $2 \times 16K \times 4 = 128K$ of 1 bit for parity.

**TABLE 4.5 SRAM designations**

| Capacity in Bits | Addressable Locations × Bits per Address | Designations | | | | |
|---|---|---|---|---|---|---|
| 16K | $16K \times 1$ | LH5167 | IDT6167 | MB81C67 | | |
| 64K | $8K \times 8$ | LH5164 | 7C185 | IDT7164 | MT5C6408 | MSM5165 | MB81C78 |
| 64K | $16K \times 4$ | LH5262 | P4C188 | 6288 | CY74S189 | IDT7188 | MT5C6405 |
| 256K | $32K \times 8$ | LH52250 | HM62832 | IDT71256 | µPD43256 | MSM51256 |
| 256K | $32K \times 8$ | TC55328P | OPTI66C256 | | | |
| 256K | $64K \times 4$ | LH52252 | P4C187 | CY7M194 | MT5C2564 | µPD43254 | MB81C86 |
| 256K | $128K \times 2$ | TE11256A | | | | |
| 256K | $256K \times 1$ | LH52251 | P4C1257 | IDT71257 | MT5C2561 | µPD43251 |
| 512K | $64K \times 8$ | MB85410 | MCM67B618FN | | | |
| 1 MB | $128K \times 8$ | 7C1024 | CY7C1009 | µPD431000 | IDT7MP4008 | |
| 1 MB | $256K \times 4$ | P4C422 | CY93422 | MT5C1005 | µPD431004 | |
| 2 MB | $256K \times 8$ | LH51256 | MB85420 | | | |

**TABLE 4.6   SRAM configurations for 32-bit boards**

| Two Banks | Address Locations × Bits per Address | Total |
|---|---|---|
| 1 chip in each bank | 32K × 8 | 64K |
| 2 chips in each bank | 32K × 8 | 128K |
| 4 chips in each bank | 8K × 8 | 64K |
| 4 chips in one bank | 32K × 8 | 128K |
| 4 chips in each bank | 32K × 8 | 256K |
| 8 chips in each bank | 32K × 8 | 512K |

**FIGURE 4.17   SRAM banks**

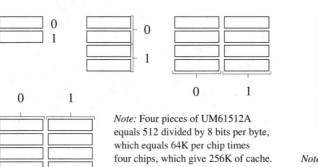

*Note:* Four pieces of UM61512A equals 512 divided by 8 bits per byte, which equals 64K per chip times four chips, which give 256K of cache.

*Note:* The above is in a 486DX. It represents eight pieces of W24257AK-15 and 1 MB RAM of same. Each is 256 bits divided by 8 bits per byte, which give 32K per chip. Four chips give 256K of cache.

**FIGURE 4.18   SRAM cache on motherboard**

2. Another example: a 386DX has eight pieces of 64K × 1-bit DIP SRAM and one more piece of 64K × 1 bit. This is 64K × 8 pieces for 512 total bits, ignoring the parity bit. Dividing by 8 bits per byte gives 64K of cache RAM.

3. In another case, a 386DX has four pieces of 256K × 1-bit memory. 4 × 256K is 1024K or 1 megabit. Dividing by 8 bits per byte gives a 128K cache.

## SRAM Pinouts

Figure 4.19 gives typical SRAM pinouts. On the 64K × 1-bit SRAM, note the address lines from 0–15, giving sixteen address lines, which is $2^{16}$ or 65,536 or 64K. However, the Din and Dout show this is only 1 bit wide. The 16K × 4-bit has A0–A13 for fourteen address lines for $2^{14}$ or 16,384 or 16K. It also has I/O lines 1–4, which shows that it is 4 bits wide or can store 4 bits of information at each of the 16K locations.

Similarly, the 32K × 8-bit has one more address line, A14, for 32K instead of 16K. Also it has four more I/O lines, I/O 5–8, making it 8 bits wide at each location. Finally, the 256K × 4-bit chip has eighteen address lines, A0–A17, giving it 262,144 or 256K locations each 4 bits wide.

Note the difference from DRAM, which has half the address lines but RAS and CAS lines to double them. Also note that while there are two 22-pin chips and two 28-pin chips shown, they are *not* pin compatible and will not work if installed in a system designed for a different chip, even though they physically fit.

## CACHE RAM

Cache RAM is not a type of RAM but a *use* of a type of RAM. As system CPU speed has increased over the years, the cost of filling the expanding RAM memory with 4, 8, or more megabytes of fast DRAM is getting more and more expensive. The faster the DRAM, the

**FIGURE 4.19  Typical SRAM pinouts**

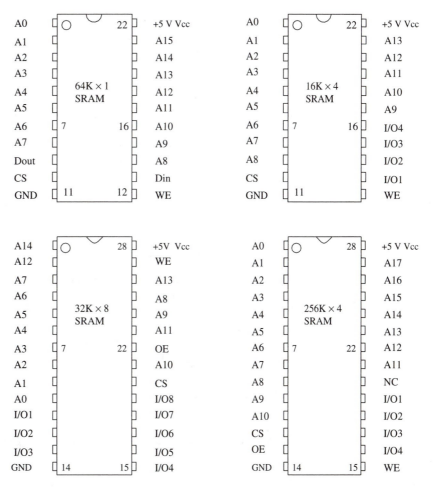

*Note:* I/O is also called DQ. CS is also called EN. WE is also called W. OE is also called G.

more it costs. To stem this expanding expense, the inherently faster SRAM is used as a small holding area for instructions (code) and data used by the microprocessor. If the cache controller is efficient, the fast cache SRAM is the only memory the CPU will see 90 percent of the time. The controlling circuits shuffle information between the cache and the slower DRAM of main memory, saving the processor this time-consuming task.

## Level 1 (L1) Cache

An internal cache is built into CPUs from the 486SX up. The 80486 family all have an 8K internal cache. The Pentium processor has two 8K caches, and most clone 80586s have the same. The Intel P6 (Pentium Pro) has the two 8K caches and a 256K cache inside the processor. On the Pentium Pro the 256K cache inside the processor is a level 2 cache, which is usually external.

## Level 2 (L2) Cache

Newer motherboards also provide an external cache, with sockets to be filled or expanded by the user. The external cache is not for use by the user, but the CPU, so no discussion of addressing it is necessary. The cache is made of SRAM, so be sure to see the section on SRAM. L2 cache is memory that is usually not in the CPU but on the motherboard. This is SRAM that is much faster than DRAM.

The Pentium Pro was the first to actually have the L2 cache in another cavity of the chip package for faster access. This made the motherboard cache on Pentium Pro systems an L3 cache. Intel Pentium II processors come with 512 KB of pipelined burst L2 cache on the SEC cartridge. I have not seen a motherboard L3 cache for this processor.

## TAG RAM

Part of the cache is used to identify which data from main memory is currently stored in the cache. This is called the TAG RAM, and it is the values stored in this area of the cache that determines if a lookup in the cache will be a hit or a miss. A hit means that the information is stored in the cache by the cache controller. A miss means the information is not stored and must be looked up in slower main DRAM memory. The higher the hit ratio, the fewer times the CPU has to look at the slower main memory and the faster the system throughput. The other area of the cache is the area to store data. This is where the data is actually stored. If increasing the cache on your motherboard, you must check your documentation to see if more cache requires more TAG RAM. Some motherboards do not socket their TAG RAM, limiting the cache you can have.

In a "write-through cache," the CPU writes the cache and the main memory with the same information, as though the information just passes through the cache. In a "write-back cache," the CPU writes to the cache only, increasing its processing speed. The cache controller then writes the information to the main memory. This increases the throughput of information and, therefore, enhances the CPU's overall performance.

## Calculating Cache Size

Often you need to know what size external cache the installed chips give, yet it is sometimes difficult to find the manufacturer's data sheet for those particular chips. Luckily, most BIOS programs tell the size of the SRAM cache upon booting. Try it and watch for the size on the system you are working on. If that fails, you can calculate the size from the number of chips and the number of pins each has. Referring to Figure 4.18, convince yourself of the following logic.

1. Four pins that all SRAM chips must have are
   a. power/VCC
   b. ground/VSS
   c. Write Enable/SE
   d. Chip Select/CE
2. The chip must have enough address and data lines for the capacity and width.

As you can see, some chips cannot be manufactured with fewer than a certain number of pins. The most common pin counts in use are 20, 22, and 28. Table 4.7 shows SRAM pinout possibilities.

**TABLE 4.7    SRAM pinout possibilities**

| Chip | Standard | Address | Data | Total | Package |
|------|----------|---------|------|-------|---------|
| 4K × 4 | 4 | 12 | 4 | 20 | 20 |
| 16K × 1 | 4 | 14 | 1 | 19 | 20 |
| 32K × 1 | 4 | 15 | 1 | 20 | 20 |
| 8K × 4 | 4 | 13 | 4 | 21 | 22 |
| 16K × 4 | 4 | 14 | 4 | 22 | 22 |
| 64K × 1 | 4 | 16 | 1 | 21 | 22 |
| 8K × 8 | 4 | 13 | 8 | 25 | 28 |
| 16K × 8 | 4 | 14 | 8 | 26 | 28 |
| 32K × 4 | 4 | 15 | 4 | 23 | 28 |
| 32K × 8 | 4 | 15 | 8 | 27 | 28 |
| 64K × 4 | 4 | 16 | 4 | 24 | 28 |
| 256K × 4 | 4 | 18 | 4 | 26 | 28 |
| 1 MB × 1 | 4 | 20 | 1 | 25 | 28 |

## VRAM

With the increasing demand for very high resolution graphics, video display adapters are carrying increasing amounts of memory. This onboard DRAM enhances speed of the graphics. For even faster, more efficient video, a new type of RAM, called VRAM (Video RAM), has been developed. With separate input and output buffers, this RAM can be written and read at the same time, speeding access. It is also called Dual-Ported RAM.

VRAM is available in 1-, 2-, and 4-megabit (Mbit) chips, with access times down to 20 ns, serial data access, and block write for faster data transfer from the system board. The 4-Mbit SMT (Surface Mount Technology) SSOPs (Super Small Outline Packages) consume half the power of four 1-Mbit chips, reduce board size, and increase reliability.

## WRAM

Window RAM is another video memory. It is dual-ported like VRAM, but gives faster video by addressing large windows (blocks) of video memory. WRAM is bound by the system bus speed, and a newer technology has taken over.

## SGRAM

The default on 100-MHz bus systems, Synchronous Graphic RAM (SGRAM) synchronizes its cycle with the system CPU bus clock, like SDRAM, up to 100 MHz. This is single ported (not dual) but performs block writes like WRAM. Together with synchronizing, this gives the fastest video available at this time.

## CMOS RAM

This complementary metal oxide semiconductor (CMOS) RAM uses very little power and is easily powered for long terms by batteries. It is used to hold the setup configuration data for your 286 or higher system. This information is changed by the BIOS SETUP program, usually entered by pressing the DELETE key on bootup. While the POST usually tests the first 1 MB of RAM, on some systems you may set the CMOS in "advanced setup" to test all RAM.

## SHADOW RAM

ROM holds the BIOS, POST, and the program to boot the drives. ROM has a much longer access time than RAM, slowing down the system. If, on bootup, the ROM is copied into RAM, the system now can access this fast RAM that is *shadowing* the slow ROM, considerably speeding up the overall system operation.

## RAM SPEED

Besides choosing the correct package for your motherboard, you must also choose the correct speed, which ranges from 200 ns to 53 ns for dynamic RAM. The memory must be fast enough for the motherboard. Figuring the speed of RAM required in a system is not as straightforward as it may at first seem.

Table 4.8 lists RAM speeds that some manufacturers are using in their systems. A safe decision is to never mix speeds in the same machine and never use slower than the system was delivered with. Adding faster memory does *not* speed up a system's ability to access that memory. Adding slower memory causes errors.

A 4.77 MHz original IBM machine could apparently use 200 ns memory. Inverting the clock speed of 4.77 MHz gives a period of 209.6 ns. Because the RAM cannot be accessed in one cycle (a 33-MHz 386 can do about 12 MIPS), 200 ns was fast enough. The question is, could you have used slower, cheaper memory? A 20-MHz 286 may have 32K of SRAM for a high-speed cache memory controller to use. In less expensive systems, "wait states" were used to make the CPU wait and do nothing, giving the slow memory time to catch up. These, of course, slowed down the overall performance of the system.

As you can see, at about 20 MHz, manufacturers gave up trying to fill the system with 4 or 8 MB of expensive, fast DRAM and went to caches of SRAM. The 80386SL, 80486DX, and 80486SX all have internal 8K instruction and data caches built into the CPU, and compatible controllers offer external caching with superfast SRAM.

**TABLE 4.8   RAM speeds**

| CPU Speed (MHz) | DRAM Speed (ns) | SRAM Speed | CPU Period (ns) |
|---|---|---|---|
| 4.77 | 150 | | 210.0 |
| 8.0 | 120 | | 125.0 |
| 10.0 | 100 | | 100.0 |
| 12.0 | 80 | | 83.0 |
| 16.0 | 80 | | 62.5 |
| 20.0 | 80 | 1 wait state | 50.0 |
| 20.0 | 53 | 0 wait state | 50.0 |
| 25.0 | 70 | | 50.0 |
| 33.0 | 70 | 25 ns—128K external and 8K internal cache | 30.3 |
| 40.0 | 70 | 20 ns—128K external and 8K internal cache | 25.0 |
| 50.0 | 70 | 20 ns—256K external and 8K internal cache | 20.0 |
| 66.0 | 60 | | 15.15 |
| 100.0 | 60 | | 10.0 |
| 120.0 | 60 | | 8.33 |
| 133.0 | 60 | | 7.52 |

## 4.2 GENERIC SYSTEM OPERATION

The POST video error code may be preceded by a four-digit code. If so, this may pin down the bank of memory the failed chip is in, or the chip itself. The memory of early systems was laid out in four banks, labeled 0 to 3, and nine rows—nine because of eight data bits (0–7) and one parity bit (labeled P). The decoding scheme shown in Table 4.9 may help. It is for the early 256K systems, which could be expanded to 640K with a memory expansion board in an expansion slot.

A POST video error code of 2008 201 on a 256K system means a failure in bank 2 and row 3 (which is the fifth chip over, starting at the parity chip).

You may ask who uses 256K anymore. The point of the above was to show a logical pattern to the error code and the location. On a system with 640K or more having a BIOS that shows an error code, you can remove one of the memory chips, note the error code, replace it and remove another, note the code, and learn to interpret the code. Of course you know to turn the system off before removing or replacing any chip and to watch for leads that are bent under.

If the error is in the first bank, the system may not operate at all. As an example of figuring out your system's code (after you have read the documentation to be sure it's not deciphered there), pull the first chip in bank 1 (leaving bank 0 alone). Turn on the system. If the error code has not changed, the failure is before this, in bank 0. If it *has* changed, write it down, power-down, replace the chip, and remove the first chip in the next bank (bank 2 if you have four banks). Turn on the system. Note the error code. It should be changed for a different bank.

Once you have the codes for the first chip in each bank, try the second and third in one bank only. This should give you a pattern. Change the hex code to binary, and see whether that gives a count or just ripples a 1 through a list of 0s, showing location by picture instead of count. Note all of this in your system's documentation (where you are keeping the CMOS SETUP and COM: interrupt settings and so on).

With newer systems using SIPs and SIMMs, there may be nine chips each on two or four removable boards, and trial and error is either as fast as or faster than deciphering bank and chip identification schemes.

### ADDRESSING RAM

The total system memory space or area is shared with ROM. Table 4.10 gives the total system memory map, showing areas used by each type of memory and typical uses. Addressing RAM is done in the same manner as ROM.

**TABLE 4.9   RAM error codes**

| Bank |
|---|
| 00 = Bank 0 |
| 10 = Bank 1 |
| 20 = Bank 2 |
| 30 = Bank 3 |

| Row |
|---|
| 00 = Parity |
| 01 = Row 0 |
| 02 = Row 1 |
| 04 = Row 2 |
| 08 = Row 3 |
| 10 = Row 4 |
| 20 = Row 5 |
| 40 = Row 6 |
| 80 = Row 7 |

*Note:* BBRR 201 means two digits for the bank and two digits for the row on the 256K system board.

**TABLE 4.10  RAM memory map (including ROM)**

| Address* | | | Usage | Usage | Usage |
|---|---|---|---|---|---|
| Hex | Dec | Type | ISA/8-Bit | ISA/16-Bit | EISA/32-Bit |
| 00000 | 0K | RAM | Conventional | Conventional | Conventional |
| 10000 | 64K | RAM | Conventional | Conventional | Conventional |
| 20000 | 128K | RAM | Conventional | Conventional | Conventional |
| 30000 | 192K | RAM | Conventional | Conventional | Conventional |
| 40000 | 256K | RAM | Conventional | Conventional | Conventional |
| 50000 | 320K | RAM | Conventional | Conventional | Conventional |
| 60000 | 384K | RAM | Conventional | Conventional | Conventional |
| 70000 | 448K | RAM | Conventional | Conventional | Conventional |
| 80000 | 512K | RAM | Conventional | Conventional | Conventional |
| 90000 | 576K | RAM | Conventional | Conventional | Conventional |
| A0000 | 640K | RAM | Video | VGA | VGA |
| B0000 | 704K | RAM | Mono 4K | VGA | VGA |
| –B1000 | 708K | | | VGA | VGA |
| B8000 | 736K | RAM | CGA 16K | VGA | VGA |
| –BC000 | 752K | | | VGA | VGA |
| C0000 | 768K | ROM | Expansion | Adapter ROM | Adapter ROM |
| –D0000 | 832K | | Expansion | EMS window | EMS window |
| –E0000 | 896K | | Expansion | BIOS 128K | |
| F0000 | 960K | ROM | Expansion | BIOS 128K | |
| FE000 | 1116K | ROM | BIOS 8K | BIOS 128K | |
| –FFFFF | 1024K | | BIOS 8K | BIOS 128K | |
| | 1 MB | RAM | n.a. | Extended RAM | Extended RAM |
| –FFFFFF | 16 MB | RAM | n.a. | Extended RAM | Extended RAM |
| 3FFFFFF | 64 MB | RAM | n.a. | n.a. | Extended RAM |

**Breakdown of Lowest 64K of RAM in an 8088 System**

| | | |
|---|---|---|
| 00000–0001F | BIOS interrupt vectors | 00–1F |
| 00080–000FF | DOS interrupt vectors | 20–3F |
| 00100–001FF | User interrupt vectors | 40–7F |
| 00200–003FF | BASIC interrupt vectors | 80–FF |
| 00400–004FF | BIOS data area | |
| 00500–005FF | BASIC and DOS data areas | |
| 00600–0FFFF | User RAM | |

*Addresses are in 64K blocks except where indented. An indented address is the end of the range starting in the line above it.

## Accessing Memory from BASIC

Both RAM and ROM can be read from the BASIC programming language, and RAM can be written to. First, review computer numbering systems in Chapter 1. Accessing memory locations from the BASIC programming language (which cannot handle addresses larger than 64K), requires you to use the SEGMENT, OFFSET technique.

BASIC automatically shifts the segment one hex position left (multiplying by 16). Then the offset is added to it, from 0–64K (0–65535). This gives the absolute address. Then you must step the segment up 64K (add 1000H to it). Since BASIC automatically shifts this left 1 bit, that is 10000H, which is 65536D, for 0–65535 or 64K locations. Therefore, to address the VGA adapter RAM, which starts at absolute address A0000H, you define the segment to A000 in basic with the command DEF SEG = &HA000.

Now, looping from 0–65535 with an address will access the next 64K. To access more, you must DEF SEG again, stepping &H1000. (For an example, see Chapter 6 on testing video RAM.)

## DESIGNATION OF BLOCKS OF RAM

Different address ranges or blocks of memory have been given different designations or names. Table 4.11 and Figure 4.20 indicate the names given to these blocks.

The expanded memory manager (EMM) program uses a 64K page frame to access the extended memory. The amount of extended memory accessed is no longer called extended memory addressed by the extended memory specification (XMS); it is called expanded memory and uses the expanded memory specification (EMS).

## USING NONCONVENTIONAL MEMORY IN 386 SYSTEMS

All the RAM you add to a system is not automatically used by all the programs you run. Following are some ways DOS 5.0 has included to access that memory.

On a system with no memory manager available in the CONFIG.SYS file, the MEM command gives the following:

```
655360 bytes total conventional memory
655360 bytes available to MS-DOS
```

**TABLE 4.11 Designation of blocks of RAM**

> Conventional Memory (0–640K) is ram that all programs can use.
>
> CMOS Memory (Complementary Metal Oxide Semiconductor) is RAM that requires very little power to retain data. It is powered by a lithium battery and holds the system configuration (number and type of drives, etc.).
>
> Expanded Memory (EMS) is memory above 640K that conforms to the Lotus/Intel/Microsoft (LIM) standard. Programs must be written to conform to this standard if they use this ram.
>
> Extended Memory is a block of RAM starting at 1 MB going to 16 MB on 286 systems and 4 GB (4080 MB) on a 386 or 486. In practice, the 386 or 486 motherboard will have eight sockets for SIMM, which, with 4 MB SIMMs, may hold up to 32 MB of RAM.
>
> Extended Memory Specification (XMS) defines HMA (below) and allows the extended memory to be used by programs for storage.
>
> High Memory Area (HMA) is a 64K block of RAM starting at 1 MB.
>
> Reserved Memory is a block of RAM from 640K to 1 MB.
>
> Upper Memory Block (UMB) is a block of RAM from 640K to 1 MB.

**FIGURE 4.20 RAM memory blocks**

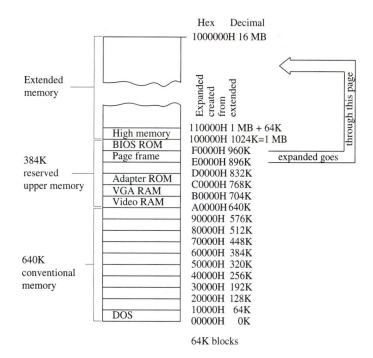

139

```
  573632 largest executable program size
1048576 bytes total contiguous extended memory
1048576 bytes available contiguous extended memory
```

From the first of the five lines, note 640K conventional (655,360 divided by 1024 converts bytes to kilobytes) and 1 MB extended memory. That 1 MB is not accessible by most programs.

## HIMEM.SYS

Setting HIMEM.SYS in your CONFIG.SYS file (and rebooting to activate it) prepares your system to use extended memory XMS including HMA (High Memory Area). This must come before other memory commands. HIMEM tests your system's extended memory when booting. It is better than the POST memory test, but not as good as some diagnostic programs test.

To set HIMEM.SYS, use the command

```
DEVICE = C:\DOS\HIMEM.SYS
```

Using the MEM command now gives the following:

```
 655360 bytes total conventional memory
 655360 bytes available to MS-DOS
 573632 largest executable program size
1048576 bytes total contiguous extended memory
1048576 bytes available contiguous extended memory
```

Note, there is no change.

The following two lines use the first 64K of extended memory to map the rest of it and load most of DOS up in HMA.

```
DEVICE = C:\DOS\HIMEM.SYS
DOS = HIGH
```

On a system with 2 MB installed, the MEM command gave the following display. (The third line depends on what is loaded.)

```
 655360 bytes total conventional memory
 655360 bytes available to MS-DOS
 622048 largest executable program size
1048576 bytes total contiguous extended memory
      0 bytes available contiguous extended memory
 983040 bytes available XMS memory
        MS-DOS resident in High Memory Area
```

Note we have increased from 573,632 in line 3 of the second example, to 622,048 in line 3 of the third example. This is about a 47K savings of conventional memory. Note in line 6, with the loss of 64K, most of the 1 MB of extended memory is now available. Line 5 is telling us we designated *all* extended memory as XMS memory; there is none left over.

## DEVICEHIGH, LOADHIGH, and INSTALL

LOADHIGH FILENAME will put DOS or terminate stay resident (TSR) programs or drivers above 640K in high memory, leaving more room for programs that run in the base or conventional memory range. DEVICEHIGH=FILENAME loads device drivers in UMA. INSTALL FILENAME loads a memory resident program on booting.

## RAMDRIVE.SYS

In DOS 5.0 the RAMDRIVE.SYS file (VDISK.SYS in DOS 3 and 4) allows you to designate your own area of RAM to use as an electronic disk drive. This is called a ramdrive, an edisk (electronic disk), a vdisk (virtual disk), or a ramdisk. Because it has no mechanical parts, it is orders of magnitude faster than a floppy or hard disk. Being DRAM, it loses all information when the system is powered-down.

The syntax is:

```
DEVICE = DRIVE:PATH RAMDRIVE.SYS BYTES SECTORSIZE NUMBEROFDIRECTORIES /E
```

The /E parameter places the ramdrive in extended memory. Of course, you must have activated the extended memory. The following two lines for the CONFIG.SYS file are typical. Don't forget to reboot to implement the new CONFIG.SYS file.

```
DEVICE = C:\DOS\HIMEM.SYS
DEVICE = C:\DOS\RAMDRIVE.SYS 960/E
```

The MEM command gives the following:

```
 655360 bytes total conventional memory
 655360 bytes available to MS-DOS
 572432 largest executable program size
1048576 bytes total contiguous extended memory
      0 bytes available contiguous extended memory
      0 bytes available XMS memory
        64K High Memory Area available
```

Note from lines 4 to 7 that 1 MB minus 64K leaves our 960K for the ramdrive. To save 47K of conventional memory as above, we add the DOS = HIGH line.

```
DEVICE = C:\DOS\HIMEM.SYS
DOS = HIGH
DEVICE=C:\DOS\RAMDRIVE.SYS 960/E
```

The MEM command gives the following:

```
 655360 bytes total conventional memory
 655360 bytes available to MS-DOS
 615344 largest executable program size
1048576 bytes total contiguous extended memory
      0 bytes available contiguous extended memory
      0 bytes available XMS memory
        MS-DOS resident in High Memory Area
```

## The MEM Command

The MEM command has three switches: /D for debug, /C for classify, and /P for program. The example of /C given below uses exactly the same CONFIG.SYS as above. The /C switch shows the quantity of RAM consumed by each program. Note that 14K of DOS is still in conventional memory, even though we loaded most of it in extended memory.

Issuing the command MEM/C gives the following screen display:

```
Conventional Memory:
Name            Size in Decimal      Size in Hex
-------         ---------------      -----------
MSDOS           14144    (13.8K)         3740
ANSI             4192    (4.1K)          1060
HIMEM            1184    (1.2K)           4A0
RAMDRIVE         1184    (1.2K)           4A0
COMMAND          2624    (2.6K)           A40
MSCMOUSE        12112    (11.8K)         2F50
DOSKEY           4128    (4.0K)          1020
FREE               64    (0.1K)            40
FREE               96    (0.1K)            60
FREE           615344    (600.9K)       963B0
Total FREE:    615504    (601.1K)
Total bytes available to programs : 615504   (601.1K)
Largest executable program size :   615344   (600.9K)
1048576 bytes total contiguous extended memory
      0 bytes available contiguous extended memory
      0 bytes available XMS memory
        MS-DOS resident in High Memory Area
```

An example using the /P switch follows. Piping through MORE.COM is necessary because there is more than one screen of information.

Issuing the command MEM/P ¦ MORE gives the following screen display:

```
Address     Name        Size        Type
-------     --------    ------      -------
000000                  000400      Interrupt Vector
000400                  000100      ROM Communication Area
000500                  000200      DOS Communication Area
000700      IO          000A60      System Data
001160      MSDOS       0013D0      System Data
002530      IO          002BF0      System Data
            ANSI        001060      DEVICE=
            HIMEM       0004A0      DEVICE=
            RAMDRIVE    0004A0      DEVICE=
                        0005D0      FILES=
                        000100      FCBS=
                        000200      BUFFERS=
                        0001C0      LASTDRIVE=
                        000740      STACKS=
005130      MSDOS       000040      System Program
005180      COMMAND     000940      Program
005AD0      MSDOS       000040      -- Free --
005B20      COMMAND     000100      Environment
005C30      MSCMOUSE    000070      Environment
005CB0      MSCMOUSE    002EE0      Program
008BA0      MEM         000050      Environment
008C00      MSDOS       000000      -- Free --
008C10      DOSKEY      001020      Program
009C40      MEM         0176F0      Program
021340      MSDOS       07ECB0      -- Free --
  655360 bytes total conventional memory
  655360 bytes available to MS-DOS
  615344 largest executable program size
 1048576 bytes total contiguous extended memory
       0 bytes available contiguous extended memory
       0 bytes available XMS memory
         MS-DOS resident in High Memory Area
```

## The EMM386.EXE Program

The EMM386.EXE program of DOS 5.0 (XMAEM.SYS and XMA2EMS.SYS in DOS 4.0) enables the use of upper (UMB) and extended (XMS) memory as expanded memory (EMS). The following two lines show the usage and are added to the CONFIG.SYS file. The default is 256K. Note that HIMEM.SYS must come before EMM386.

```
DEVICE = C:\DOS\HIMEM.SYS
DEVICE = C:\DOS\EMM386.EXE
```

The MEM command gives the following:

```
  655360 bytes total conventional memory
  655360 bytes available to MS-DOS
  565216 largest executable program size
  655360 bytes total EMS memory
  262144 bytes free EMS memory
 1048576 bytes total contiguous extended memory
       0 bytes available contiguous extended memory
  622592 bytes available XMS memory
         64K High Memory Area available
```

Line 5 shows the EMS 256K to be free, out of the 640K total in line 4. To increase the amount of EMS memory, just add the size desired as shown below.

```
DEVICE = C:\DOS\HIMEM.SYS
DOS = HIGH
DEVICE = C:\DOS\EMM386.EXE 864
```

The MEM command gives the following:

```
 655360 bytes total conventional memory
 655360 bytes available to MS-DOS
 608128 largest executable program size
1277952 bytes total EMS memory
 884736 bytes free EMS memory
1048576 bytes total contiguous extended memory
      0 bytes available contiguous extended memory
      0 bytes available XMS memory
        MS-DOS resident in High Memory Area
```

Line 5 shows that we now have 864K (884,736/1,024 = 864K) of EMS memory. Lines 1, 4, and 9 (64K) add up to 1952K, just 48K less than the full 2 MB. (Just in passing, WordPerfect uses EMS, so adding these lines to your CONFIG.SYS file will minimize hard drive access and therefore speed your program.

## MEMMAKER.EXE

DOS 6.x has a program called MEMMAKER.EXE that automatically configures your memory using the above command. It can probably do a much better job than you can.

## READING AND WRITING THE CMOS MEMORY

(**Note:** This section is recommended for technicians.)

The CMOS memory is 64 bytes labeled 0–63. This is on a chip such as the 146818 real-time clock and CMOS RAM. It is accessed through I/O addresses 70 and 71 hex. Address 70 is loaded with the byte to be read from or written to 71. Table 4.12 gives the CMOS memory map. Program Listing 4.1 shows how the information in this table can be used.

**TABLE 4.12   CMOS memory map**

| Byte (Hex) | Data Stored |
|---|---|
| 0 | Seconds |
| 2 | Minutes |
| 4 | Hours |
| 7 | Day |
| 8 | Month |
| 9 | Year, last two decimal digits (xx92) |
| E | Diagnostic status byte |
| F | Shutdown status byte |
| 10 | First hex digit drive A: type, second digit is B: type |
| 12 | First hex digit drive C: type, if 14 or below; second digit is D: type, if 14 or below |
| 14 | Equipment configuration byte |
| 15 | Low base memory byte |
| 16 | High base memory byte |
| 17 | Low expansion memory byte |
| 18 | High expansion memory byte |

TABLE 4.12 *(continued)*

| Byte (Hex) | Data Stored |
|---|---|
| 19 | Drive C: type, if above 15 |
| 1A | Drive D: type, if above 15 |
| 2E | CMOS CHECKSUM high byte |
| 2F | CMOS CHECKSUM low byte |
| 30 | Low expansion memory byte |
| 31 | High expansion memory byte |
| 32 | Century, year first two decimal digits (19xx) |
| 33 | Information flags set at power-up |

**PROGRAM LISTING 4.1**
**CMOS Setup**

```
10    REM SETUP.BAS sets up the CMOS memory
20    DIM DT$(54): REM to hold five types of floppy drives
30    DIM A(64): REM to hold bytes 0-63 of CMOS memory
40    DT$(0) = "Not installed": DT$(1) = "360K 4.25 inch": DT$(2)
      = "1.2 M 5.25 inch": DT$(3) = "720K 3.5 inch": DT$(4) =
      "1.44 M 3.5 inch": DT$(5) = "2.88 M 3.5 inch"
50    KEY OFF
60    WHILE CHO <> 3
70    CLS
80    PRINT"1. Display current CMOS settings"
90    PRINT"2. Change or Setup CMOS settings"
100   PRINT"3. Quit this program"
110   INPUT "Choice ",CHO
120   ON CHO GOSUB 480,160
130   WEND
140   CLS:PRINT "If changes were made, you must reboot to implement
      them."
150   END
160   REM change CMOS subroutine
170   RES$=""
180   WHILE RES$<> "Y" AND RES$<>"y"
190   CLS
200   INPUT "Day xx ",DY$
210   INPUT "Month xx ",MO$
220   INPUT "Year xxxx ",YEAR$
230   CN$ = LEFT$(YEAR$,2)
240   YR$ = RIGHT$(YEAR$,2)
250   INPUT "Hour ",HR$
260   INPUT "Minute ",MN$
270   INPUT "Second ",SC$
280   GOSUB 950 :REM show floppy types
290   INPUT "Type of floppy drive A: (0-4) ",FDA$
300   INPUT "Type of floppy drive B: (0-4) ",FDB$
310   PRINT "Date = ";MO$+"/"+DY$+"/"+CN$+YR$
320   PRINT "Time = ";HR$+":"+MN$+":"+SC$
330   GOSUB 950 :REM show floppy drive types
340   PRINT "Floppy A: is type ";FDA$
350   PRINT "Floppy B: is type ";FDB$
360   INPUT "Is the above correct y/n ";RES$:RES$=LEFT$(RES$,1)
370   WEND
380   FD$=FDA$ + FDB$
390   OUT &H70,7 :OUT &H71,ASC(CHR$(VAL("&H" + DY$)))
400   OUT &H70,8 :OUT &H71,ASC(CHR$(VAL("&H" + MO$)))
410   OUT &H70,9  :OUT &H71,ASC(CHR$(VAL("&H" + YR$)))
```

```
420    OUT &H70,50 :OUT &H71,ASC(CHR$(VAL("&H" + CN$)))
430    OUT &H70,0  :OUT &H71,ASC(CHR$(VAL("&H" + SC$)))
440    OUT &H70,2  :OUT &H71,ASC(CHR$(VAL("&H" + MN$)))
450    OUT &H70,4  :OUT &H71,ASC(CHR$(VAL("&H" + HR$)))
460    OUT &H70,16 :OUT &H71,ASC(CHR$(VAL("&H" + FD$)))
470    RETURN
480    REM CMOSLOOK.BAS shows CMOS contents and system configuration
490    FOR B = 0 TO 63: REM 64 bytes of CMOS
500    OUT &H70, B: REM I/O address 112D or 70H
510    A(B) = INP(&H71): REM I/O address 113D or 71H
520    NEXT B
530    REM to see raw data, remove "REM" from next line
540    REM FOR B = 0 TO 63: PRINT HEX$(A(B)); " "; : NEXT B:
       INPUT "Press Return to continue ", X
550    CLS : PRINT "Data in CMOS memory gives this configuration ":
       PRINT
560    REM
570    REM examine date, byte 50=first 2 digits year, 9=last 2 dig
       year
580    REM byte 8=month, 7=day
590    FDY$ = HEX$(A(50))
600    SDY$ = HEX$(A(9)): IF LEN(SDY$) < 2 THEN SDY$ = "0" + SDY$
610    MO$ = HEX$(A(8)): IF LEN(MO$) < 2 THEN MO$ = "0" + MO$
620    DA$ = HEX$(A(7)): IF LEN(DA$) < 2 THEN DA$ = "0" + DA$
630    PRINT "Date = "; MO$; "/"; DA$; "/"; FDY$; SDY$
640    REM
650    REM examine time, byte 4=hr,2=min,0=sec
660    PRINT "Time = "; HEX$(A(4)); ":"; HEX$(A(2)); ":"; HEX$(A(0))
670    REM
680    REM examine base memory, bytes 22=high, 21=low
690    REM high first then low, change hex to dec
700    PRINT USING "Base memory = ##### K "; (256 * A(22)) +
       A(21)
710    REM
720    REM examine extended memory, bytes 24=high, 23=low
730    PRINT USING "Extended memory = ##### K "; (256 * A(24)) +
       A(23)
740    REM
750    REM examine floppies, byte 16, first digit=A, 2nd=B
760    DA =INT((VAL(HEX$(A(16))) / 10))
770    DB = VAL(HEX$(A(16))) - (DA * 10)
780    PRINT "Floppy drive A: = "; DT$(DA)
790    PRINT "Floppy drive B: = "; DT$(DB)
800    REM
810    REM examine hard drives, byte 18 first digit = C if under 16
820    REM byte 18 second digit is D if under 16. If 16 or over,
830    REM byte 25 = C, and byte 26 = D
840    DC = A(25): DD = A(26)
850    DC = INT(A(18) / 16)
860    DD = A(18) - (DC * 16)
870    IF DC = 15 AND A(25) > 0 THEN DC = A(25)
880    IF DD = 15 AND A(26) > 0 THEN DD = A(26)
890    PRINT "Hard drive C: is ";
900    IF DC = 0 THEN PRINT "Not installed" ELSE PRINT "Type "; DC
910    PRINT "Hard drive D: is ";
920    IF DD = 0 THEN PRINT "Not installed" ELSE PRINT "Type "; DD
930    INPUT "Press Enter to Continue ",RES
940    RETURN
950    REM floppy type subroutine
```

```
960    PRINT "Floppy drive types"
970    PRINT " 0 = not installed "
980    PRINT " 1 = 360K 5.25 inch"
990    PRINT " 2 = 1.2M 5.25 inch"
1000   PRINT " 3 = 720K 3.5 inch"
1010   PRINT " 4 = 1.4M 3.5 inch"
1020   RETURN
```

## WORKING WITH MEMORY IN WINDOWS 98

The first test of the system memory is still done by the BIOS, and the CMOS SETUP program often allows you to select if you want to test all memory on booting.

In Windows 98, there are a number of ways to see how much system memory is installed. The simplest may be by clicking on the following sequence: double-click on My Computer, then click on Help, About Windows 98. This will show physical memory available to Windows 98.

It is best to let Windows 98 manage its own virtual memory settings. However, to set the amount of virtual memory allocated click on Start, Settings, Control Panel, System, Performance (here you may again see the amount of total memory installed), and Virtual.

To see I/O memory and driver address space, click on Start, Settings, Control Panel, System, Device Manager, Computer, Properties, and I/O or Memory.

In Windows 98, start troubleshooting memory problems by clicking on the following sequence: Help, Contents, Troubleshooting, Windows 98 Troubleshooters, Memory. This will open the window shown in Figure 4.21 asking three questions, and leading you to more questions when you click on "Next." Your answers cause Windows to lead you through a flowchart narrowing down the possible causes of the problem and describing how to solve them. It is really quite handy. Of course this help is limited in scope, because

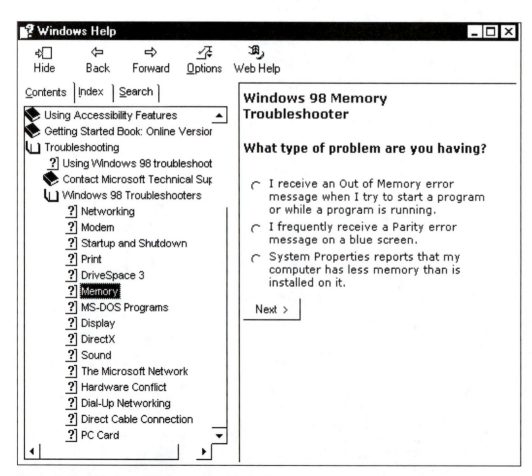

**Figure 4.21    Windows 98 Memory Troubleshooter**

146

if the RAM is not installed correctly, the system is not going to run at all. If one SIMM or DIMM is failing in any bank other than 0, then the system may boot and just have less memory than you think you installed.

## 4.3 ASSEMBLING MEMORY

Install your RAM in the system board before you install your system board in the case. Check the direction (polarity) of the DIPS or SIPS, and search for any motherboard jumpers concerning RAM. Check the system board documentation to determine proper settings. SIMM is much simpler to install than the old DIPS, so many of the headaches, such as bent or sprung leads, are eliminated. Some system boards allow DIPS or SIMMs, or a combination. Shunt jumpers are sometimes required to be set so the system knows which is installed. Some motherboards have both SIMM and DIMM sockets, and the latest boards have only DIMM sockets. This is because of the market takeover by 64-bit processors.

To install SIMMs, note Figure 4.22a, b, and c. First the SIMM is inserted in the socket at a slight angle to vertical, as in Figure 4.22a. They are polarized and difficult to get backward. The SIMM is then leaned forward as shown in Figure 4.22b. The side retention springs snap around the SIMM, holding it securely.

Because of the tilt, it is important to install the SIMMs in order. The direction you push in is pointing to the socket to fill first. Figure 4.22c shows the proper order. Another order makes the leaning SIMM hit the one before it.

Installing DIMMs is even easier than SIMMs, but DIMMS require firm seating. They slide into their socket staight up and down, no leaning. They are polarized by connector notches (keys) and go in one way only. The retention clips hold a DIMM up slightly out of the socket, as shown in Figure 4.23.

It is important to press the DIMM down while pulling the retention clips up and in. A fully seated, installed DIMM was shown in Figure 4.16B. Note the retention clips are snapped in the DIMM's side notches.

Figure 4.17 shows typical choices on filling your banks of DIP RAM. At least one whole bank must be filled. With 80286 and 80386SX CPUs, the data bus is 16 bits wide, so you must use a minimum of 2 of the same capacity 8-bit SIMMs (9 bits, with parity). With 80386DX, 80486SX, and 80486DX CPUs, the data bus is 32 bits wide, so you must use a minimum of 4 of the same capacity 8-bit SIMMs or one 32-bit SIMM. With 80586 and 80686 CPUs, the data bus width is 64 bits, so you must use a minimum of 8 of the same capacity 8-bit SIMMs or two of the same capacity 32-bit SIMMs. 64-bit DIMMs need only one module installed in a 64-bit system such as the 686s and Pentium IIs.

**Figure 4.22 (a) Installing the SIMM at a tilt**

**Figure 4.22** *(continued)*
**(b) leaning or snapping the
SIMM forward; (c) installing
SIMMs in order**

Run through the following steps to check your newly installed memory.

1. Confirm that the RAM is fast enough for your system board. Remember, 70 ns is usually fast enough for all boards, because they use a cache of static RAM with 20 ns chips. Although the SETUP program usually automatically adjusts for memory size, you should check the CMOS setup to confirm proper settings for RAM.
2. The POST will test the first 1 MB of RAM memory. On some systems, the advanced CMOS setup will allow you to choose to test all of the main RAM.
3. You may test the full memory from DOS with versions 5.0 and up, if you have a bootable floppy with a CONFIG.SYS file containing the line:

```
DEVICE = C:\DOS\HIMEM.SYS
```

The HIMEM tests the extended memory on booting. SIMM testers are available to test your SIMM before you install it or to test suspicious or intermittent SIMM.

**Figure 4.23  A DIMM inserted but not seated**

## 4.4  BASIC TROUBLESHOOTING

As you can see, there is no POST audio code to consider in troubleshooting memory, but, as discussed earlier, you must consider any POST video error code. Then you can proceed to note any visual clues and take appropriate measurements. Details of these steps are as follows.

a. POST audio error code: none
b. POST video error CODE of 2xx. (See Table 4.7 and discussion.)
c. Measurements
   1. Check that RAM speeds are sufficient for the system. Users play with memory, so check polarity in socket, and look for pins bent under or sprung out. Keep only same speed in bank.
   2. Check jumpers per documentation. If no documentation is available, carefully note jumper positions in writing. This will be your future documentation. Using a binary count, change memory jumpers to all possible configurations, powering and booting each time.
   3. If system has numerous banks, remove one chip, note error code, repeat, decipher code, find failing chip from code.
   4. If system has only a few banks of SIP or SIMM, replace one at a time, powering-down, swapping, and powering-up. Watch pin 1 with SIP! Some system boards have no silk screen designating pin 1. Look at the bottom of the board for a square copper foil pad on one end of the row of pads for the chip. That will be pin 1.
   5. Intermittent memory failures can be extremely frustrating. Try to isolate the problem by determining whether the system fails after a certain time (indicating heat failure), always fails with the same program (corrupted but seems to run), or always fails after the same *sequence* of programs (memory conflicts). There are SIMM testers that test the memory without putting it in the computer.
   6. Diagnostic programs allow you to run memory tests repeatedly and log any failures to a file. This lets you "burn in" memory or run for extended periods to catch an intermittent problem. If you have no diagnostic program, put the HIMEM.SYS command in your CONFIG.SYS file as described in section 4.2, and reboot to test extended memory.

## 4.5  CHAPTER REVIEW/QUIZ QUESTIONS

1. Three possible random access memory configurations are
   a. RAM, ROM, and CMOS.
   b. Slow, medium, and fast.
   c. DIP, SIP, and SIMM.
   d. Static, dynamic, and ROM.

2. Touching the pins or connections on memory with your fingers may cause
    a. Oxidation or tarnishing and intermittent failures.
    b. You to receive a shock.
    c. ROM failure.
    d. Power supply failure.

3. RAM should be installed immediately
    a. After the motherboard is safely in place in the case.
    b. Before the motherboard is in the case.
    c. Before powering the system up.
    d. After powering the system up.

4. Intermittent RAM failures may be most easily checked by
    a. A diagnostic program.
    b. A digital logic probe.
    c. A multimeter.
    d. Swapping chips.

5. The RAM can be
    a. Read and written.
    b. Read but not written.
    c. Written but not read.
    d. Neither read nor written.

6. The ROM can be
    a. Read and written.
    b. Read but not written.
    c. Written but not read.
    d. Neither read nor written.

7. The memory from 0 to 640K (or 0–A0000) is used for
    a. BIOS ROM.
    b. Video RAM.
    c. Conventional RAM.
    d. Extended RAM.

8. On a 286 or above, the last 128K of memory, from 896K to 1 MB (or E0000–FFFFF), is used for

    a. BIOS ROM.
    b. Video RAM.
    c. Conventional RAM.
    d. Extended RAM.

9. The memory above 1 MB (or 100,000H) is used for
    a. BIOS ROM.
    b. Video RAM.
    c. Conventional RAM.
    d. Extended RAM.

10. SRAM
    a. Needs to be refreshed periodically but consumes less power.
    b. Is fast but consumes more power.
    c. Maintains information when the computer is turned off.
    d. Is the Short Read Address Memory.

11. DRAM
    a. Needs to be refreshed periodically but consumes less power.
    b. Is fast but consumes more power.
    c. Is Darlington Rom And Memory.
    d. Is the RAM supported by a battery.

12. CMOS
    a. Needs to be refreshed periodically but consumes less power.
    b. Is fast but consumes more power.
    c. Loses information when the computer is turned off.
    d. Is the RAM supported by a battery.

13. Shadow RAM
    a. Holds the BIOS program copied from ROM for speed.
    b. Holds the video information.
    c. Maintains information when the computer is turned off.
    d. Is the RAM supported by a battery.

14. The cache is
    a. An area on the hard drive to hold information from memory.
    b. SRAM holding data copied from DRAM for speed.
    c. Part of the ROM.
    d. Saved on the floppy drive.

15. The RAM speed must
    a. Match the CPU speed exactly.
    b. Be slower than the CPU speed.
    c. Be equal to or faster than the CPU speed.
    d. Match the ROM speed.

16. Two types of error codes the POST attempts to give are
    a. RAM and ROM.
    b. Exact error code and approximate error code.
    c. CPU and SYSTEM error codes.
    d. Audio error codes and video error codes.

17. The physical memory type that is most convenient to install is
    a. DIP.
    b. SIP.
    c. SIMM.
    d. ROM.

18. Once you have done the CMOS SETUP
    a. Memorize it.
    b. Write it down or print it out.
    c. Change it often to keep in practice.
    d. Set all future systems up exactly the same way.

19. How many 30-pin 8-bit SIMMs are required as the minimum installed in a 286?
    a. 1
    b. 2
    c. 4
    d. 8

20. How many 30-pin 8-bit SIMMs are required as the minimum installed in a 386?
    a. 1
    b. 2
    c. 4
    d. 8

21. How many 30-pin 8-bit SIMMs are required as the minimum installed in a 486?
    a. 1
    b. 2
    c. 4
    d. 8

22. How many 30-pin 8-bit SIMMs are required as the minimum installed in a 586?
    a. 1
    b. 2
    c. 4
    d. 8

23. How many 72-pin 36-bit SIMMs are required as the minimum installed in a 386?
    a. 1
    b. 2
    c. 4
    d. 8

24. How many 72-pin 36-bit SIMMs are required as the minimum installed in a 486?
    a. 1
    b. 2
    c. 4
    d. 8

25. How many 72-pin 36-bit SIMMs are required as the minimum installed in a 586?
    a. 1
    b. 2
    c. 4
    d. 8

26. How many 168-pin 64-bit DIMMs are required as the minimum installed in a 586?
    a. 1
    b. 2
    c. 3
    d. 4

27. How many 168-pin 64-bit DIMMs are required as the minimum installed in a 686?
    a. 1
    b. 2
    c. 3
    d. 4

28. SDRAM is
    a. Standard Dynamic Random Access Memory.
    b. Used in slower systems only.
    c. Able to synchronize with the motherboard system bus.
    d. Used exclusively for video memory.

29. SGRAM is
    a. Standard Grade Random Access Memory.
    b. Used in slower systems only.
    c. Video memory able to synchronize with the motherboard system bus.
    d. Used exclusively for cache memory.

30. WRAM is
    a. Write-only Random Access Memory.
    b. Windows Memory, so it is used exclusively on systems running Windows.
    c. Used exclusively for cache memory.
    d. A dual-port, block write memory used for video.

# 5 Keyboards

**OBJECTIVES**

After completing this chapter you should be able to

- Identify PC/XT or AT keyboards
- Specify the keyboard needed for a system
- Test keyboards
- Troubleshoot and repair or replace keyboards

## 5.1 GENERAL DISCUSSION

The keyboard is the most common interface from user to computer. The computer system must know which key is pushed.

With 101 or more keys on a modern keyboard, if each key switch were wired directly to the computer, that would be 101 wires plus one for ground. A 102-wire cable is rather fat and awkward. On the other hand, $101_{10}$ (decimal) is $110\ 0101_2$ (binary), so 101 choices sent as a binary number would require only 7 bits. This changing of decimal to binary is called **encoding.**

Sending these 7 bits in parallel would require seven data wires plus one each for power and ground—nine total wires. However, if it is sent in a *serial* bit stream, the 7 bits could all go on one wire, plus the two for power and ground—a total of three wires. Obviously this is the way to go. Two more wires were added for clock and reset, making the clone keyboard cable a five-conductor cable, as shown by the jack pinouts in Figure 5.1.

Besides sending a binary number to represent each key, keyboard efficiency is increased in yet another way. Instead of individual wires for each key, the keys are wired in a matrix (rows and columns), as shown in Figure 5.2. Note that an array of eight rows by eight columns can access sixty-four keys. By noting the row and column in which a pressed key is located, the keyboard circuitry identifies the key with a 6-bit binary code. This is in parallel. It changes this to send the binary code for that key out in serial format, one bit after another. In the computer that code is intercepted, and the character is recognized.

Although numerous keyboard configurations exist, the three main types we consider are the 83-key for the PC or XT; the 101-key for the AT, 286, 386, and 486 systems; and the 104-key for Windows 95 and 98 or 586 and higher processors. Because a keyboard is a complex electromechanical device, integration has simplified it into an easily repairable subassembly. Simple troubleshooting should result in a simple repair.

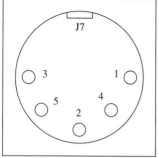

Standard 5 Pin DIN

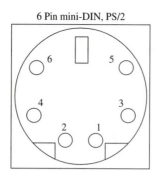

6 Pin mini-DIN, PS/2

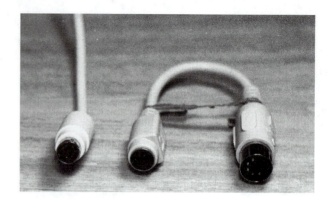

**FIGURE 5.1    Keyboard jack**

Table 5.1 shows the voltages to expect from the keyboard jack (J7) on the motherboard. You measure with the keyboard disconnected. (Of course the system must be off when you disconnect the keyboard, then you power-up and measure.) If the voltages are correct, the motherboard is doing its part, so the keyboard and cable are suspect.

The two keyboard connectors in popular use in personal computers today are standard 5-pin DIN jack and the 6-pin mini-DIN or PS/2 connectors shown in Figure 5.1.

All keyboards directly interface with the motherboard without an intervening circuit card. Almost all problems I have seen with keyboards are dust or dirt under the keys in the

**FIGURE 5.2    Keyboard matrix**

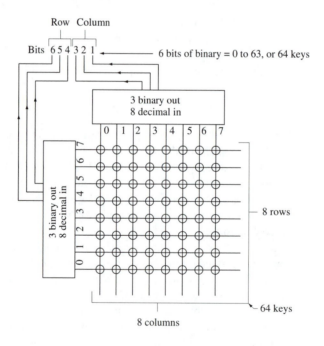

**TABLE 5.1    Keyboard signals on motherboard jack (female)**

| 5-Pin DIN | | Keyboard Signals | 6-Pin Mini-DIN |
|---|---|---|---|
| 1 | 5 V | Keyboard clock, pulled high | 5 |
| 2 | 5 V | Keyboard data out, pulled high | 1 |
| 3 | 5 V | Keyboard reset, pulled high | |
| 4 | 0 V | Ground | 3 |
| 5 | 5 V | Power from motherboard | 4 |
| | | Reserved | 2 |
| | | No connection | 6 |

switches or broken copper traces on the printed circuit board. If one key is out, it is probably dust. If many are out, it may be a trace. Of course the cable itself gets rough use and is suspect if the failure is major. A simple continuity check will prove it.

Be careful when opening any keyboard, because some almost fly apart with numerous little springs and key caps falling all over. On most, the separate key caps can be popped up and off.

Dirt and dust problems can be cured by compressed air or careful manual cleaning. When performing continuity tests to check for a cable problem, carefully document the matrixing of the keys in columns and rows. These columns and rows are not always physically clear.

## 5.2 GENERIC SYSTEM OPERATION

## 83-KEY KEYBOARD FOR PC, XT, AND 8088 SYSTEMS

The original IBM clone PC keyboard had 83 keys (later 84) in two blocks, ten in two columns of five on the left for the ten special function keys, with all other keys in a block to the right. The numeric keypad keys were dual function, with the 2, 4, 6, and 8 keys being direction arrows and the 1, 3, 7, and 9 keys being End, PageDown, Home, and PageUp editing keys. The Num Lock key toggles the keypad between these functions.

Figure 5.3 shows an example of the 84-key keyboard, and Figure 5.4 shows the key scan codes, which are the numbers sent as serial data in signals from the keyboard to the motherboard. The motherboard interprets these signals by the key scan code to recognize which key was pressed.

(**Note:** The remainder of this section is recommended for technicians.)

Figure 5.5 shows how a keyboard matrix can be "scanned," creating a scan code for each key. The clock sends a squarewave clock signal to the counter, which repeatedly counts from 0 to 7 in binary. This sends 3 bits to the parallel register to hold when loaded and to the binary-to-decimal decoder. Thus rows 0–7 (normally pulled high) are "scanned" repeatedly with a low signal. When a key is pressed, the scanning low pulls the column input to the encoder low,

**FIGURE 5.3  84-key keyboard**

**FIGURE 5.4  83-key keyboard scan codes**

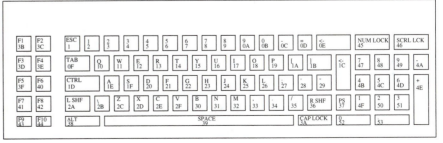

83-key keyboard, 8088 systems, PC, XT

*Note:* Key name on top of key is in decimal alphanumeric. Key scan code on bottom is in hexadecimal.

FIGURE 5.5   Scanned keyboard

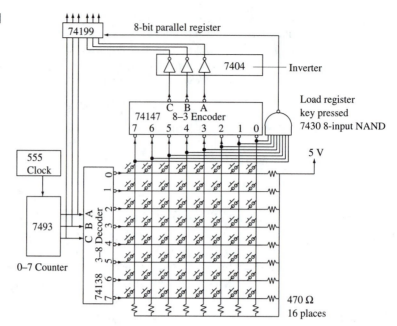

doing two things. First, the encoder now puts the other 3 bits of the code out (because it is active low, they need inverting), completing the 6-bit scan code. Second, the NAND notes any keystroke and loads the 6-bit code into the register. For serial output, a serial shift register would load this and shift it out serially.

By removing the back of the keyboard, you can examine the signals at the other end of the cable from the system board keyboard jack. On the keyboard end, they will be routed to a standard 5-pin .1 header (the pins are 0.1 inch apart) and numbered in a row, with the same numbers as the jumbled numbering on the motherboard jack. Figure 5.6 shows these signals.

Some 286 and higher boards have the standard five-pin .1 header with numbers in order, with a jack and cable going to the jack on the case. On pin 1 the strobe is about 5 microseconds (µs) long followed after about 80 µs by eight more pulses, each about 5 µs wide and about 52 µs apart.

FIGURE 5.6   83-key keyboard signals, pins 1 and 2

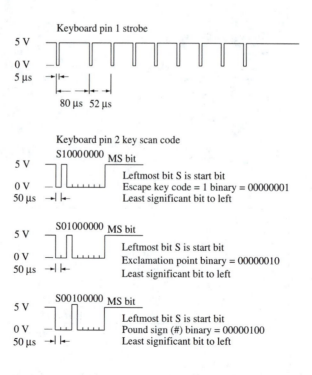

On pin 2, the data for the key scan code appears. Three examples are given in Figure 5.6: the Esc key, the exclamation point (!), and the pound sign (#). Note that there is a start bit, then the bits follow left to right from least to most significant bit. This may seem backward to binary, but it is the same as RS-232 serial, except not inverted.

If the keyboard is receiving its 5 V supply from the system board on pin 5 and the ground is good on pin 4, and if the strobe and data are not seen on the scope when you press the above keys, the problem lies in the keyboard. Check for the common problems listed above.

When you release any key that you have been holding down, a key break signal is sent that is 128 higher than the original scan code. This can be seen in passing on a scope, but it is not repeatable for extended viewing.

The key scan codes for the 83-key keyboard common on 8088 systems, such as the PC and XT, are given here in two formats. Use Figure 5.4 if you want to know the scan code for a certain key. Use Table 5.2 if you are examining scan codes with the oscilloscope and want to find the key that was pressed.

In general, the 8088 system keyboard has a 8048 microcomputer in it. Pinouts for the 8048 are given in Figure 5.7. The 8048 is an 8-bit CPU with a 1K ROM and a 64-byte RAM. It is supported by three or four chips, one of which is usually a 7404 or 7405 used as an output buffer. Pin 17 of the 8048 is the serial out pin, and it is there that you should see the data signal with the oscilloscope, then through a buffer to the 5-pin header and out the

**TABLE 5.2  83-key keyboard scan codes**

| Scan Code | Key Name | Scan Code | Key Name |
|---|---|---|---|
| 1 | Esc | 2B | \ |
| 2 | 1 | 2C | Z |
| 3 | 2 | 2D | X |
| 4 | 3 | 2E | C |
| 5 | 4 | 2F | V |
| 6 | 5 | 30 | B |
| 7 | 6 | 31 | N |
| 8 | 7 | 32 | M |
| 9 | 8 | 33 | , |
| A | 9 | 34 | . |
| B | 0 | 35 | / |
| C | - | 36 | Right Shift |
| D | = | 37 | * |
| E | Backspace | 38 | Alt |
| F | Tab | 39 | Space |
| 10 | Q | 3A | Caps Lock |
| 11 | W | 3B | F1 |
| 12 | E | 3C | F2 |
| 13 | R | 3D | F3 |
| 14 | T | 3E | F4 |
| 15 | Y | 3F | F5 |
| 16 | U | 40 | F6 |
| 17 | I | 41 | F7 |
| 18 | O | 42 | F8 |
| 19 | P | 43 | F9 |

**TABLE 5.2** *(continued)*

| Scan Code | Key Name | Scan Code | Key Name |
|-----------|----------|-----------|----------|
| 1A | [ | 44 | F10 |
| 1B | ] | 45 | Num Lock |
| 1C | Enter | 46 | Scroll Lock |
| 1D | Ctrl | 47 | 7 |
| 1E | A | 48 | 8 |
| 1F | S | 49 | 9 |
| 20 | D | 4A | - |
| 21 | F | 4B | 4 |
| 22 | G | 4C | 5 |
| 23 | H | 4D | 6 |
| 24 | J | 4E | + |
| 25 | K | 4F | 1 |
| 26 | L | 50 | 2 |
| 27 | ; | 51 | 3 |
| 28 | ' | 52 | 0 |
| 29 | ` | 53 | . |
| 2A | Left Shift | | |

**FIGURE 5.7  8048 keyboard controller pinout**

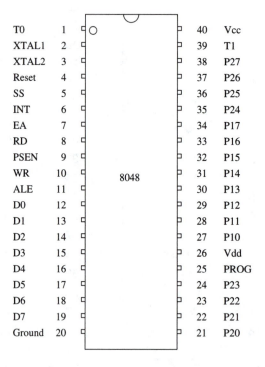

cable. The serial signal is received in the motherboard by a 74322, a serial-to-parallel register, on pin 17 also. It is output on lines 4–7 and 13–16 in parallel format. The pinouts for the 74322 are given in Figure 5.8.

A generic block diagram of the keyboard-to-motherboard interface is given in Figure 5.9. On the left is circuitry in the keyboard, on the right is circuitry on the motherboard. In the keyboard, note the oscillator giving clock pulses to the 8048 microcomputer. The microcomputer scans the matrix and receives the binary data for which key

FIGURE 5.8    74322 serial-to-parallel register

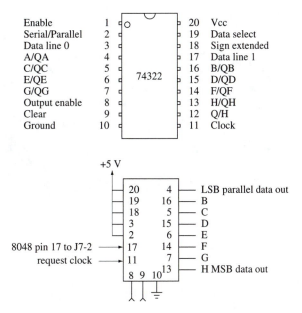

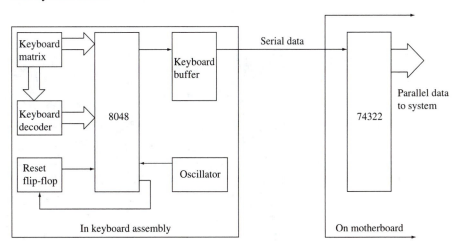

*Note:* Typical usage in 8088 system; keyboard in serial-to-parallel register.

was pressed. This is sent out serially (shifted) to a buffer, then out of the keyboard to the motherboard.

On the motherboard the 74322 parallel-to-serial shift register returns the data to parallel form and sends it to an 8255 PPI (Programmable Peripheral Interface) that is used to interface I/O to the system bus.

## 101-KEY KEYBOARD FOR 286/386/486 SYSTEMS

The newer 101-key keyboards include all the old keys and add separate editing keys. These are often laid out in one line across the top and in three blocks of keys. The line across the top starts at the left with the Esc key, followed by twelve function keys. The keypad is moved to the right, making room for the added editing keys. These keys can be ignored and the Num Lock key used with the keypad as before, or you may now have separate editing and numeric keypad keys.

It is important to note that these newer keyboards usually have a switch on the bottom to set the keyboard for an 8088 or 286/386/486 mode. Some automatically sense the mode, but you should check for the switch. Some keyboards have a DIP switch on the bottom with two switches or bits. Counting in binary, 00 is not set, 01 is for XT, 10 is for AT, and 11 is for autosensing. Figure 5.10 is a photo of a typical 101-key keyboard, and Figure 5.11 gives the key scan codes.

**FIGURE 5.9    PC keyboard block diagram**

159

**FIGURE 5.10    Typical 101-key keyboard**

**FIGURE 5.11    101-key keyboard key scan codes**

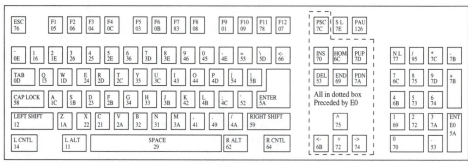

286/386/486 keyboard type 101

*Note:* Key name on top of key is in decimal alphanumeric. Key scan code on bottom is in hexadecimal.

## 104-KEY KEYBOARDS FOR WINDOWS 95 AND 98 SYSTEMS

With Windows 95 and 98 taking over as the standard operating systems, more and more systems are using keyboards with a 6-pin mini-DIN (PS/2) male plug. Adapters are available to go either direction, from a 5 pin to a 6 pin, or vice versa. This plug is the same as the PS/2 mouse plug, so the jacks on the rear of the case are usually labeled. Be careful not to plug the keyboard into the mouse port, or vice versa. Keyboards made especially for Windows have a key with the Windows logo, which pops up the Start Menu from any program that is running. Pressing the Windows and Break together pops up the System Properties dialog box. The Windows and F keys open the "FIND" program. The Windows and F1 keys give Windows Help (F1 alone gives help for the active window). Windows and E opens Explorer. Windows and R opens the RUN program. Windows and TAB cycle through open programs with buttons on the taskbar. These keyboards have 104 keys, the same as the 101 key keyboards, with function keys 11 and 12 and the Windows key added. In most other ways, it is the same as the 101-key keyboard. Note the Windows key in Figure 5.12, one key right of the bottom left corner.

**FIGURE 5.12    The 104-key Windows keyboard**

Windows key

## ERGONOMIC KEYBOARDS

Anyone who has typed for any period on a computer keyboard knows that it puts a strain on the wrists. Just lean forward and put your arms on a table and you should note that they fall naturally at an angle almost 45 degrees inward. To get the fingers back in a straight line for lying on the "home" keys (ASDF JKL;) you need to twist your wrists outward about that same amount. This leads to cumulative trauma disorders (CTD) or repetitive strain injuries (RSI) such as carpal tunnel syndrome for those who spend their days working on a keyboard.

What is proper keyboard placement? The following excerpt is quoted from the Typing Injury FAQ (frequently asked questions) at www.tifaq.com, associated with the Cumulative Trauma Disorder Resource Network.

> Keyboards should be located to allow a comfortable, neutral posture during use. This is typically considered to be directly in front of you, at seated elbow height. Proper keyboard placement assists in keeping a neutral keyboarding posture, which is generally stated as: shoulders back and relaxed; the upper arms resting down, to the side of your body; the elbows making an approximate 90 degree bend; the forearms horizontal, parallel to the floor; and the wrists being in-line with the forearms with minimal bending up or down, left or right.

The FAQ sheet adds that current research shows that a slight backward tilt of the keyboard conforms more to the natural position of the wrists. One product that helps reduce wrist strain is a soft wrist support, such as the one shown in Figure 5.13.

Some keyboard manufacturers make alternative keyboards that are ergonomically designed to allow the wrists and hands to be in a more natural position. The basic idea is to break the keyboard down the middle and bend it to an angle instead of your wrists. These are called split keyboards. Figure 5.14 shows one popular model. Those that allow the user to change the angle are called adjustable split keyboards. The fanciest of these allow you to change the horizontal and vertical angle, and even raise or lower parts of the keyboard.

Trays for shelves to attach to a desk and lower the keyboard to a more natural height are also available. They also allow sliding the keyboard back out of the way when not in use (Figure 5.15).

The following URLs (Universal Resource Locators) may be useful:

> www.adessonic.com
> www.acercomponents.com/psckeyboards.html
> www.cirque.com
> www.darwinkeyboards.com
> www.ergopro.com
> www.ids2.com/pace
> www.intr.net/mykey

**FIGURE 5.13    A wrist support**

**FIGURE 5.14    An ergonomic keyboard**

**FIGURE 5.15    A keyboard tray to save wrists and room**

www.kinesis.ergo.com
www.microsoft.com/products/hardware/natkeybd
www.sni.de/pc/tastatur/ur/kbpce.htm
www.wpdesigns.com

## SIGNALS

(**Note:** The remainder of this section is recommended for technicians.)

Table 5.3 shows the scan codes that an oscilloscope can read at the jack for a 101-key keyboard. With the scope on pin 2 from the keyboard, you can again see the binary bit patterns being sent serially over the line from the keyboard to the system. On the 101-key keyboard, a start bit, eight data bits, an odd parity bit, and a stop bit make 11. The stop bit is high, so on the scope you see 10 bit positions. Set the scope at 0.1 milliseconds/ division, and decalibrate until you can see the entire pattern. This will give you 1 bit per division. The keys shown in the dotted lines in Figure 5.11 actually have two patterns sent, an E0 followed by another. The second pattern is the same as the keys on the keypad with the same functions. The first is to differentiate them. Setting the scope on 0.2 ms/div and decalibrating as above will allow you to use the horizontal position control to slide back and forth between the two patterns. Holding a key down causes it to autorepeat, so a stable trace is seen. Pause does not repeat, and it is hard to catch without a storage scope.

**TABLE 5.3    101-key keyboard
scan codes on scope at jack**

| Scan Code | Key Name | Scan Code | Key Name |
|---|---|---|---|
| 1 | F9 | 4B | L |
| 3 | F5 | 4C | : |
| 4 | F3 | 4D | P |
| 5 | F1 | 4E | - |
| 6 | F2 | 52 | ' |
| 7 | F12 | 53 | Del (keypad) |
| 8 | F8 | 54 | [ |
| 9 | F10 | 55 | = |
| B | F6 | 58 | Cap Lock |
| C | F4 | 59 | Right Shift |
| D | TAB | 5A | Enter |
| E | ` | 5B | ] |
| 11 | Left Alt | 5D | \ |
| 12 | Left Shift | 62 | Right Alt |
| 14 | Left Ctrl | 64 | Right Ctrl |
| 15 | Q | 66 | Backspace |
| 16 | 1 | 69 | 1 (keypad) |
| 1A | Z | 6B | 4 (keypad) |
| 1B | S | 6C | 7 (keypad) |
| 1C | A | 72 | 2 (keypad) |
| 1D | W | 73 | 5 (keypad) |
| 1E | 2 | 74 | 6 (keypad) |
| 21 | C | 75 | 8 (keypad) |
| 22 | X | 76 | Esc |
| 23 | D | 77 | Num Lock |
| 24 | E | 78 | F11 |
| 25 | 4 | 79 | + (keypad) |
| 26 | 3 | 7A | 3 (keypad) |
| 29 | Space | 7B | – (keypad) |
| 2A | V | 7C | * (keypad) |
| 2B | F | 7D | 9 (keypad) |
| 2C | T | 7E | Scroll Lock |
| 2D | R | 83 | F7 |
| 2E | 5 | 126 ? | Pause |
| 31 | N | E0 4A | / (keypad) |
| 32 | B | E0 53 | Del |
| 33 | H | E0 69 | End |
| 34 | G | E0 6B | Left Arrow |
| 35 | Y | E0 6C | Home |
| 36 | 6 | E0 70 | Insert |
| 3A | M | E0 72 | Down Arrow |
| 3B | J | E0 74 | Right Arrow |
| 3C | U | E0 75 | Up Arrow |
| 3D | 7 | E0 7A | Page Down |
| 3E | 8 | E0 7C | Print Screen |
| 41 | , | E0 7D | Page Up |
| 42 | K |  |  |
| 43 | I |  |  |
| 44 | O |  |  |
| 45 | 0 |  |  |
| 46 | 9 |  |  |
| 49 | . |  |  |
| 4A | / |  |  |

The 286/386/486 system keyboard usually is based on an 8049. A little more sophisticated than the 8048 used in the 8088 systems, the 8049 has a 2K ROM and a 128-byte RAM. It is one of only two chips in the keyboard; the other is a 7407 buffer.

This data is received in the motherboard by an 8042 or 8742 microcomputer. It is because of the ROM BIOS programming in this microcomputer that not all keyboard controller chips will work in all systems. You must thoroughly check out any system's keyboard after replacing the keyboard BIOS chip. The system BIOS ROMs are usually sold in a set with a keyboard BIOS ROM, to overcome this possible incompatibility. Figure 5.16 gives the block diagram of the 101-key keyboard and motherboard circuitry.

Figure 5.17 shows the 5-pin DIN jack and the single header it comes into on the keyboard. Note that the pins are not in the same order of 1–5 as the DIN jack. Also note the key with the cap removed, exposing the switch for cleaning.

Figure 5.18 shows the copper stripes or traces on the keyboard PCB that may be broken. This failure usually creates a pattern of keys failing at the same time.

Program Listing 5.1 is a program that flashes the LEDs on the keyboard (on most systems) and gives the scan codes of a key that is pressed. Press ENTER to quit the program. For an 8088 system, the scan codes are also the scope trace. For 286 systems, the scope trace is as given in Table 5.3.

**PROGRAM LISTING 5.1    AT keyboard scan code test**

```
CLS : REM The program is for an 80X86 AT class computer.
PRINT "This flashes the 3 LEDS for Num Lock, Caps Lock, and Scroll
Lock"
INPUT "Press Enter to continue !", X
```

**FIGURE 5.16    AT keyboard block diagram**

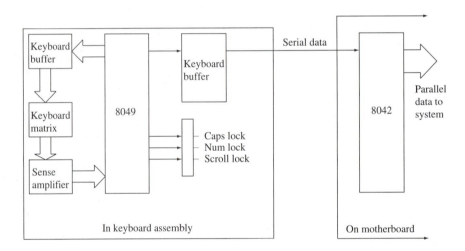

**FIGURE 5.17    DIN and single header keyboard jacks**

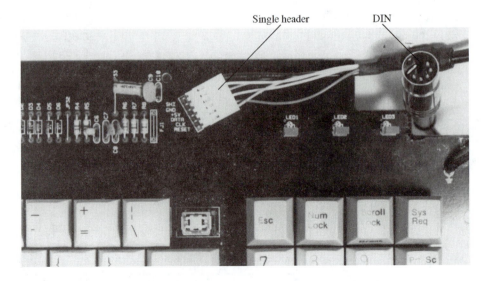

**FIGURE 5.18 Keyboard copper stripes or traces**

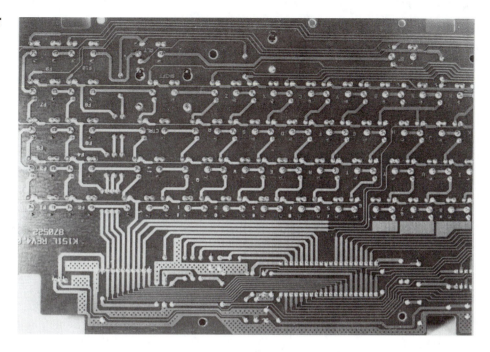

```
FOR flash = 7 TO 0 STEP -1
   OUT &H60, &HED
   OUT &H60, flash: REM turn on three LEDs
   FOR DELAY = 1 TO 3000: NEXT DELAY
   PRINT "Flash "; 8 - flash
NEXT flash
PRINT "Press any key for 8088 system (83-key) scan code of that key
!"
PRINT "This will be seen on scope with start bit, LSB first, MSB
last"
PRINT "On 80286/386/486 system, code will be same as 8088"
PRINT "but scope trace at keyboard will be as shown in 286 scan
codes"
PRINT "(101-key) table in this manual."
PRINT "Press any key for code, or Enter to quit"
WHILE CODE <> 28
   KEEY$ = INPUT$(1)
   CODE = INP(&H60)
   PRINT "Key = "; KEEY$; " Scan Code = "; HEX$(CODE); " Hex."
WEND
```

## 5.3 INSTALLING KEYBOARDS

The keyboard is the simplest component to install. A PC/XT needs a matching PC/XT keyboard; an AT system needs an AT keyboard. If there is an XT/AT switch on the bottom of the keyboard, set it appropriately. The connector must match the motherboard jack, either 5-pin DIN or 6-pin mini-DIN/PS2. Adapters are available to convert in either direction. If using Windows 95 or 98, use a keyboard with the Windows logo key. With the power to the system off, just plug in the keyboard. Confirm the polarity of the DIN plug, and do not force it.

The BIOS SETUP program has a choice to select "installed" or "not installed," but it is always defaulted to "installed" and is not a concern. In Windows 95 or 98, click Start, Settings, Control Panel, Keyboard. This opens a window allowing you to set the Repeat Delay, the Repeat Rate, and the Cursor Blink Rate or change the Language and enable an indicator on the task bar. Of course Help, Search, "keyboard" gives information on these and changing the layout and using the Windows key.

## 5.4 BASIC TROUBLESHOOTING

As you can see, there is no POST audio code to consider in troubleshooting keyboards, but you may have a POST video error code to consider. Then you can proceed to noting any visual clues and taking appropriate measurements. Details of these steps are as follows:

a. POST audio error code: none
b. POST video error code of 3xx
c. Measurements
   1. Check switch on bottom for XT/AT setting.
   2. Check that 6-pin mini-DIN is not in mouse jack.
   3. Keyboard dead. Remove, power up, check system board keyboard plug per Table 5.1 to see if system board keyboard plug is functioning.
   4. **Technicians only:** Keyboard connected, back off. Power-up, scope check the signals per Figure 5.5. Cables are abused! Are voltages from system board present?
   5. Check continuity of keyboard cable.
   6. If a few keys are out, is it dirt in the switch or a stripe break? If the keys are in a pattern, suspect a stripe break and make continuity checks, soldering broken lines. If keys are random, try blowing out dirt with compressed air.
   7. Replace keyboard.

## 5.5 CHAPTER REVIEW/QUIZ QUESTIONS

1. When you connect or disconnect the keyboard to the system,
   a. The system must be turned on.
   b. The system must be turned off.
   c. It does not matter whether the system is on or off.
   d. Use an antistatic wrist strap to protect circuitry.

2. The plug on the keyboard is a
   a. 9-pin D-shell connector.
   b. 15-pin D-shell connector.
   c. 5-pin DIN connector or 6-pin mini-DIN.
   d. RCA phono connector or 9-pin D-shell.

3. In general, the two types of keyboards are the
   a. Large and small type.
   b. 8088 and 80286 types.
   c. Integrated and nonintegrated.
   d. High-speed and normal types.

4. If connecting to a 286 or higher, check the keyboard for
   a. Enough memory.
   b. High enough speed.
   c. 83-key or 101-key.
   d. A switch set to AT.

5. If the keyboard appears dead, check
   a. Each key, one at a time.
   b. The motherboard keyboard jack for proper voltages.
   c. Printed circuit board foil traces or stripes.
   d. For dust under the keys.

6. If the keyboard appears dead, check
   a. For broken printed circuit board traces.
   b. Keyboard cable continuity.
   c. For dust under the keys.
   d. Oscilloscope signals on pin 2.

7. If one or two random keys are failing, check
   a. For broken printed circuit board traces.
   b. Keyboard cable continuity.

    c. For dust under the keys.
    d. The voltages on the motherboard keyboard jack.

8. If a pattern of keys are failing, check
    a. For broken printed circuit board traces.
    b. Keyboard cable continuity.
    c. For dust under the keys.
    d. The voltages on the motherboard keyboard jack.

9. If the keyboard is totally dead, the problem is
    a. Probably the cable or chips in the keyboard.
    b. Dust in the keyboard.
    c. A printed circuit board stripe broken.
    d. Definitely in the motherboard.

10. The keyboard is
    a. Not easily repairable.
    b. Never worth repairing.
    c. Fail-safe.
    d. Directly connected to the motherboard.

11. The keyboard sends out information to the motherboard
    a. In parallel along eight lines.
    b. In serial along one line.
    c. In octal along four lines.
    d. In decimal along ten lines.

12. Ninety-five percent of keyboard failures are due to
    a. Electronic chip malfunction.
    b. Static discharge destruction.
    c. Coca-Cola.
    d. Dust, broken stripes on the PCB, or broken cable.

13. The XT and AT keyboards are
    a. Not really different.
    b. Nine bits and 11 bits and therefore incompatible.
    c. Interchangeable.
    d. Not functional on 386s.

14. The switch found on the bottom of some keyboards selects
    a. 8088 or 286/386/486.
    b. PC or XT.
    c. Serial or parallel data transmission.
    d. 80386 or 80386SX.

15. Standard straight keyboards may cause
    a. Cumulative trauma disorders (CTD).
    b. Repetitive strain injuries (RSI).
    c. Carpal tunnel syndrome (CTS).
    d. All of the above.

16. Split, curved, angled, and adjustable keyboards may be classified as
    a. Windows 95 keyboards.
    b. Legacy keyboards.
    c. Ergonomic keyboards.
    d. Parallel keyboards.

17. The 6-pin mini-DIN or PS/2 keyboard plug may be incorrectly inserted in the
    a. Parallel port.
    b. Serial port.
    c. Video port.
    d. Mouse port.

18. The 6-pin mini DIN or PS/2 keyboard may be made to work with a motherboard with a standard 5-pin DIN jack by
    a. Cutting wire number 6.
    b. An adapter.
    c. Breaking off pin 3.
    d. Nothing will make it compatible.

19. Pressing the Windows logo key on the 104-key keyboard will
    a. Pop up the task bar.
    b. Reboot the system.
    c. Print the screen to the clipboard.
    d. Give the help screen.

# 6

# Monitors, Video Displays

**OBJECTIVES**

After completing this chapter you should be able to

- Describe operation of displays
- Identify display controller card type
- List specifications of different types of displays
- Install memory in display adapter card
- Select and install display card and display
- Install display software
- Troubleshoot and repair or replace adapter or display

## 6.1 GENERAL DISCUSSION

Most computer systems require a card plugged into an expansion slot to control the video. The card drives a monitor or display that connects to the card. Figure 6.1 shows the standard jacks that are on the back of the card, which shows at the back of the computer. Composite uses a standard female RCA phono jack like the one on the back of an audio tape deck. In general, the 9-pin D-shell indicates a monochrome, CGA, or EGA type of monitor. A high-density 15-pin D-shell indicates a VGA or better display monitor.

The overall operation of the computer monitor or display is very similar to a television set. The final output is on a screen called a picture tube or a CRT (Cathode-Ray Tube). Like all picture tubes, it is a sealed, evacuated, glass container having no air inside to interfere with the electronics. At the end of the small neck are the pins that fit into a socket.

A monochrome or one-color tube is the simplest, so we will examine that first. A typical monochrome CRT is shown in Figure 6.1. The elements in the CRT make up a device known as an electron gun, because it "shoots" electrons. The electron gun is made up of a filament just like the filament in a standard incandescent lightbulb. Its purpose is to heat the next element, so it is sometimes called the **heater.** It heats the cathode, which has voltage applied to it to deliver electrons to be "shot." Next come two or three control grids to precisely control the beam. Finally, the front face of the CRT has phosphors that glow when energized. A very high positive voltage, from 15,000 to 35,000 volts, is applied to this anode (positive terminal). Positive charges attract negative charges, so this strongly attracts electrons from the cathode. They break away from the hold of the atom's nucleus, leaving orbit and flying through the tube in the form of a beam or ray, hence the term, cathode-ray tube. The electrons hit the phosphors at the other end, causing a dot to glow.

**FIGURE 6.1  Video jacks and CRT**

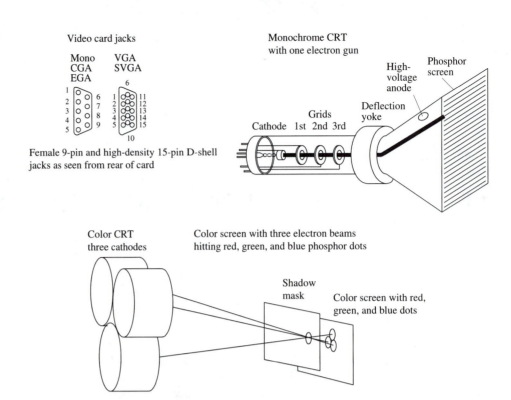

Video card jacks

Mono
CGA
EGA

VGA
SVGA

Female 9-pin and high-density 15-pin D-shell jacks as seen from rear of card

Monochrome CRT with one electron gun

High-voltage anode

Phosphor screen

Deflection yoke

Grids
Cathode  1st  2nd  3rd

Color CRT three cathodes

Color screen with three electron beams hitting red, green, and blue phosphor dots

Shadow mask

Color screen with red, green, and blue dots

To form a picture of dots, the beam is pulled quickly back and forth horizontally to make the phosphors glow in a line across the screen from side to side. Additionally, the beam is pulled up and down slowly, causing the horizontal line to move from the top of the screen to the bottom. This is accomplished by electromagnets in a "yoke." Because the phosphors glow for a short time after they have been hit, and because the eye retains the image of each dot for a short time, the effect is now a white square or rectangle called a **raster.** Now, by varying the video signal to the cathode, the beam is either turned off or turned on to varying degrees, giving different energy to different dots it strikes. This gives the black, gray, and white dots that make up the picture we see.

The color picture tube has three identical electron guns shooting three beams. The phosphors on the screen are not all white but are in triangular groups of three dots, one each of red, green, and blue. These three dots make up one bigger dot from which the picture is made. Each gun shoots at one of the colors, so the guns are named the red, green, and blue guns, even though the beams are identical. If all three color dots are on to the same degree, the overall color our eye perceives is white. If only green is on, the color is green. If blue and green are on, the color is yellow. By controlling the degree to which each of the three basic colors is on, a vast spectrum of colors can be shown.

## MONITOR SPECIFICATIONS

Text/mono displays have one color, usually green or amber. They print only preformed characters on the screen. These characters are of higher resolution ($9 \times 14$ array of pixels or dots) than "graphics" displays, and they are easier on the eyes for those using the computer for nongraphics work. They are also the least expensive displays to be found. Some of the better high-resolution graphics displays have a text mode that is comparable to text/mono displays. Some companies also produce monographics adapter cards that will do graphics on text/mono displays (in one color, of course). The mono card can be identified by the 9-pin DB jack and one crystal, perhaps 16 MHz.

Composite displays are very similar to televisions and may give color graphics. These displays can be identified by a one-conductor (and ground shield) RCA phono jack for video from the computer. The adapter card is usually a CGA and must have the same jack. A cable with an RCA phono plug on each end connects the display to the adapter card.

Color graphics array (CGA) monitors have three color signals and an intensity signal. The three lines for signals are equivalent to three binary bits, or $2^3$. The maximum different states or combinations of 3 bits is eight states, represented by the numbers 0–7 in binary. So eight colors are possible. The fourth line, intensity, is for brightness of the color. The original eight colors are dark; with the intensity bit they are light. This gives the appearance of sixteen colors maximum. Unfortunately, this is possible only in text mode. Four colors are available in low-resolution graphics mode ($320 \times 200$ pixels); in high-resolution graphics ($640 \times 200$ pixels or dots) the maximum number of colors is only two, black and white. The CGA card can be identified by the 9-pin female D-shell connector (DB9 jack) and the lack of a crystal—it gets its 14.31818-MHz clock from the motherboard through pin B30 of the I/O connector.

Enhanced graphics array (EGA) monitors improved resolution to $640 \times 350$ pixels in high resolution. At the same time, the number of signal lines doubled from three to six, with a primary and a secondary line for each of the basic colors red, green, and blue. This gives six lines or $2^6$ states, for sixty-four maximum possible combinations. However, these are in a palette, and only sixteen can be used at one time. If the controller card has more memory, more colors are allowed. The same jack is used as on the CGA card.

Video graphics array (VGA) monitors again improved resolution to $640 \times 480$ pixels in high resolution. Colors were increased to 256 at the low resolution of $320 \times 200$ and to 16 for the high resolution. These were the most popular in the late 1980s. The VGA and SVGA adapter cards can be identified by the high-density 15-pin female D-shell connector and from one to four crystals at 40, 14.31818, 32.5, 25.17, or 28.32 MHz, or almost any value.

The SVGA adapter was most popular on systems in the early 1990s. These super VGA monitors were originally $800 \times 600$ with 256 colors in low resolution and 16 in high resolution. SVGAs went through a stage with a resolution of $1024 \times 768$ and 1 million colors, with dot pitches of less than 0.28. A typical SVGA may give the specifications shown in Table 6.1.

Note that more memory gives more colors, up to the maximum ability of the monitor. These high-end monitors handle graphical user interfaces (GUI), painting, and drawing programs with photographic quality.

## Pixels

The picture on the screen of the monitor is made up of small dots. The more dots and the smaller each is, the better the picture. The dots are called picture elements or **pixels.** Standard displays now range from 640 pixels horizontally and 200 vertically, to 1024 dots both horizontally and vertically. In text or alphanumeric (A/N) mode, only characters defined in a "character box" of a given size can be displayed. A character box is typically $8 \times 8$ or $8 \times 14$ pixels. In graphics mode, each pixel can be used to create any design you wish. This is known as all-points-addressable (APA) mode.

## Dot Pitch

The picture quality is controlled not only by the number of pixels but by the **dot pitch.** This is the distance between pixels. If they are far apart, the picture is not "crisp." If they are close together, the clarity of the picture in tiny details is improved. The original IBM CGA

**TABLE 6.1  Typical SVGA specifications**

| Pixels | Memory – Colors | Memory – Colors | Memory – Colors |
|--------|-----------------|-----------------|-----------------|
| $640 \times 480$ | 512K – 256 | 1 MB – 16,000,000 | Not needed |
| $800 \times 600$ | 512K – 256 | 1 MB – 64,000 | 2 MB – 16,000,000 |
| $1024 \times 768$ | 1 MB – 256 | 2 MB – 64,000 | 3 MB – 16,000,000 |
| $1280 \times 1024$ | 2 MB – 256 | 3 MB – 64,000 | 4 MB – 16,000,000 |
| $1600 \times 1200$ | 2 MB – 256 | 4 MB – 64,000 | 8 MB – 16,000,000 |
| $1800 \times 1440$ | 4 MB – 256 | 8 MB – 64,000 | 8 MB – 16,000,000 |

*Note:* Horizontal pixels × vertical pixels × bytes color = total video card RAM:
1 byte = 8 bits = 256 colors
2 bytes = 16 bits = 64,000 colors
3 bytes = 24 bits = 16,000,000 colors

had a dot pitch of 0.43 mm, which is considered very poor. A dot pitch of 0.31 mm is standard, while 0.28 mm is considered good and 0.26 mm excellent.

## Bandwidth

The bandwidth of a display monitor is the range of frequencies to which the circuitry is sensitive. Considering that the circuitry must respond to changes fast enough to change each pixel, you can rough out bandwidth by multiplying the number of horizontal times vertical pixels times the number of times the screen is written in one second. For overhead, a 1.3 fudge factor (1.9 for CGA) brings most display calculations close to reality. Rough bandwidths are given in Table 6.2.

## Interlaced/Noninterlaced

As the number of horizontal lines on displays rose, the bandwidth and cost of displays rose also. One way to hold the bandwidth down was to display only half the number of lines at a time. This was done by displaying the odd-numbered lines, then the even-numbered lines, in other words, *interlacing* them. As you can see, all the lines get displayed but at a time offset. Some people notice flicker when watching an interlaced display. Noninterlaced displays show the lines sequentially, that is, in order. This requires a faster, higher-quality monitor and is the better choice, if you can afford it.

Table 6.3 shows the relative quality of today's video displays.

To calculate these resolutions, just multiply the highest number of characters horizontally by the horizontal character pixels, and do the same for vertical.

| | | |
|---|---|---|
| Mono | $80 \times 9 = 720$ horizontally | $25 \times 14 = 350$ vertically |
| CGA | $80 \times 8 = 640$ horizontally | $25 \times 8 = 200$ vertically |
| EGA | $80 \times 8 = 640$ horizontally | $25 \times 14 = 350$ vertically |
| VGA | $80 \times 8 = 640$ horizontally | $30 \times 16 = 480$ vertically |
| SVGA | $50 \times 16 = 800$ horizontally | $46 \times 13 = 598$ vertically |

## TYPES OF MONITORS

All color monitors use the CRT technology that we described earlier, but they differ in the technology they use to handle signals. Thus, we have digital, analog, and multisync displays.

### Digital Monitors

The original monitors were digital. Digital monitors use the binary counting system. Their integrated circuits use transistor–transistor logic (TTL), so these monitors are also called TTL monitors. Their signals are either *on* (2.0–5.0 V DC) or *off* (less than 0.8 V DC). Digital monitors use a 9-pin D-shell connector. CGA and EGA monitors are digital.

### Analog Monitors

The other monitors are analog. Analog monitors employ signals that vary between 0.0 and 1.0 V DC. This allows graduated levels of brightness of each of the three basic colors, so

**TABLE 6.2  Video adapter frequencies**

| Monitor Type | Horizontal (KHz) | Vertical (Hz) | Rough Bandwidth (MHz) |
|---|---|---|---|
| MDA | 18.43 | 50 | $720 \times 350 \times 50 \times 1.3 = 16.4$ MHz |
| CGA | 15.75 | 60 | $640 \times 200 \times 60 \times 1.9 = 14.6$ MHz |
| EGA | 21.85 | 60 | $640 \times 350 \times 60 \times 1.3 = 17.5$ MHz |
| VGA | 31.47 | 70 | $640 \times 480 \times 60 \times 1.3 = 24.0$ MHz |
| SVGA | 35.16 | 60 | $800 \times 600 \times 60 \times 1.3 = 37.4$ MHz |

**TABLE 6.3 Video monitor comparisons**

| Type | | Pixels | Character Box | Colors |
|------|------|--------|---------------|--------|
| Mono/Text | | 720 × 350 | 9 × 14 | 1 |
| CGA | Text | 640 × 200 | 8 × 8 | 16 |
| | Low | 320 × 200 | 8 × 8 | 4 |
| | High | 640 × 200 | 8 × 8 | 2 |
| EGA | Text | 640 × 350 | 8 × 14 | 16 |
| | Low | 320 × 200 | 8 × 8 | 16 |
| | High | 640 × 350 | 8 × 14 | 4* |
| | | | | 16** |
| VGA | Text | 640 × 480 | 8 × 16 | 16 |
| | Low | 320 × 200 | 8 × 8 | 256 |
| | High | 640 × 480 | 8 × 16 | 16 |
| SVGA | | 800 × 600 | | 256 |
| VGA1024+ | | 1024 × 768 | | 256 |
| High-End | | 648 × 480 | | 16,000,000 |
| | | 1280 × 1024 | | 256 |
| | | 1600 × 1200 | | 16,000,000 |
| | | 1800 × 1440 | | 16,000,000 |

\* 64K RAM.
\*\* 128K RAM.

users see many more shades of color. Analog monitors use a high-density 15-pin D-shell connector. VGA and SVGA monitors are analog.

## Multisync Monitors

Some monitors automatically determine the frequency of the incoming signals and shift to the proper mode. They are able to *sync*hronize with *multi*ple frequencies, hence the name multisync. These are more universal, working with many types of display adapters. They are usually very high quality and very expensive.

## VIDEO ADAPTERS

The circuit card that plugs into the motherboard and prepares the signals for the monitor is called a video adapter card. These must be matched to the correct motherboard connector, whether 8-bit ISA, 16-bit ISA, EISA, VESA, PCI, or AGP. To handle the video information more quickly and burden the system less, video adapter cards have their own video memory. In the beginning this was standard DRAM. Then came the dual-ported Video RAM (VRAM), which could be read and written at the same time for higher speed. Window RAM (WRAM), also dual ported, boosted speed above VRAM by addressing large windows (blocks) of video memory. Synchronous Graphic RAM (SGRAM) synchronizes with the system CPU bus clock, like SDRAM, up to 100 MHz, for the fastest video RAM in use today.

## Composite Monitor Adapter

Composite video is exactly the video that is used in television. The waveform is shown in Figure 6.2. It can be accessed on most CGA cards from a standard 4-pin .1-inch header.

**FIGURE 6.2    Composite video waveform**

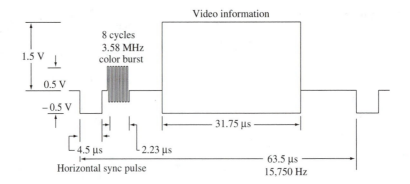

This signal will run a composite monitor directly, and can be connected to an RF modulator that can be connected to a standard TV. Although it gives 640 horizontal by 200 vertical pixels black and white, or 320 by 200 in four colors, the modulation and subsequent demodulation by the TV tuner make it much less crisp than CGA. Both signals are available on the CGA adapter. Color tint can be adjusted by a trimmer capacitor on the motherboard. Video adapter cards are controlled by accessing their I/O addresses, given in Table 6.4.

## MDA Monochrome Monitor Adapter

The monochrome display gives text only (with some new MonoGraphics Cards giving graphics) in a 720 × 348 pixel format. It uses a 9-pin female D-shell on the monochrome monitor adapter (MMA) card. Pin 1 is ground on the MDA, CGA, and EGA adapters. Pin 1 is *not* ground on the high-density 15-pin VGA. The monochrome card usually has a parallel port also, to conserve expansion slots. It has a single 16.257 MHz crystal oscillator. Many early cards used the MC6845 CRT controller. Table 6.5 gives the MDA signal names and connector pinout, and Figure 6.3 shows signals from the card with no monitor attached.

The card has a character-generator ROM (usually 24-pin, usually 8K) and 4K of RAM (using eight 2114s). The first four 2114s hold the characters on the display at that time, and the second four hold the attributes of those characters.

A simple block diagram of a TTL monochrome monitor is shown in Figure 6.4. Note that the input signals from the display adapter card need no color signals, and the high-voltage section need develop only about 15,000 volts, instead of the 30,000 volts common in color displays.

**TABLE 6.4    Video I/O port addresses**

| Address | Device |
|---|---|
| 3B4H, 3B5H, 3B8H, 3BAH, 3BCH, 3BDH, 3BEH, 3D1H–3D8H | Mono (MDA) |
| 2B0H TO 2DFH | Alternate Enhanced Graphic Adapter (EGA) |
| 3B0H TO 3BFH | Monochrome Display and Printer Adapter |
| 3C0H TO 3CFH | Enhanced Graphic Adapter (EGA) |
| 3D0H TO 3DFH | Color Graphics Adapter (CGA) |

**TABLE 6.5    Mono/text signals and MMA pinout**

| Pin | Signal |
|---|---|
| 1, 2 | Ground |
| 3, 4, 5 | None |
| 6 | Intensity |
| 7 | Video |
| 8 | Horizontal |
| 9 | Vertical |

**FIGURE 6.3   Monochrome video signals**

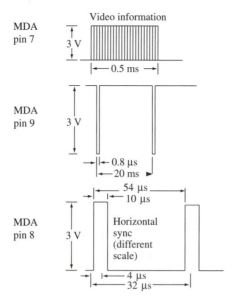

## CGA Monitor Adapter

CGA uses a 9-pin D-shell female on the card and male on the monitor cord for RGB (Red, Green, Blue) video, and a 4-pin standard 1-inch header for the composite video. The names of the signals are shown in Table 6.6 and the signal waveforms are shown in Figure 6.5 for RGB. (Waveforms for composites were shown in Figure 6.2.)

The CGA can often be recognized by its lack of a crystal oscillator; it uses the 14.31818-MHz OSC signal from B30 of the expansion slot bus. This signal can be adjusted to control the color tint by adjusting the trimmer capacitor on the motherboard for 14.31818 MHz. (If the card is a color/mono card, it will have the 16-MHz crystal for mono.) It also uses the 6845 and will have a character-generator ROM. It has 16K of RAM, perhaps in two 4164s. The 4-pin 0.1 inch header for composite is not on all cards, but it is on most.

This card can give twenty-five lines of eighty characters (or forty characters) of sixteen colors in text mode, and in graphics mode, 640 horizontal by 200 vertical pixels of two colors, or 320 × 200 pixels of four colors. It's no wonder everybody switched so quickly to VGA. CGA cards were mostly 8-bit ISA bus expansion cards.

**FIGURE 6.4   Simple TTL monochrome monitor block diagram**

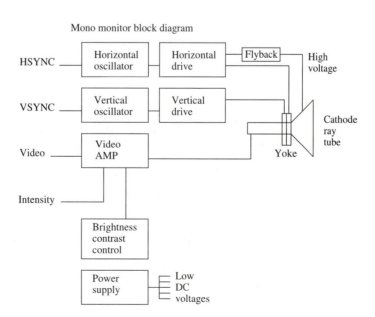

**TABLE 6.6   CGA signal names and CGA pinout**

| 9-Pin D-Shell Pin | Signal (RGB) |
|---|---|
| 1, 2 | Ground |
| 3 | Red video info |
| 4 | Green video info |
| 5 | Blue video info |
| 6 | Intensity |
| 7 | None |
| 8 | Horizontal |
| 9 | Vertical |

| 4-Pin Header Pin | Signal (Composite) |
|---|---|
| 1 | +12 V |
| 2 | |
| 3 | Video |
| 4 | Ground |

**FIGURE 6.5   CGA waveforms**

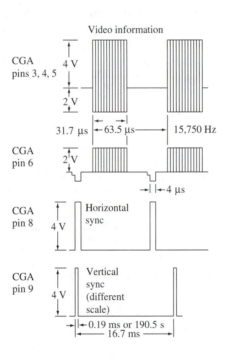

Figure 6.6 is a simple block diagram of a CGA monitor. Note the separate red, green, and blue inputs. Recall that the CRT has three electron guns and that each gun is aimed at dots of one of the three colors, red, green, and blue. Because this monitor is digital or TTL, the guns must either be on or off. With three guns it should give only $2^3$ or eight colors. However, note from Table 6.6 that pin 6 is intensity. With pins 3, 4, and 5, this gives us four control lines, so the CGA actually provides $2^4$ or 16 colors as the absolute maximum.

## Testing CGA Video RAM

Table 6.7 gives the video RAM and ROM locations for various monitor types. Note that the CGA card has onboard RAM at addresses B8000H–BBFFFH. This RAM is often in the physical form of two 4416s, which are 16K by 4 bits wide. This RAM can be tested using the program given in Program Listing 6.1.

**FIGURE 6.6   Simple CGA color monitor block diagram**

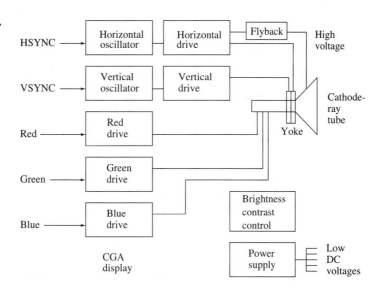

**TABLE 6.7   Video RAM and ROM locations**

| Type | RAM | | ROM | |
|------|-----|---|-----|---|
| MDA | B0000H–B1000H | 4K | In BIOS | |
| CGA | B8000H–BBFFFH | 16K | In BIOS | |
| EGA | A0000H–BFFFFH | 64K | C0000H–C3FFFH | 16K |
| VGA | A0000H–BFFFFH | 128K | C0000H–C5FFFH | 24K |

**PROGRAM LISTING 6.1
Testing CGA video RAM**

```
10   REM Testing CGA video RAM. Name of program is CGRAMTST.BAS
20   ON ERROR GOTO 200
30   DIM NEXBIT(8)
40   FOR I = 0 TO 7 :REM get 1,2,4,8,16,32,64,128 for bit
     positions
50   NEXBIT(I)= 2^I
60   NEXT I
70   DEF SEG = &HB800 :REM the beginning of 16K of RAM on adapter
     card. Remember this is shifted left one hex digit to = HB8000.
80   CLS:PRINT "This test takes up to 4 minutes. Please stand by!"
90   FOR I = 1 TO 5000:NEXT I :REM time delay
100  SCREEN 2 :KEY OFF:CLS :REM high res
110  FOR BYTE = 0 TO 16383 :REM 16K :REM Move through 16K
120  FOR BIT = 0 TO 7 :REM Move through 8 bits (0-7)
130  POKE BYTE,NEXBIT(BIT):REM Place a 1 in bit position
140  IF PEEK(BYTE) = NEXBIT(BIT) THEN POKE BYTE,0 ELSE
     RAMERR = BYTE:GOSUB 200
150  IF PEEK(BYTE)<>0 THEN RAMERR=BYTE:GOSUB 200
160  NEXT BIT
170  NEXT BYTE
180  CLS:BEEP:BEEP:PRINT "RAM Tests Good!"
190  END :REM main
200  REM ERROR ROUTINE
210  CLS:IF RAMERR >0 THEN PRINT "RAM ERROR AT SEGMENT B800H,
     OFFSET ";BYTE:END
220  IF ERR = 5 AND ERL = 100 THEN PRINT "CGA COLOR monitor not
     responding, test stopped!":END
230  RETURN
```

## EGA Monitor Adapter

While the CGA monitor gave reasonable high-resolution graphics (at least horizontally), the text character was made of an 8 × 8 pixel array and was therefore poorly formed. A CGA monitor can cause eye strain for those reading text for long periods. Some users overcame this problem by using an MDA for text and a CGA for graphics.

The EGA adapter and monitor solved this problem with an 8 × 14 character array. The graphics was improved to 640 × 350 with four colors. Table 6.8 gives the EGA signal names and connector pinout. The EGA card has a 9-pin female D-shell jack, so it is recognizable as a digital or TTL monitor. Note, however, that although you can only have a high or a low on any particular line, there are now two lines for each electron gun, which allows each of the three guns to have two intensities. (The CGA intensity controlled all three guns at once.) This gives six control lines for $2^6$ or sixty-four possible color combinations. Most EGA cards were made for the 8- or 16-bit ISA bus.

With the advent of VGA displays, EGAs are no longer constructed. If an EGA display is beyond practical repair, it may need to be replaced with a VGA display. A new VGA display adapter to drive the display to its maximum capabilities is certainly worth the investment.

## VGA Monitor Adapter

With the higher resolution displays available in the 1990s, a new, simpler human/computer interface was possible. Instead of memorizing and typing in dozens of DOS commands, a number of GUIs (Graphical User Interfaces) developed, one of which is WINDOWS. With a GUI the user sees graphical "Windows" (different sections of the screen used for different functions or programs), icons (small action-oriented pictures, such as a printer), menus, and pointing devices (mice, trackballs, touchscreens). GUIs are supposed to make it easier for the user to get real work done. Some old DOS diehards prefer the keyed commands and call a GUI a "WIMP" (Windows, Icons, Menus, Pointers) interface.

A drawback of GUIs is that they load down the CPU with huge amounts of video data. To speed the rate at which the video data is displayed on the screen and to relieve the main CPU of this time-consuming task, video accelerator cards have been designed. Video accelerator cards have circuits and whole chips specifically designed to handle the graphics commands and even memory designed just for video cards (VRAM).

VGA monitors have the resolution (640 × 480) and colors (sixteen) necessary for GUIs. The card has a high-density 15-pin D-shell connector, showing they are analog monitors. While this seems to allow infinite color possibilities, memory is still digital, and the color information must be stored in memory. With more memory, VGA cards can show more colors. VGA adapter cards were made for the 16-bit ISA bus, the VESA local bus, and the EISA bus. You must match the card you buy to the proper expansion bus on your system motherboard. VGA waveforms are shown in Figure 6.7. Table 6.9 gives the VGA signal names and pinout.

**TABLE 6.8   EGA signal names and EGA pinout**

| 9-Pin D-Shell | EGA |
|---|---|
| 1 | Ground |
| 2 | Secondary red |
| 3 | Primary red |
| 4 | Primary green |
| 5 | Primary blue |
| 6 | Secondary green |
| 7 | Secondary blue |
| 8 | Horizontal |
| 9 | Vertical |

**FIGURE 6.7 VGA waveforms**

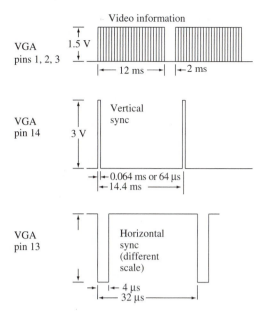

**TABLE 6.9 VGA signal names and VGA pinout**

| Pin | Signal |
|-----|--------|
| 1 | Red |
| 2 | Green |
| 3 | Blue |
| 4 | None |
| 5 | Self-test |
| 6 | Red ground |
| 7 | Green ground |
| 8 | Blue ground |
| 9 | None |
| 10 | Digital ground |
| 11 | None |
| 12 | None |
| 13 | Horizontal sync |
| 14 | Vertical sync |
| 15 | None |

**Testing VGA Video RAM** Program Listing 6.2 is a program that can be run in BASIC (make sure your version handles video mode 12, as QBasic does). It tests 128K of RAM on the VGA adapter, starting at absolute address A0000H or 640K.

**PROGRAM LISTING 6.2**
**Testing VGA Video RAM**

```
10    REM Testing VGA video RAM. Name of program is VGRAMTST.BAS
15    ON ERROR GOTO 1000: REM if not VGA, bail out
20    DIM NEXBIT(8)
25    DEFLNG B
30    FOR i = 0 TO 7: REM get 1,2,4,8,16,32,64,128 for bit positions
40    NEXBIT(i) = 2 ^ i
50    NEXT i
60    DEF SEG = &HA000: REM the beginning of 128K of RAM on adapter
      card
```

```
65    CLS : PRINT "This test takes up to 20 minutes. Please stand by!"
66    FOR i = 1 TO 5000: NEXT i: CLS
70    SCREEN 12: KEY OFF: CLS : REM high res
75    FOR sgmnt = 1 TO 2
76    IF sgmnt = 1 THEN DEF SEG = &HA000: REM first 64K bytes
77    IF sgmnt = 2 THEN DEF SEG = &HB000: REM second 64K bytes
80    FOR byte = 0 TO 65533: REM Move through 64K at a time
90    FOR BIT = 0 TO 7: REM Move through 8 bits (0-7)
100   POKE byte, NEXBIT(BIT): REM Place a 1 in bit position
110   IF PEEK(byte) = NEXBIT(BIT) THEN POKE byte, 0 ELSE
      ramerr = sgmnt: ofsterr = byte
112   POKE byte, 0: REM Place all 0's in byte
114   IF PEEK(byte) <> 0 THEN ramerr = sgmnt: ofsterr = byte
120   NEXT BIT
125   IF byte MOD 2000 = 0 THEN PRINT "."; : REM show test is
      progressing
130   NEXT byte
135   NEXT sgmnt
138   CLS : FOR i = 1 TO 5: BEEP: NEXT i: REM Notify user of end of
      test.
140   IF ramerr = 1 THEN PRINT "RAM Error, Segment = &HA000 ,
      Offset = "; ofsterr ELSE PRINT "RAM Tests Good!"
150   IF ramerr = 2 THEN PRINT "RAM Error, Segment = &HB000 ,
      Offset = "; ofsterr ELSE PRINT "RAM Tests Good!"
160   END: REM main
1000  REM ERROR subroutine
1010  IF ERR = 5 AND ERL = 70 THEN PRINT "VGA Not Responding, test
      stopped!": END
```

The program in Program Listing 6.3 must be run in a BASIC that can handle screen modes to 13, such as QBasic. It can use some modifications and enhancements, but as is it does a basic screen mode test of modes 1 (CGA low resolution), 2 (CGA high resolution), 9 (EGA high resolution), 12 (VGA high resolution), and 13 (VGA low resolution). Table 6.10 gives the QBasic video screen modes used in this program. Table 6.11 gives the screen modes in DOS.

**PROGRAM LISTING 6.3**
**Testing screen graphics modes to VGA**

```
4     ON ERROR GOTO 1000
5     KEY OFF: CLS
9     SCREEN 1: REM CGA 320 x 200, 4 color ************************
10    COLOR 0, 0
11    C = 0
12    LINE (1, 1)-(319, 199), , B
13    FOR I = 0 TO 6.28 STEP .3
14    Y = 40 * SIN(I) + 75: X = 40 * COS(I) + 160
15    C = ((C + 1) MOD 3) + 1
16    CIRCLE (X, Y), 10, C
17    NEXT I
19    X = 110: Y = 130
21    FOR CLR = 1 TO 3
22    X = X + 20
23    LINE (X, Y)-(X + 20, Y + 50), CLR, BF
26    NEXT CLR
27    LOCATE 2, 10: PRINT "CGA 320 x 200, 4 color"
28    LOCATE 3, 10: INPUT "Press ENTER to continue", A$
29    SCREEN 2: REM CGA 640 x 200, 2 color ************************
30    C = 0
40    LINE (1, 1)-(639, 199), , B
41    FOR I = 0 TO 6.28 STEP .3
```

```
42    Y = 40 * SIN(I) + 75: X = 40 * COS(I) + 320
43    C = C + 1
44    CIRCLE (X, Y), 10, C
45    NEXT I
49    LOCATE 2, 30: PRINT "CGA 640 x 200, 2 color"
50    LOCATE 3, 30: INPUT "Press ENTER to continue", A$
110   SCREEN 9: REM 640*350 16 colors 0-15 ***********************
111   LINE (1, 1)-(639, 349), , B
115   RANDOMIZE TIMER
120   FOR I = 0 TO 6.28 STEP .5
130   Y = 50 * SIN(I) + 120: X = 50 * COS(I) + 320
135   C = INT(15 * RND) + 1
140   CIRCLE (X, Y), 20, C
150   NEXT I
153   R = 15: C = 30
154   LOCATE R, C
155   FOR j = 1 TO 10
160   FOR I = 1 TO 15
170   COLOR I
180   PRINT CHR$(219); CHR$(219);
190   NEXT I
200   R = R + 1: LOCATE R, C
209   NEXT j
210   LOCATE 2, 30: PRINT "EGA 9 640 x 350, 16 colors"
211   LOCATE 3, 30: INPUT "Press ENTER to continue", A$
410   SCREEN 12: REM 640*480 16 colors 0-15 *********************
411   LINE (1, 1)-(639, 479), , B
415   RANDOMIZE TIMER
420   FOR I = 0 TO 6.28 STEP .5
430   Y = 50 * SIN(I) + 200: X = 50 * COS(I) + 320
435   C = INT(15 * RND) + 1
440   CIRCLE (X, Y), 20, C
450   NEXT I
453   R = 20: C = 30
454   LOCATE R, C
455   FOR j = 1 TO 10
460   FOR I = 1 TO 15
470   COLOR I
480   PRINT CHR$(219); CHR$(219);
490   NEXT I
500   R = R + 1: LOCATE R, C
509   NEXT j
510   LOCATE 2, 30: PRINT "VGA 12 640 x 480, 16 colors"
511   LOCATE 3, 30: INPUT "Press ENTER to continue", A$
512   SCREEN 13: REM 320*200 256 colors 0-255 *********************
513   LINE (1, 1)-(319, 199), , B
520   FOR I = 0 TO 6.28 STEP .3
530   Y = 40 * SIN(I) + 75: X = 40 * COS(I) + 160
535   C = INT(I * 3): IF C < 1 OR C > 15 THEN C = C + 40
540   CIRCLE (X, Y), 10, C
550   NEXT I
552   LOCATE 18, 1
555   REM FOR j = 1 TO 12
560   FOR I = 0 TO 239
570   COLOR I
580   PRINT CHR$(219);
590   NEXT I
600   PRINT
610   REM NEXT j
```

```
611   COLOR 7
615   LOCATE 2, 2: PRINT "VGA MODE 13 320 x 200, 256 colors"
620   LOCATE 3, 2: INPUT "Press ENTER to continue", A$
900   CLS : PRINT "LAST MODE SHOWN WAS HIGHEST VIDEO MODE"
905   PRINT "SUPPORTED ( TO 13)"
910   END
1000  REM ERROR ROUTINE  ******************************************
1010  IF ERR = 5 AND ERL = 9 THEN CLS : PRINT "CGA MODE 1 320 x 200
      NOT SUPPORTED"
1020  IF ERR = 5 AND ERL = 29 THEN CLS : PRINT "CGA MODE 2 640 x
      200 NOT SUPPORTED"
1030  IF ERR = 5 AND ERL = 110 THEN CLS : PRINT "EGA MODE 9 640 x
      350 NOT SUPPORTED"
1040  IF ERR = 5 AND ERL = 410 THEN CLS : PRINT "VGA MODE 12 640 x
      480 NOT SUPPORTED"
1050  IF ERR = 5 AND ERL = 512 THEN CLS : PRINT "VGA MODE 13 320 x
      200 NOT SUPPORTED"
1060  RESUME 900
```

**TABLE 6.10  QBasic video modes (screen modes)**

| | |
|---|---|
| SCREEN 0: | Text mode only<br>$40 \times 25$ or $80 \times 25$<br>16 colors assigned to any of 16 attributes (with CGA or EGA)<br>64 colors assigned to any of 16 attributes (with EGA or VGA) |
| SCREEN 1: | $320 \times 200$ graphics<br>16 background colors and one of two sets of 3 foreground<br>   colors assigned using COLOR statement with CGA<br>16 colors assigned to 4 attributes with EGA or VGA |
| SCREEN 2: | $640 \times 200$ graphics<br>16 colors assigned to 2 attributes with EGA or VGA |
| SCREEN 3: | Hercules adapter required, monochrome monitor only<br>$720 \times 348$ graphics |
| SCREEN 4: | Supports Olivetti Personal Computers<br>Supports AT&T Personal Computers 6300 series<br>$640 \times 400$ graphics<br>1 of 16 colors assigned as the foreground color (selected by the COLOR<br>   statement); background is fixed at black, PALETTE statement not<br>   supported |
| SCREEN 7: | $320 \times 200$ graphics<br>$40 \times 25$ text format, $8 \times 8$ character box<br>Assignment of 16 colors to any of 16 attributes |
| SCREEN 8: | $640 \times 200$ graphics<br>$80 \times 25$ text format, $8 \times 8$ character box<br>Assignment of 16 colors to any of 16 attributes |
| SCREEN 9: | $640 \times 350$ graphics<br>$80 \times 25$ or $80 \times 43$ text format, $8 \times 14$ or $8 \times 8$ character box<br>16 colors assigned to 4 attributes (64K adapter memory), or<br>64 colors assigned to 16 attributes (more than 64K adapter memory) |
| SCREEN 10: | $640 \times 350$ graphics, monochrome monitor only<br>$80 \times 25$ or $80 \times 43$ text format, $8 \times 14$ or $8 \times 8$ character box<br>Up to 9 pseudocolors assigned to 4 attributes |

**TABLE 6.10** *(continued)*

SCREEN 11: VGA or MCGA
640 × 480 graphics
80 × 30 or 80 × 60 text format, 8 × 16 or 8 × 8 character box
Assignment of up to 256K colors to 2 attributes

SCREEN 12: VGA
640 × 480 graphics
80 × 30 or 80 × 60 text format, 8 × 16 or 8 × 8 character box
Assignment of up to 256K colors to 16 attributes

SCREEN 13: VGA or MCGA
320 × 200 graphics
40 × 25 text format, 8 × 8 character box
Assignment of up to 256K colors to 256 attributes

**TABLE 6.11  DOS video screen modes**

| Mode | Adapter | Pixels | | Characters | | Colors |
| --- | --- | --- | --- | --- | --- | --- |
| | | Horizontal | Vertical | Row | Column | |
| 0 Text | CGA | 320 | 200 | 25 | 40 | 16 |
| | EGA | 320 | 350 | 25 | 40 | 16 |
| | VGA | 360 | 400 | 25 | 40 | 16 |
| 1 Text | CGA | 320 | 200 | 25 | 40 | 16 |
| | EGA | 320 | 350 | 25 | 40 | 16 |
| | VGA | 360 | 400 | 25 | 40 | 16 |
| 2 Text | CGA | 640 | 200 | 25 | 80 | 16 |
| | EGA | 640 | 350 | 25 | 80 | 16 |
| | VGA | 720 | 400 | 25 | 80 | 16 |
| 3 Text | CGA | 640 | 200 | 25 | 80 | 16 |
| | EGA | 640 | 350 | 25 | 80 | 16 |
| | VGA | 640 | 400 | 25 | 80 | 16 |
| 4 Graphic | All | 320 | 200 | 25 | 40 | 4 |
| 5 Graphic | All | 320 | 200 | 25 | 40 | 4 |
| 6 Graphic | All | 640 | 200 | 25 | 80 | 2 |
| 7 Text | EGA | 720 | 350 | 25 | 80 | 2 |
| | VGA | 720 | 400 | 25 | 80 | 2 |
| D Graphic | EGA | 320 | 200 | 25 | 40 | 16 |
| | VGA | 320 | 200 | 25 | 40 | 256 |
| E Graphic | EGA | 640 | 200 | 25 | 80 | 16 |
| | VGA | 640 | 200 | 25 | 80 | 256 |
| F Graphic | EGA | 640 | 350 | 25 | 80 | 2 |
| | VGA | 640 | 350 | 25 | 80 | 2 |
| 10 Graphic | EGA | 640 | 350 | 25 | 80 | 16 |
| | VGA | 640 | 350 | 25 | 80 | 256 |
| 11 Graphic | VGA | 640 | 480 | 30 | 80 | 2 |
| 12 Graphic | VGA | 640 | 480 | 30 | 80 | 16 |
| 13 Graphic | VGA | 320 | 200 | 25 | 40 | 256 |

## SUPER VGA (SVGA) MONITOR ADAPTERS

Following VGA, the Super Video Graphics Array (SVGA) monitor adapter cards started at 800 × 600 resolution and have kept rising in resolution. They use the same high-density 15-pin D-shell connector. With higher resolution comes the cost of more video memory. It takes 8 MB of video memory to have 1280 × 1024 true color resolution (16.8 million colors) with a refresh rate of 85 Hz. At 1600 × 1200, a card like this can handle only 65536 colors and a 75 Hz refresh rate. 1800 × 1440 pixel resolution is available for a price. Some newer video cards have 32-bit, full-motion, full-duplex video for video conferencing. This supports two-way video phone communication for home or business. A card with an NTCS output jack will connect directly to your TV set for presentations; and cards are made with the television tuner on board for watching TV on your computer or capturing frames or video from a TV broadcast.

## ACCELERATED GRAPHICS PORT (AGP)

As the need for speed has increased, video cards have been produced for the PCI bus for higher overall speed. A new bus was designed just for the video. The Accelerated Graphics Port (AGP) is based on extensions and enhancements of the PCI bus. With its own bus, the AGP bus has direct memory access to the main memory, thereby not slowing down the CPU with video tasks. The AGP can double or triple three-dimensional performance. There is a special display adapter connector on the motherboard. In Figure 6.8a note the dark AGP motherboard connector just above the PCI connectors. In Figure 6.8b note an AGP adapter card.

There is some controversy about this port. Some think this will overburden the main CPU and system RAM, slowing overall system response. The marketplace will determine the success of AGP. Intel's 440LX, BX, and EX chipsets support the AGP bus, so it may well be the new standard for graphics video cards.

## WINDOWS 95 AND 98

Click on Start, Settings, Control Panel, Display. The Background tab allows you to select a wallpaper. The Screen Saver tab allows you to choose from a few screen savers and set energy-saving features of monitors that support those features. You can set the system to automatically turn the monitor and hard drive off after selected times. The Appearance tab lets you set the color and scheme of your Windows. The Effects tab lets you set a number of visual effects, such as large or small icons, animated windows, and smoothing the edges of the screen fonts. The Web tab allows you to view your desktop as a Web page. Perhaps the most important is the

**Figure 6.8  (a) The AGP motherboard connector**

**Figure 6.8** (*continued*) (b) an
AGP adapter card

Settings tab. Here you choose the number of colors and resolution your monitor will display. Of course your monitor and video adapter card must be capable of displaying the settings you choose. In case you wonder why anyone would have the settings less than maximum, consider that a 1600 × 1200 resolution makes small, hard-to-read print, icons, and windows. You may have an Advanced button on this Settings window. This gives a General tab that lets you change font size and some compatibility parameters. The Adapter tab gives you the type of video card installed. The Monitor tab shows the type of monitor installed. The Performance tab allows you to set the graphics hardware acceleration anywhere from none to full. The Color Management tab lets you select a color profile to fine-tune your monitor's color display.

## 6.2 GENERIC SYSTEM OPERATION

(**Note:** This section is recommended for technicians.)

### VIDEO ADAPTER CARD

When there is no video, a quick oscilloscope check of the waveforms on the jack on the rear of the video card tells you whether the card is working. If it is not, try cleaning the edge connector contacts and reseating it. If it still is not working, replace it. If the signals appear fine, the problem is in the monitor. Since the general scope of this book is failed systems that once worked, you know the monitor matched the card. However, some cards have a DIP switch accessible through the rear mounting plate. Someone may have changed these. Find documentation and confirm settings. If no documentation is available, there are only sixteen possible combinations of a four-pole switch, so take good notes and try them all. Of course, no one needs to tell you to power-down first, change the switches, and power back up.

Someone also may have changed the motherboard video switch or jumper. Check those also. If installing a new monitor, you must confirm that the card and the monitor are compatible.

### VIDEO DISPLAY OR MONITOR

**DANGER:** If you are uncomfortable around high voltage, don't even open the cover of a display. The high voltage can be from 12,000 to 28,000 V DC, and the boost voltage can range from 200 to 850 V DC. The CRT anode is usually at the top of the tube, with a fat (1/4-inch diameter) red wire going to it. It has a large (2-inch diameter) round rubber boot over it for your protection. A charge can remain on this line for days after the unit is turned off. *Never* *assume* that it is *not* hot—*always assume that it is hot*. Another danger is implosion of the tube. The neck is the weakest point. A third danger is that some units have a *hot* chassis. Besides frying yourself, you can destroy your scope if you are not using an isolation transformer.

Following a block diagram for a general monitor (Figure 6.4 or 6.6), note the DC power supply. On some systems, this supply is almost completely separate from the video board. While a few power supplies are still linear, many are now switching supplies, similar to the PC's supply. Measure the output of the DC supply, and compare it to your documentation. Since you're usually flying blind, be prepared for DC voltages up to 300 V. Table 6.12 gives typical DC voltages.

If you can disconnect the DC supply from the rest of the system, that is a good point to start the divide-and-conquer troubleshooting technique. Figure 6.9 shows a typical low-voltage power supply for a display. This is very similiar to the system power supply discussed in Chapter 2. Note the small transformers for the oscillator at the top right and the main transformer at the top left, the size of which tells us this is a switcher. The large electrolytics in the center hold more than 100 V DC, so beware. At the bottom left are two heat sinks for the power transistors. The small electrolytics at the top left are for final filtering. Except for the transformer, generic parts are available, so it is reasonable to attempt the repair of the switching low-voltage power supply.

The horizontal sync controls timing of a horizontal sweep sawtooth signal, which goes to a horizontal output driver. This is usually a TO-3 or similar high-power transistor mounted on a heat sink 2 inches square. One or two of these will drive the flyback. Check for a 5–15 V signal on the base. Figure 6.10 shows two large heat sinks with the horizontal driver and output mounted.

**TABLE 6.12  Video display typical DC voltages**

| Monitor 1 | +230 | +85 | +33 | +12 |
|---|---|---|---|---|
| Monitor 2 | +148 | +109 | | |
| Monitor 3 | +12.07 | +12 | +11.84 | |

**FIGURE 6.9  Low-voltage switching supply for a display**

**FIGURE 6.10  Horizontal driver and output with heat sinks**

The high voltage is usually developed from the horizontal output going through a **flyback transformer.** The flyback develops a number of voltages, and changing it is the pits. A cheap flyback costs $21, and changing it will take you more than an hour. What may be out, however, is the rectifier. Some of these high-voltage rectifiers are monolithic and easy to change. Just make sure you've discharged everything before testing. Table 6.13 gives typical yoke voltages, and Table 6.14 gives typical resistances. Figure 6.11 shows the overall view, with the anode lead and rubber boot. The low-voltage power supply is in the cage to the left. Video circuits are in the cage to the right. Figure 6.12 shows the flyback transformer on the top and the high-voltage tripler diodes and caps to the bottom.

**TABLE 6.13  Video display typical yoke AC voltages (pulses or spikes)**

| Horizontal | +200 V peak |
|---|---|
| Vertical | +12 V peak |

**TABLE 6.14  Video display typical resistance values**

| Location | From + (Red) | To − (Black) | Resistance (VΩ) | |
|---|---|---|---|---|
| Yoke horizontal | | Across | 2 | |
| Yoke vertical | | Across | 13 | |
| Horizontal drive transistor | C | B | 120 | |
| | C | E | 400 | |
| | B | C | 60K | electrolytic charging |
| | E | C | 50K | electrolytic charging |
| Horizontal transformer primary | | Across | 70 | |
| Horizontal transformer secondary | | Across | 0.2 | |
| Horizontal output transistor | C | B | 10 | in circuit |
| | E | B | 0.1 | in circuit |
| | B | C | 60K | in circuit cap charging |
| | B | E | 0.1 | in circuit |
| Flyback (10-pin) | | | | |
| Pin 1 | | 2, 3, 4, 5, 7, 9 | <1 | |
| Pin 10 | | 5, 7, 9 | 10 | |
| Pin 10 | | Horizontal yoke 1 | 0 | |
| Pin 10 | | Horizontal yoke 2 | 2 | |

**FIGURE 6.11  High-voltage
anode and boot**

**FIGURE 6.12  Flyback and
high-voltage rectifiers**

Flyback

High-voltage rectifiers

If you have a number of identical monitors to repair, you can isolate the problem with each one to power supply, logic board, yoke, flyback, or CRT. In that case, cannibalization can bring a few systems back to life quickly.

Figure 6.13 gives a basic generic schematic diagram of the horizontal output section of a typical display with the flyback transformer. Because of the high voltages, it is quite reasonable that this is the section that seems to fail most often in TVs and monitors. The CRT with pinouts is also shown.

**Warning:** Because the CRT is a sealed, evacuated, glass container, having no air inside to interfere with the electron beam or ray, if it is broken, it will implode and send shards of glass all over. The thin neck is the most likely point to break, so use much caution when touching the picture tube. Also, there are up to 35,000 V on the anode and the flyback transformer. At the end of the small neck of the CRT are the pins that fit into a socket; one of these pins has up to 1000 V. A typical color CRT with voltages is shown in Figure 6.13. The whole purpose of the display is to make the CRT work and show video. The function of the tube can best be explained by examining the elements involved. The elements in the CRT make up a device known as an electron gun because it "shoots" electrons.

The first element is a filament or heater. The filament has from 5 to 12 V AC or DC across it. It glows with heat and heats up the next element, the cathode. This energy input makes the cathode electrons more energetic, raising them to a higher energy level, ready to leave orbit and be free electrons.

The cathode (terminal for a negative charge) is the second element in the CRT. A positive voltage of 40 to 150 V DC is applied to the cathode. Here the actual video information in volts AC is applied to vary the energy of the electron beam. This signal is controlled by the red, blue, and green video signals.

The next element is the first grid, or grid 1. This can be thought of as a disk with a hole in the middle, a washer, or a flat donut. Its voltage varies from 0 V (grounded) to the cathode voltage, but it is usually less than the cathode voltage. This means that the grid 1 voltage is negative with respect to the cathode voltage, suppressing the electron flow when the grid 1 voltage is lower, and increasing electron flow when it is higher. The brightness control is sometimes applied to this element.

The next element is grid 2, called the first anode on mono CRTs and the screen on color CRTs. Voltage is from 300 to 500 V and is usually controlled by a screen control, sometimes located on the small PCB connected to the socket for the CRT and sometimes located on the flyback transformer.

**FIGURE 6.13 Horizontal output and flyback transformer**

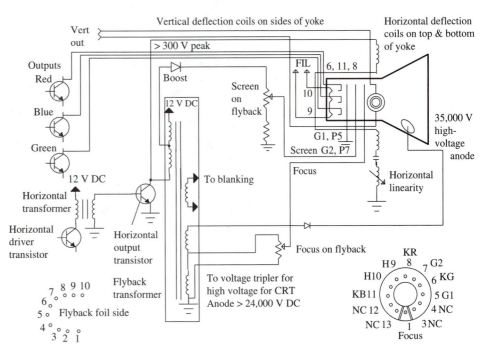

The next element is grid 3, the focus grid electrode. Its control usually is mounted on the flyback transformer next to the screen control; it is typically 3 to 7 kV. For safety, this pin (usually pin 1) is separated from the others in the socket "key" by plastic dividers. Also, the red wire coming from the flyback transformer usually goes directly into the plastic socket, not to a pin accessible from the back of the CRT PCB. To prevent radio frequency interference (RFI), the CRT PCB often has a metal plate shield with tabs soldered to ground on the board. This of course makes it much more difficult to examine these voltages. This element is sometimes labeled G4 for grid 4, and grid 3 is connected to the high voltage. If that is the configuration, the voltage on G3 will be 0 to 500 V. G3 would then *not* be accessible at the socket.

The final element is the second anode, sometimes called the accelerating anode. This is the high-voltage anode, and it is connected to a conductive aquadag coating on the inner walls of the CRT from this point up to the phosphors on the face. The voltage is 15 kV or so on a mono CRT, and 25 to 35 kV on a color CRT. **Caution:** Never measure this with anything other than a special high-voltage probe.

Checking these voltages will tell you if the CRT is receiving the voltages necessary to operate properly.

## 6.3 ASSEMBLING VIDEO

Carefully plug the video adapter card into an expansion slot on the system board that matches the bus type of the card. Use 8-bit cards in 8-bit slots, if possible. Eight-bit cards usually work in all other slots, except PCI and AGP, but that keeps those slots from being available for higher-level cards. The leftmost slot (slot 0) facing the system as you would use it is most common, but any slot will work (Figure 6.14). Confirm the system board video jumper settings (mono/color), and check adapter documentation for video adapter card configuration jumper settings. Set rear adapter card bracket switches (if any) per documentation for matching display. On systems with CMOS memory (286 and higher), do the CMOS SETUP to set the display type. Eight-bit adapter cards may not fit in 16-bit slots because of the rear "skirt" that hangs down behind the 8-bit slot. What seems almost amazing is that *some* 16-bit cards are designed to be able to function in 8-bit slots, even though the 16-bit extension connector dangles in space.

**FIGURE 6.14  Installing an AGP video adapter card**

A quick test of video can be made with only the motherboard, the power supply, the video card, and the display. If any video appears on the screen, the video is probably OK. On higher end cards, a diskette or CD of driver software is often included to maximize the card's performance. This software often automatically installs itself. Just read the manual, follow the installation instructions, and everything will go nicely and easily. Just in case that doesn't work, I offer the following.

Here are two lists showing you how to install the drivers. The first list is for automatic installation. (This may or may not be included with your software.) The second list is for manual installation.

For automatic device driver installation,

1. Change the active drive to A:

```
A:
```

2. Find the name of the program.

```
DIR *.EXE
DIR *.COM
DIR *.SYS
```

One of these should give a program with a recognizable name, such as INSTALL.EXE or SETUP.EXE.

3. Run that automatic installation program.

```
INSTALL (or SETUP)
```

For Windows 95 or 98, click Start, Run, type in your CD drive letter, such as D:, then Browse. Click on a file named SETUP or INSTALL and click OK.

For manual device driver installation,

1. Create a subdirectory with an appropriate name.

```
C:
MD\displayd
```

2. Change to that subdirectory.

```
CD\displayd
```

3. Copy all the files from the floppy to the hard drive.

```
COPY A:*.* C: /V
```

4. Identify the driver.

```
DIR *.EXE
DIR *.COM
DIR *.SYS
```

The name should be recognizable, though it may be a bit cryptic, such as VID1024.SYS.

5. Files ending in .COM or .EXE are run in the AUTOEXEC.BAT file. Files ending in .SYS are loaded in the CONFIG.SYS file. Edit each file as required. If you have no instructions from the documentation, try DIR *.DOC and DIR *.TXT and DIR READ*.*. If you find a file with a similar name, such as VIDEO.DOC, use the type command to read it: TYPE VIDEO.DOC ¦ C:\DOS\MORE. This will display the file on the screen page by page, allowing you to browse through it.

6. In Windows 95 and 98, click on Start, Settings, Control Panel, Display, Settings. Choose the number of colors and the resolution you want, up to the limit your monitor will display. If you have an Advanced button on this Settings window, click the Adapter tab to see if the type of video card installed is correct and the Monitor tab to see if the type of monitor installed is correct.

## 6.4 BASIC TROUBLESHOOTING

Troubleshooting a video problem begins by noting any POST audio or video error codes. Then you can proceed to noting any visual clues and taking appropriate measurements. Details of these steps are as follows. **Warnings:** Never use freeze spray around the CRT and use extreme caution around the high-voltage circuits. Also, wear safety glasses!

a. POST audio error code of one long and two short beeps, or one long and three short beeps, or one short and one long beep with bad video, or one short beep with bad video

b. POST video error code of 4xx for mono or 5xx for color

c. Measurements

1. Is the monitor/display plugged in? Is the power light on? If so, check brightness and contrast settings on monitor. Rub your arm on CRT face. Does the hair on your arm bristle from the high voltage? (Or does the CRT hold a small scrap of paper pressed on it?) If so, suspect adapter card. If not, it doesn't prove there is no high voltage, because some monitors are well shielded.

2. Check connections, switches, and jumpers on the system board and display adapter. EGA/VGA cards may have a 4PST DIP switch on the rear. Try all sixteen possible combinations by changing the switch to each and rebooting each time. Document your results.

3. Reseat or replace video RAM. Use a pencil eraser to clean the edge connector contacts, and reseat the video adapter board.

4. Power-down, remove all other cards, power-up. If the display now has video, repeat this step, adding one card each time, until the conflicting card is found.

5. When appropriate and possible, swap the adapter card for a known good video card, then swap the monitor for a known good monitor.

6. If your video is working enough to see writing on the screen, in Windows 95 or 98, click on Start, Help, Index, Display Troubleshooter. For other information, click on Start, Help, Index. Typing Display, Monitor, Video separately for dozens of topics of interest about your systems video.

(**Note:** Steps 7–14 are recommended for experienced technicians.)

7. Disconnect the monitor, power-up system, use scope to check signals on adapter jack per appropriate figure or table in this chapter. Good signals indicate a monitor problem, improper signals indicate an adapter card problem.

8. Disconnect monitor from system. Remove case, and bleed high-voltage anode to ground. **Caution:** Use your best safety techniques. Locate the outputs of the low-voltage DC supply. Plug into power. Measure the outputs using Table 6.7 as a guideline. If one or more is missing, you're in luck—it's one of the few blocks you can repair without good documentation. If all voltages are missing, the fuse is a good place to begin, then check the power switch.

9. Once low voltages are established, check high voltage *only* with a proper high-voltage meter. If you don't have one, you can usually tell if high voltage is there by the whining of the horizontal (15–18 kHz) and the crackling sound of high-voltage buildup on the CRT. Full high voltage should hold small scraps of paper to the CRT.

10. If DC voltages are good and you have high voltage but still no video (when connected to the adapter), try cleaning the brightness and contrast control potentiometers with a tuner cleaner.

11. Get a good TV repair manual. Note the meanings of typical problems.

> Horizontal white line, vertical drive out. Vertical white line, horizontal yoke out (if horizontal driver were out, you would not have high voltage, so no brightness). Rolling picture, vertical hold. Leaning picture or torn horizontally, horizontal hold.

12. If colors are wrong, there is a trimmer cap on the system board to adjust color.

13. Check horizontal driver and output transistors, and vertical driver transistor.

14. Measure resistances of flyback. Typical in-circuit values are from less than 1 Ω to 15 Ω.

1. The two basic categories of display monitor are
   a. Digital and analog.
   b. High and low voltage.
   c. Amber and green screens.
   d. High speed and low speed.

2. A CGA display has a resolution of
   a. $720 \times 350$.
   b. $640 \times 200$.
   c. $640 \times 350$.
   d. $640 \times 480$.

3. An SVGA monitor at minimum has a highest resolution of
   a. $720 \times 350$.
   b. $640 \times 200$.
   c. $640 \times 480$.
   d. $800 \times 600$.

4. Graphics mode in a display is
   a. A/N.
   b. APA.
   c. RGB.
   d. PDQ.

5. The maximum number of colors in graphics mode on a CGA display is
   a. 2.
   b. 4.
   c. 16.
   d. 256.

6. The maximum number of colors in graphics mode on a VGA display is
   a. 2.
   b. 4.
   c. 16.
   d. 256.

7. A 9-pin D-shell connector is used on what displays?
   a. Mono only
   b. Color only
   c. Digital only
   d. Analog only

8. A high-density 15-pin D-shell connector is used on what displays?
   a. Mono only
   b. Color only
   c. Digital only
   d. Analog only

9. An interlaced display shows
   a. Every line sequentially.
   b. Every odd line, then every even line.
   c. Only color lines.
   d. Only mono lines.

10. A multisync display
    a. Only works on digital adapter cards.
    b. Only works on analog adapter cards.
    c. Only works on mono adapter cards.
    d. Works on all adapter cards.

11. VRAM
    a. Can be used instead of dynamic RAM on the motherboard.
    b. Is virtual RAM.
    c. Is used on the motherboard for a video cache.
    d. Is special RAM for use on the video adapter card.

12. Who should open a display monitor case?
    a. Only highly experienced users
    b. Only electronic technicians
    c. Only electrical engineers
    d. Abe Lincoln

13. Displays that have signals with voltages of either 0 or 5 V are
    a. Digital.
    b. Analog.
    c. RGB.
    d. Noninterlaced.

14. Displays that have many more color choices are
    a. Digital.
    b. Analog.
    c. RGB.
    d. Noninterlaced.

15. Displays that are so fast they can show every line in order are
    a. Digital.
    b. Analog.
    c. RGB.
    d. Noninterlaced.

16. Displays with 15 pins in a high-density 15-pin D-shell connector are
    a. Digital.
    b. Analog.
    c. Interlaced.
    d. Noninterlaced.

17. Displays with a maximum graphics resolution of $640 \times 200$ are
    a. VGA.
    b. EGA.
    c. SVGA.
    d. CGA.

18. Displays with a maximum graphics resolution of $640 \times 480$ are
    a. VGA.
    b. EGA.
    c. SVGA.
    d. CGA.

19. The order of increasing resolution and number of colors is
    a. EGA, VGA, SVGA, CGA, SVGA+.
    b. CGA, EGA, VGA, SVGA, SVGA+.
    c. Mono, EGA, CGA, VGA, SVGA+.
    d. Green screen, amber, RGB, VGA.

20. Video display adapter cards all have
    a. 9-pin D-shell connectors.
    b. High-density 15-pin D-shell connectors.
    c. 32-bit edge card connectors.
    d. Random access memory on board.

21. Other than set external controls on the display, what can the microcomputer technician do to repair monitors?
    a. Replace the high-voltage flyback transformer
    b. Change the CRT for a new one
    c. Repair switching, low-voltage power supply
    d. Nothing

22. Video failure (not system) can be isolated to what two blocks?
    a. CPU or display
    b. Video adapter card or display monitor
    c. CMOS SETUP or display cable
    d. External controls or internal failures

23. If the failure is traced to the video adapter card, try
    a. Replacing RAM and cleaning edge card connector.
    b. Changing all jumpers on card.
    c. CMOS SETUP for video.
    d. Not to touch anything, just change the card.

24. If the failure is traced to the display monitor, try
    a. Adjusting all external controls, checking continuity of the wires in the cable from the adapter card, then replace display.
    b. Adjusting high voltage, moving purity magnets.
    c. Replacing horizontal output driver transistor.
    d. Replacing flyback transformer.

25. A quick check of a display for high voltage is
    a. Beyond a user's capability.
    b. Done with a multimeter.
    c. Done with a small bit of paper on the screen.
    d. Done by drawing an arc with a screwdriver from the CRT anode.

26. The best tool in diagnosing a "blank" display is
    a. A multimeter.
    b. A diagnostic program.
    c. Careful observation of symptoms.
    d. An oscilloscope.

27. If the display pilot is on but the screen is "blank," the speaker makes no beeps, the drives make no sound, and the power supply fan is running, then
    a. The display is definitely bad.
    b. The video adapter card is probably bad.
    c. The motherboard has a problem.
    d. The drives have a problem.

28. If the display pilot is on, but the screen is "blank," the speaker beeps once, and the drives make sounds, then
    a. The motherboard is definitely bad.
    b. The video adapter card or display probably has a problem.
    c. The system power supply definitely has a problem.
    d. The video adapter card definitely has a problem.

29. A PCI video card will fit in an AGP slot.
    a. True
    b. False

30. An AGP video adapter is mainly designed for
    a. More colors.
    b. Higher resolution.
    c. Higher-speed graphics.
    d. Faster math display.

31. What RAM was designed specifically for video adapter?
    a. DRAM
    b. SGRAM
    c. SDRAM
    d. SRAM

# 7

# Floppy Drives

**OBJECTIVES**

After completing this chapter you should be able to

- List differences between sizes of drives
- Identify the four common sizes by sight
- Identify a floppy controller card by sight
- List steps of physical preparation to install each drive (jumpers, terminators)
- Perform physical installation of each of the four types
- Perform system floppy setup
- Manage your floppies from Windows 95 and 98
- Perform tests of floppy drives
- Troubleshoot and correct problems

## 7.1 GENERAL DISCUSSION

The main memory (RAM) of a computer loses data when the system is turned off. A "permanent" memory was needed, and many of the earliest personal computers used a standard audio-cassette tape recorder. As tape streamed past the recording heads, data was recorded, one program or file after the other. This gave only sequential, or serial, access to the data. To look up the last file on the tape, you had to scan through the entire tape. Floppy disk drives were designed to overcome this problem and speed up the accessing of data and programs.

The floppy disk is a disk-shaped piece of plastic with an oxide coating very similar to the coating on a standard audiotape. Rather than being written sequentially, data is written in concentric rings or tracks. As the disk spins, the data in one track passes the recording head and can be written or read. The head itself also moves along the radius of the disk, from track to track. Therefore, the drive can jump quickly to any track, accessing information quickly. Modern drives have two heads, one for the top of the disk, one for the bottom.

All floppy drives have a door latch or button. Opening the door (which is usually just the latch) or pushing the door button physically lifts the heads from the floppy. Do not use the floppy door while the system is turned off—you should turn on the system before you close the floppy door or insert the floppy, and you should open the floppy door or push the button before you turn off the system. This keeps accidental spikes or pulses from damaging the information on your disk.

The capacity of your drive must be set up in the CMOS memory. This is usually done at boot up by pressing the DELETE key. Once in the CMOS SETUP, you can set the capacities of drives A: and B:. If these are set incorrectly, they either will not work or, worse, they will work halfway. You may get the directory but be unable to read a file, because the directory is on track 0, so no head movement is needed to read it. DOS also has a means of doing this in the CONFIG.SYS file. The command DRIVPARM = /D:0 /F:2 will set up drive A: as a 720K floppy drive. For the D: parameter, use 0 for A: and 1 for B:. For the F: parameter, use 0 for 360K 5.25", 1 for 1.2 MB 5.25", 2 for 720K 3.5", and 3 for 1.44 MB 3.5" drives.

Figure 7.1 shows the physical appearance of floppy drives. The drive on the right is a full-height drive, not available any longer.

The drive on the left is the 3.5" drive. To be mounted conveniently in your system, it may need a 5.25" frame for 3.5" drives.

## 360K 5.25" FLOPPY DRIVES

Almost all early ISA systems had full-height 360K floppy drives that used double-sided double-density 5.25" disks (DSDD). With forty tracks, nine sectors, 512 bytes per sector, and two sides, they had a capacity of 368,640 bytes or (divided by 1024) 360K. These soon gave way to the half-high drives using the same media. The spindle speed of all 360K drives is 300 Revolutions Per Minute (RMP) or 5 Revolutions Per Second (RPS).

## 1.2 MB 5.25" FLOPPY DRIVES

The first 286 systems were upgraded to use 1.2 MB floppies. These need a higher density disk, such as the Maxell MD2-HD, a double-sided *high*-density disk. With eighty tracks, fifteen sectors, 512 bytes per sector, and two sides, they have a capacity of 1,228,800 bytes, or 1.2 MB. The spindle speed of all 1.2 MB drives is 360 RPM or 6 RPS.

Figure 7.2 shows the track and sector arrangement of both sizes of high-density disks, and Figure 7.3 identifies physical characteristics of both sizes of floppy disks.

## 720K 3.5" FLOPPY DRIVES

This drive is a double-density drive. It can use double-density 3.5" disks. With eighty tracks, nine sectors, 512 bytes per sector, and two sides, these disks have a capacity of 737,280 bytes,

**FIGURE 7.1 Three types of floppy drives**

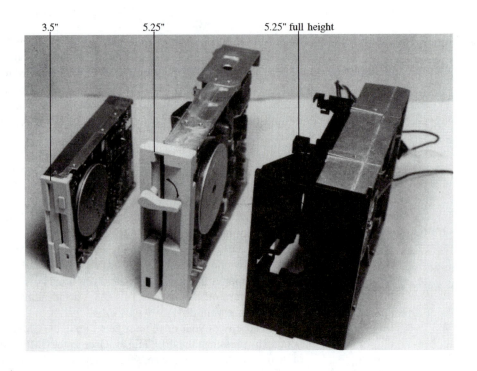

**FIGURE 7.2    Floppy sectors
and tracks**

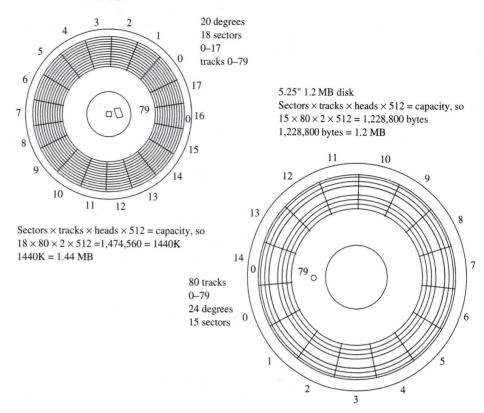

or (divided by 1024) 720K. The 3.5" drives often have a 34-pin dual header (two rows of 17 pins, 1-inch on center) instead of an edge connector. Note from Figure 7.3 that the 5.25" floppies are write protected when the notch is *covered* or closed, whereas the 3.5" disks are write protected when the hole is *open*. They are *opposite!* The speed of the spindle is 300 RPM or 5 RPS.

## 1.44 MB 3.5" FLOPPY DRIVES

The 3.5" 720K drive capacity was soon doubled to 1.44 MB. The 1.44 was the most common in the late 1980s and early 1990s. It requires high-density disks. With eighty tracks, eighteen sectors, 512 bytes per sector, and two sides, these disks have a capacity of 1,474,560 bytes or 1.44 MB. Note in Figure 7.3 the extra hole just opposite the write protect hole. Some technicians argue you can punch a hole in a 720K floppy case and format it at the higher capacity. In fact, a punch for this purpose is currently being marketed. The hole indeed tricks the drive,

**FIGURE 7.3    Parts of the 3.5"
and 5.25" floppy disk**

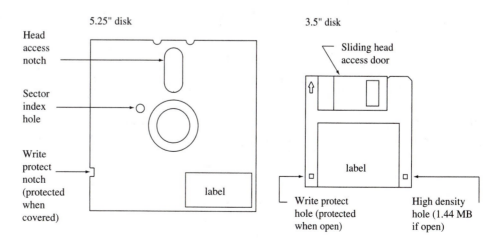

which senses a 1.44-MB disk. Many argue against it, and for the trivial difference in price, it does not seem worth it. The speed of the spindle is 300 RPM or 5 RPS.

## 2.88 MB 3.5" FLOPPY DRIVES

Toshiba first produced and licensed the 2.88 MB 3.5" drive. These disks use barium ferrite platelets that are magnetized vertically, allowing information to be squeezed more than on the horizontally magnetized floppies. The drives are backward compatible.

Table 7.1 summarizes the configurations of the five types of floppy disks.

## FLOPTICAL DRIVES

The largest capacity floppy drive available is the **floptical drive.** Using optical recording, a 21 MB drive is said to be three times faster than a standard floppy drive, and it can also read and write standard disks. Externally, the floptical is identical to a 3.5" standard 1.44 MB magnetic media disk. The cost is less than $600 for the drive and controller. You could install it in place of your 3.5" 1.44 MB drive.

## IDENTIFYING FLOPPY DRIVES

While the physical size of your drive is obvious from the external size and width of the disk insertion slot, determining whether the drive you have is high density may require a phone call to the manufacturer. A 5.25" high-density drive has a jumper selection not available on the low-density drive. This is the DC or DCH jumper (Disk Change line jumper). If present, it indicates a high-density drive. A high-density 3.5" drive has a media sensor on the right, just inside the insertion slot, to feel for or see the high-density hole (Figure 7.2). You may need to remove the cover to confirm the presence of this sensor.

Some manufacturers identify their drives by the model number. TEAC, a major manufacturer of drives, uses label color and a suffix on the model number (Table 7.2).

### Sides

Almost all modern disks are double sided, meaning both sides are used. This requires two heads on the drive, a top and a bottom head. It is also the reason for the difference of tracks and cylinders.

**TABLE 7.1    Floppy tracks and sectors.**

| Diameter | Capacity | Tracks | Sectors | Bytes/Sector | Sides |
|----------|----------|--------|---------|--------------|-------|
| 5.25" | 360K | 40 | 9 | 512 | 2 |
|  | 1.2 MB | 80 | 15 | 512 | 2 |
| 3.5" | 720K | 80 | 9 | 512 | 2 |
|  | 1.44 MB | 80 | 18 | 512 | 2 |
|  | 2.88 MB | 80 | 36 | 512 | 2 |

**TABLE 7.2    Floppy drive identification**

| Capacity | Label Color | Model Number Suffix |
|----------|-------------|---------------------|
| 360K | blue | BR |
| 720K | green | F |
| 1.2 MB | orange | GFR |
| 1.44 MB | yellow | HF |

### Tracks

Tracks are the concentric rings around the disk, similar to the track of a phonographic record, although the latter is really a spiral. The heads travel together and access the same track by moving linearly, in and out.

### Cylinders

A cylinder is all tracks of the same number. A double-sided disk has a cylinder of two tracks. A six-headed (three-platter) hard drive has six tracks in a cylinder. All of these tracks are in a vertical line, so the heads can read all tracks in a cylinder without moving.

### Sectors

A sector is a length of a track, similar to a pie slice of one ring. Each sector has a sector address and 512 8-bit bytes to hold information. This gives us the ability to calculate the storage capacity of a disk:

$$(\text{Tracks} \times \text{sectors} \times 512 \text{ bytes} \times \text{sides}) \div 1024 = \text{capacity}$$
$$(80 \times 18 \times 512 \times 2) \div 1024 = 1440K \text{ or } 1.44 \text{ MB}$$

As the disk spins at 300 or 360 RPM, all the sectors of one track pass under the heads for access without moving the head. All clone disks are **soft sectored,** meaning there is one index hole to identify position, and the 9, 15, 18, or 36 sectors are timed. The index sensor is by the spindle. **Hard-sectored** disks had an index hole for each sector.

### Clusters

Clusters are groups of sectors (usually four, for 2K total) on which one file resides. If many small files are written and a few are erased, holes of blank sectors are left between filled ones. If a large file is then written, it is placed in all empty sectors, starting at the beginning and going until it is all written on the disk. This file is said to be fragmented, having a little of its data here, a little there, and so on. This causes the head to move from track to track and sector to sector to read all the file.

Programs called defragmenters or disk optimizers go through the disk and collect all these fragments, then write them in contiguous sectors and tracks, so the drive can just cruise smoothly along, reading this file. This, of course, speeds up the data retrieval.

### File Allocation Table (FAT)

The **file allocation table** (FAT) is a table on the disk telling how each cluster is allocated. This helps keep track of where files are on the disk and how many bytes they take up. There are two copies, in case one is damaged. Some DOS programs can use this copy to re-create files that were erased but not written over. In general, use of the FAT is transparent to the user (that is, you never see or notice it). Some DOS programs use it.

Although Windows 98 can convert a 16-bit FAT table to a 32-bit FAT table, this is done for efficient storing using fewer sectors for large drives. It may be of use on a 512-MB hard drive and certainly is needed on 3 GB and above. It is of no use on the floppy drive but is just mentioned here.

## 7.2 GENERIC SYSTEM OPERATION

Table 7.3 shows the signals for the floppy controller cable, which uses a 34-pin edge connector or dual header (two rows of seventeen, red or blue stripe on pin 1).

Note that lines 10–16 (seven lines) are twisted to be reversed between the two drive connectors. It is useful to know that a cable with five twisted lines is for hard drives and one with seven twisted lines is for floppies.

Note also that this twist swaps lines 10 and 16 (motor enable A and B), and 12 and 14 (drive select A and B). Thus, the drive selects are the same pins on both drives but alternate pins on the controller card because the cable reverses them, giving alternate selections. On

**TABLE 7.3  Signals for floppy controller cable**

| Pin | Signal | Direction |
|-----|--------|-----------|
| 8 | Index | From drive |
| 10 | Motor enable A | To drive |
| 12 | Drive select B | To drive |
| 14 | Drive select A | To drive |
| 16 | Motor enable B | To drive |
| 18 | Direction stepper motor | To drive |
| 20 | Step pulse | To drive |
| 22 | Write data | To drive |
| 24 | Write enable | To drive |
| 26 | Track 0 | From drive |
| 28 | Write protect | From drive |
| 30 | Read data | From drive |
| 32 | Select head 1 | To drive |
| 34 | Disk changed | From drive |

the card edge, pin 1 is the end by the notch. On a dual header, pin 1 has a square pad. Two jacks before the twist and two after are for choice of edge or header plugs, *not* for four drives.

## CONFIGURATION JUMPERS

### Drive Select Jumper

The drive select jumper tells the drive which signal to respond to. If both controller cable drive connectors had all lines the same, each drive would need to be set differently, such as drive 1 and drive 2. However, as noted in Table 7.3, lines 10–16 are twisted to be reversed between the two drive connectors. Therefore, drive select jumpers should be set the same on both drives for the second selection. They will be labeled DS 0 1 2 3 or D0 D1, or sometimes just 00 01. If selections offered for the jumper are drives 1–4, select 2. If selections offered are drives 0–3, select 1 (the second of 0,1,2,3).

Many 360K full-height drives have an empty DIP socket for jumpers to set drive select jumper. The typical setting is jump pin 3 to pin 14 (across third position from pin 1 end) with a wire.

### Media Sensor Jumper

A 1.44 MB 3.5" floppy drive sometimes has a jumper to detect the hole on the rear of high-density floppies across from the write protect hole. It is labeled MS or just M.

### Disk Changed Jumper

Line 34 is used in some systems to tell the system if the disk in the drive has been removed and another inserted. This "disk changed" DC or DCH signal is used on some systems as the standard ready (SR or RY or RDY).

A typical drive gives a negative pulse from 1.5 to 3.0 ms long the first time the disk is accessed after a disk removal and insertion. Repeated accesses do not give the pulse again until the disk has again been changed.

Some drives hold the directory in RAM and do not reread the directory unless the disk has been changed. A 1.2 MB 5.25" drive I repaired gave the directory of the first disk inserted after booting and *always* gave that same directory no matter what disk was put in. As you can guess, it was not sending the controller the DC pulse. Pressing CTRL+BREAK did cause the drive to check the directory.

## TERMINATING RESISTORS

Electronic signals can "reflect" from the end of a wire. This interference from the reflected signal can cause errors in circuits trying to read the signal. To stop spurious signal reflections, a terminating resistor is placed at the end of the control cable. There will be a SIP or DIP socket on the 5.25" drive. With the standard two-drive cable with seven twisted lines, the drive on the end is A:, so it should have the terminating resistors. The B: drive is second and should *not* have terminating resistors.

Typical resistor values are 150 to 330 $\Omega$; you can use only regular resistors if you need to. Some 1 k$\Omega$ SIP resistor packs are permanently installed and jumper selected. Many 3.5" drives use distributed termination, meaning they have permanently installed resistors and share the job of terminating the line. These drives have no removable pack or jumper for terminating resistors.

## DOS COMMANDS FOR FLOPPIES

### ATTRIB

The ATTRIB command sets the attributes of a file. Switch choices are R (read only), A (archive, changed since last backup), S (system), H (hidden), and /S for "all subdirectories." A plus sign (+) sets the attribute, and a minus sign (-) clears it. The syntax is: ATTRIB sign switch path /S. An example is: ATTRIB +R MYFILE.TXT. The ERASE and DELETE commands will *not* remove this file from the disk now. It is read only.

### CHKDSK (and ScanDisk)

The CHKDSK (CHecK DiSK) command shows and corrects errors in the file allocation table. The syntax is: CHKDSK drive /F/V, where the /F is a switch to "fix" the table, and the /V is to show each filename as checked. The lost chains are renamed FILE0001.CHK. If this file contains data, you may recognize it and import it back into the data file. If the file contains binary data (what looks like garbage), it is likely part of a program, and one of the programs on the disk will start failing. An example is: CHKDSK A:/F.

With the arrival of DOS 6.0, its upgrades, and Windows 95 and 98, Microsoft recommends using the ScanDisk utility to detect and correct errors on both hard disks and floppies. In Windows 98, Click on Start, Programs, Accessories, System tools, ScanDisk. Start gives detailed byte information about the files on your floppy disk. A detailed discussion of ScanDisk is presented in Chapter 10.

### COPY

The COPY command copies one or more files from one location to another. The syntax is: COPY file1 file2, and an important switch is /V for verify. You can add or concatenate files by using the plus sign: COPY file1 + file2 file3. This command adds the first two files together and places the result in the last file. An example is: COPY junk1 + junk2 junkyard.

### DISKCOPY

The DISKCOPY command copies one floppy to another, making an exact duplicate of the original. This means, for example, that it copies errors on the source disk to the target disk, and if the source is fragmented, so is the target. The only useful switch is /V for verify, which checks that the target copy is correctly copied. An example is: DISKCOPY A: B:.

## DISKCOMP

The DISKCOMP (DISK COMPare) command compares two disks to see whether they are identical. Disks that were copied with the DISKCOPY command will be. Disks that were just copied with the COPY command will not be. An example is: DISKCOMP A: B:.

## DRIVPARM and DRIVER.SYS

DRIVPARM and DRIVER.SYS are two commands that cannot be entered from the DOS command line. They are loaded in the CONFIG.SYS file that you create to help configure your system. The system reads this file (and AUTOEXEC.BAT) automatically when it is cold or warm booted. Therefore, if you change either of these files with the EDIT command, you must reboot before they take effect. In the CONFIG.SYS file, programs are usually loaded with the DEVICE= or DEVICEHIGH= commands.

The DRIVPARM command sets the specifications for a physical drive. Important switches are: /D:X, where X value 0 is drive A:, 1 is B:, 2 is C:, and so on; /F:X where X value 0 is 360K, 1 is 1.2 MB 5.25", 2 is 720K 3.5", 5 is hard disk, 7 is 1.44 MB 3.5", 9 is 2.88 MB 3.5". If using 5 for F:X, use the following also: /H:X, where X is heads; /S:X, where X is sectors; /T:X, where X is tracks.

This command is used when your system is refusing to properly identify a drive. Put it in your CONFIG.SYS file. The following example sets drive B: to a 1.2 MB drive.

```
DRIVPARM=/D:1/F:1
```

DRIVER.SYS is a similar command, but it only works on floppy drives, not on hard drives. The switches are the same, but it is loaded in the CONFIG.SYS as follows.

```
DEVICE=DRIVER.SYS/D:1/F:1
```

## FC

The FC (File Compare) command compares two files and shows if they are the same or shows you the differences. An important switch is /A for abbreviate, which shows only the first and last of different lines, rather than all. An example is: FC filename filename.

## FORMAT

Formatting is done with the DOS FORMAT command. Formatting destroys all data on the disk. The low-level format actually defines the exact location of tracks and sectors on the media. The high-level format writes the file allocation table and the root directory on the disk. On the floppy, both of these are done by the DOS FORMAT command. The syntax is: FORMAT A:. Important switches are /S for transferring the system or making a bootable disk; /U for unconditional format; /F:size, where size is 360, 720, 1.2, 1.44, or 2.88. An example, which makes a bootable 720K disk, is: FORMAT A:/F:720/S/U. **Note:** Formatting a 1.44-MB disk to 720K in a 1.44-MB drive may require tape over the 1.44 MB media sensor hole.

## RECOVER

RECOVER recovers files lost due to a corrupted file allocation table. The syntax is: RECOVER filename. This is recovered in a file named FILE0001.REC without the bad sectors. You must reenter the lost data yourself.

## XCOPY

The XCOPY command is similar to the COPY command, but it includes all subdirectories. This can be a real time-saver. The syntax is: XCOPY drive1:file1 drive2:file2. Important switches are /S for subdirectories and /V for verify.

(**Note:** The remainder of this section is recommended for technicians.)
Figure 7.4 shows photos of a 5.25" drive and a 3.5" drive.

FIGURE 7.4   Inside 5.25" and 3.5" drives

5.25" drive

3.5" drive

## WINDOWS 95 AND 98

In Windows 95 and 98, you may see information on your floppy drives at least two ways. You may double-click on the My Computer icon on the desktop. Double-clicking on the Floppy A: icon gives the directory of a floppy in that drive. Windows 98 allows you to click on View and As A Webpage to see the directory shown the same as a Web page. If there is no floppy in the drive, you will get the "Drive not ready" error. Clicking once on the icon for drive A:, then on File, Properties gives a window with two tabs, General and Tools. The General tab window is quite handy, and shows a pie chart of capacity used and free including the numerical information, and type of drive. The Tools tab opens a window showing dates of the last Scandisk, Backup, and Defrag, allowing you to perform one of those functions now.

Another way to find information on your floppy drives is by clicking on Start, Settings, Control Panel, System, Device Manager, double-click on Disk Drives, and double-click on Floppy Disk. This opens a window with three tabs: General, Settings, and Drivers. From the Device Manager window, you may also double-click on Floppy Disk Controller and double-click on the line that opens up below that. This will give a window with three tabs:

General, Driver, and Resources. Click on Resources to see the DMA#, IRQ#, and I/O Address of the floppy drives.

To create a Startup Disk to use to boot the system when the hard drive has a problem, or for certain virus checks, click on Start, Settings, Control Panel, Add/Remove programs, Startup Disk.

There is no "troubleshooter" for floppy drives. Clicking on Start, Help, Search, and typing in "floppy" brings up about ten topics on floppies that may be of some interest.

## ADJUSTING THE SPEED

The spindle drive motor speed in the 1.2 MB drive is 360 RPM. In all others the speed is 300 RPM. In some of the older drives this is adjustable through a trimmer resistor. In others, it is a function of the system and not adjustable.

You can determine the drive speed with your oscilloscope; 300 RPM is 5 RPS, or a frequency of 5 Hz and a period of 0.20 seconds. With the scope time base at 20 ms (horizontal sweep) and volts/div at 1 (vertical sensitivity), connect to pin 8, index. Pin 1 will have a red or blue stripe along the ribbon cable and, usually, a small triangular arrow designating it on the side of the plastic edge connector. With the red stripe on the right, the top row is even pins. The fourth pin from the right is pin 8, index. Set trigger slope to negative, mode to normal. You should see pulses exactly ten divisions or 200 ms apart. Each pulse is the index sensor light coming through the index hole at 5 Hz, giving 300 RPM.

The pulse width is dependent on the width of the actual index hole, which is what we are looking at. With the scope at 0.5 ms/div you should see a pulse width of eight divisions or 4 ms. Of course, this may vary according to manufacturer's index hole sizes.

The above method gives the speed of the drive. Figure 7.5 shows the strobe pattern on the spindle motor. If a fluorescent lamp lights this while the spindle is turning the proper speed, the pattern appears to stand still, so you adjust speed to "freeze" the pattern for 60 Hz. Note also the separate spindle motor control board. Again, in many modern drives, the speed is not adjustable.

The drive spindle speed of the 1.2 MB 5.25" drive is 360 RPM. This gives 6 RPS, for a period of 0.167 seconds or 167 ms. With the scope horizontal sweep set at 20 ms/div, the scope trace shows negative pulses (from 4 V DC down to 0 and up to 4 V DC again) every 8.33 divisions.

**FIGURE 7.5  Strobe pattern for setting speed**

Strobe pattern

## ALIGNING THE HEADS

Figure 7.6 shows the head and head drive mechanism of a drive. Note that the top head is raised to expose the bottom head.

The **azimuth** of a head is the position of the head tangential to the track on the disk. If the head is rotated clockwise or counterclockwise on a vertical axis, it loses signal strength. On audiocassette tape recorders, you could adjust this azimuth with a screw on one side of the head. You would rotate the head for the best high-frequency response. On a floppy drive it is usually not adjustable.

The **radial alignment** of a head is the lateral position of the head (in and out from track 0 up). It must be centered over the track. Professional alignment disks are needed to adjust this correctly, and they come with directions. However, if you are working on an old drive that is grossly out of alignment (or if you are just experimenting with an old drive), you may rough it in as follows.

Insert a disk with BASIC and the program in Program Listing 7.1 on it, loosen the set screw that secures the head once aligned, and turn the adjustment screw to note the movement of the head forward or backward. To start, adjust the head back (to the outside rim of the disk). Now run the program. Note in line 20 this is for drive A:. You may change that as you please. All it does is keep doing DIRs while you adjust the head forward until it in fact prints the filenames on the screen. Once it prints the DIR, keep on adjusting forward carefully counting $\frac{1}{8}$ turns of the screw, until it stops printing DIRs. This is to purposefully go too far. Now back up the head half of the turns you counted, and the head should be in the middle of the track. Tighten the set screw, and try a few different disks. Remember, this does not take the place of a professional alignment disk and is only for "roughing in."

**PROGRAM LISTING 7.1    Radial head alignment**

```
5   KEY OFF:CLS
10  ON ERROR GOTO 40
15  TRY=TRY+1
20  CLS:PRINT "attempt ";TRY:FILES "A:*.*"
25  IF INKEY$<>"" THEN CLS:PRINT"bye":END
30  GOTO 15
35  END
40  RESUME 15
```

## THE WRITE PROTECT SWITCH

To test the write protect switch and circuitry, connect a meter to pin 28 of the controller cable connector. With the red stripe on the right, this is the fourteenth pin from the stripe on the top row. A high (3 V DC) here shows the disk is not protected; a low (0 V DC) shows it

**FIGURE 7.6    Head drive mechanism and head**

— Head

is protected. Use a write protected and a not write protected floppy to pulse this line and prove the circuitry works. In Figure 7.7, the photo on top is of a typical write protect switch (by door, on left). It may be either mechanical or optical. The photo on the bottom is an optical track zero switch. It controls pin 26 of the floppy control cable.

## SPINNING THE SPINDLE

Pin 12 (sixth on the top row from the red line on the right) is for drive select. It is high when not selected and low when selected. Pin 16 (eighth from the right) is for motor enable. High turns the spindle drive motor off, low turns it on and spins the disk.

You can test the spindle drive motor and circuitry with the drive connected only to the power supply—no controller cable. Ground both pins 12 and 16. This will select the drive and enable the motor. If the spindle does not spin, the spindle drive motor or related circuitry is not functioning.

You can use the BASIC program in Program Listing 7.2 to rotate the spindle of drive B:. The comment in line 20 shows how to modify it to rotate the spindle of drive A:.

The I/O port addresses used in this program and others that follow, as well as the interrupt request number (IRQ) for the floppy drive, are given in Table 7.4.

**FIGURE 7.7   Write protect and track 0 switches**

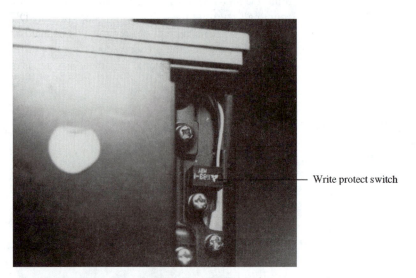

Write protect switch

Track zero switch

**TABLE 7.4  Floppy controller I/O port addresses and IRQ**

| Device | Address | Interrupt |
|---|---|---|
| Floppy disk controller | 3F0H to 3F7H | IRQ6 (8088 and AT systems) |

**PROGRAM LISTING 7.2**
**Spindle drive motor test**

```
10 REM spin_b.bas
15 FOR CYCLE=1 TO 1000
20 OUT &H3F2,&H25 :REM change second number to &H14 for drive A:
30 NEXT CYCLE
```

## STEPPING THE HEAD

Pin 20 is the step pulse line. You can test the head positioning stepper motor with the drive connected only to the power supply—no controller cable. Have a disk in and the door closed. Manually move the head to roughly halfway between track 0 (outside of disk) and the end track, then ground pin 12 and pulse pin 20 (tap with ground lead). The head will step one way or the other. You can control this movement with line 18 (direction stepper motor) if you like. There is a track 0 detect switch that you can test manually if necessary.

The programs in Program Listings 7.3 and 7.4 can be used to cycle the head of drive B: in and out to test the head stepper motor. The first is in BASIC, the second in assembly. Both tell you how to change for drive A:.

**PROGRAM LISTING 7.3  Head stepper motor test in BASIC**

```
10   REM cycle_b.bas different approach from cycle_b.com
20   FOR CYCLE = 1 TO 10 :REM cycle 10 times
30   OUT &H3F2,&H25 :REM Change to &H14 to cycle drive A:
40   OUT &H3F5,7
50   OUT &H3F5,1
60   FOR DELAY=1 TO 4000:NEXT DELAY
65   OUT &H3F2,&H25 :REM change second number to &H14 for drive A:
70   OUT &H3F5,&HF
80   OUT &H3F5,1
90   OUT &H3F5,&H10
100 FOR DELAY = 1 TO 4000:NEXT DELAY
110 NEXT CYCLE
```

The program in Program Lising 7.4 can be entered in the DOS DEBUG program. At the hyphen prompt, type "A" for assemble, then the program, with an ENTER at the end. Then "N CYCLE_B.COM" to name it. Semicolon lines (i.e., remarks) are not needed.

Type RCX to prepare the CX register, and type 2A at the colon for length of file to write. Finally, type W to write it to the disk and Q to quit debug.

**PROGRAM LISTING 7.4  Head stepper motor test in assembly**

```
DB   9
DEC  BYTE PTR [0100]
MOV  AX,0201
MOV  BX,0200
MOV  CX,0001
MOV  DX,0101
INT  13
MOV  AX,0201
MOV  BX,0200
MOV  CX,2701
MOV  DX,0101
```

**209**

```
INT   13
CMP   BYTE PTR [0100],00
JA    0101
INT   20
; first line sets 9 in location 100
; second decrements location 100 by 1
; register AH is code for operation, AL is number of sectors
; to get
; code 02 in AH means read sectors into buffer
; 01 in AL just means get 1 sector
; register BX holds address to load at
; register DH holds head #, DL holds drive ID code
; 00h=A:, 01h=B:, 80h=C:, 81h=D:
; change to MOV DX,0100 for drive A:
; INT 13 is disk I/O
; Compare the number in 100 with 0
; Jump if Above 0 to location 101, the start of action after
; data
; INT 20 is terminate and return to DOS
```

## OBSERVING DATA FLOW

You can observe data flow, with the drive fully connected to the power and control cables, by giving the system the DIR command at the DOS prompt. Line 30 (fifteenth from the right) is for read data and gives 3 Vp-p pulses about 1 μs wide. This only verifies that some kind of data is being transferred, not the validity of the data itself. Similarly, line 22 (eleventh from right) is for write data. Copying a file to another name (such as COPY junk1 junk2) will pulse this line. Realize this data is from the controller board, so this procedure verifies it but not the drive write circuits.

Simple, rough drive testing can obviously be done with the COPY command, because it requires the drive to write to the disk, and the TYPE command (as in TYPE junk2), because it requires the drive to read from the disk. If a drive has caused problems in the past, do not trust this simple test. Head movement must be checked, so DISKCOPY (of a full disk) is slightly better than just COPY.

## FORMAT/WRITE INCOMPATIBILITIES

A 5.25" floppy in 360K format has forty tracks; a 1.2-MB format has eighty. As you can guess, the tracks on the 1.2 MB are half as wide, with the forty track being approximately 0.0118" wide and eighty tracks 0.00625" wide. If you format a disk at 360K in a 1.2-MB drive, it is *readable* by a 360K drive. If you format a disk at 360K in a 360K drive, it is also *readable* in a 1.2-MB drive. The problem comes when you *write* on the disk with the opposite drive. Now the tracks are partly wide and partly thin, and you're begging for trouble.

The moral of this sad tale is to write on the same capacity drive you formatted with.

## 7.3 ASSEMBLING FLOPPY DRIVES

Newer systems have the floppy drive controller built into the motherboard, so you need only the drive and cable. Installing a floppy drive, in an older system, requires a controller card, a cable, and a drive. Ensure that the cable connector matches the drive and the controller, and that it is not a hard drive cable. The controller may be integral with the system board, or it may be a dedicated floppy-only controller, half of a hard/floppy controller, or part of a multi-I/O card. To prepare the case, remove a 3.5" front bezel as in Figure 7.8.

**FIGURE 7.8    Removing the
case's front 3.5" plastic bezel**

1. Set jumpers on the controller adapter card, if any, per documentation. Connect the cable to the card and insert in an expansion slot, or just plug the cable into the motherboard. Watch polarity. Remember pin 1 on the cable has a red or blue wire. On the connector, it may have a small raised triangle or a silk-screened "1" (Figure 7.9).

2. Set the jumpers on the drive, and physically mount the drive in the case. On PC/XTs this was directly screwed to side brackets. In the AT, it will slide in the case on slide rails screwed to the sides of the drives. In towers, it is back to direct bracket connection. You may need a 5.25" frame for 3.25" drives, disk drive rails, and possibly drive block-out bezels for empty slots. See Appendix B for sources of these supplies. (Figure 7.10)

**FIGURE 7.9    Plugging in the
floppy drive signal cable**

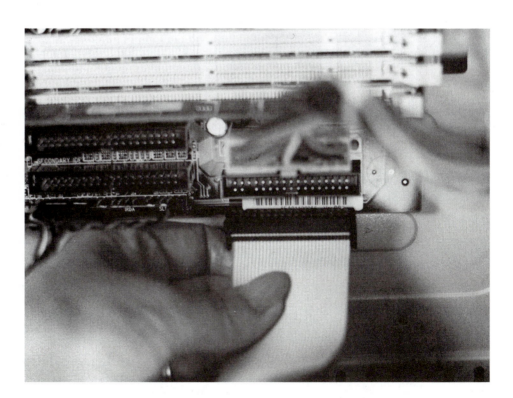

**FIGURE 7.10  Physically
installing the 3.5" floppy disk
drive in the case**

3. Connect the controller cable to the drive, as in Figure 7.11. Note the polarity of connector by colored wire and triangle, silk-screen, or square copper pad on pin 1. Connect a power cable from the supply to the drive, as in Figure 7.12. Note the mini-power plug *can* be inserted inverted despite keying. Female tabs or wings go against the male plastic. If your supply doesn't have a free power plug, a power "Y" splitter cable may be needed to change one plug into two. In Figure 7.13, note the adapter card that may be required to change the 34-pin dual header to an edge card connector, and adapt the mini-Molex power supply connector to the standard supply connector. Most newer systems have enough power connectors with the proper plugs.

4. Once the floppy drive is physically installed, the system must be informed that it is there. On a PC/XT, the motherboard switches must be set. This is switch bank 1, often labeled SW1. The settings can be found in Table 3.2.

   On AT systems, the CMOS SETUP must be done, installing the proper capacity drives. You may also choose boot sequence, booting hard drive C: before floppy drive A:, to speed system bootup. The default is floppy drive A:. The ATX system automatically sets the drive correctly.

5. Always test any device you install. In fact, it is a good idea to test all devices after any change, just to be sure there are no conflicts. With a known good DOS disk in drive A:, turn on the system (cold boot). If the error "No ROM basic, system halted" appears, the system cannot find any boot device. Check installation again. Doing a

FIGURE 7.11   Plugging in the
floppy signal cable

FIGURE 7.12   Plugging in the
floppy drive power cable

DIR command, then copying a text file to a disk in the drive with the COPY command and reading the file with the TYPE command is a simple test. Diagnostic programs can be run for a more thorough test. An example of a simple test of drive A: follows:

a. Put a known good floppy of the correct capacity in drive A:.

b. DIR A:

  If this gives the files properly, fine. If not, go to step c.

**FIGURE 7.13    Floppy connector adapter**

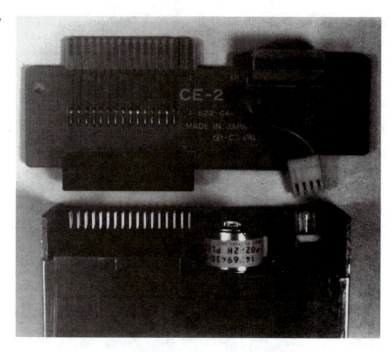

c. COPY C:\AUTOEXEC.BAT A:/V

   AUTOEXEC.BAT is just an example of a text file that is probably on C: drive. /V is for verify, which proves proper reading and writing.

d. TYPE A:\AUTOEXEC.BAT

   This should show AUTOEXEC.BAT on display if it was written correctly.

In Windows 95 and 98, check the directory on your floppy drives by double-clicking on the My Computer icon on the desktop. Double clicking on the Floppy A: icon gives the directory of a floppy in that drive. If there is no floppy in the drive, you will get the "Drive not ready" error. If there is a floppy in the drive and you get that error or another, check your cables and refer to the Windows 98 section earlier in this chapter.

If this test works, you probably have a properly functioning floppy drive. If not, proceed to Basic Troubleshooting.

## 7.4  BASIC TROUBLESHOOTING

Troubleshooting floppy drive problems begins by noting any POST audio or video error codes. Then you can proceed to noting any visual clues and taking appropriate measurements. Details of these steps are as follows.

a. POST audio error code: 1 short beep, good video, no boot with a DOS disk in drive A:
b. POST video error code of 6xx
c. Measurements
   1. Confirm that you have a floppy of compatible density with good information on it in the drive. Confirm that the drive is set up correctly in CMOS. Confirm that the cable has seven twisted wires, not five.
   2. With the DIR A: or DIR B: command, watch for the drive access light to come on, showing that the disk adapter card is in fact trying to read the correct drive.
   3. If the drive attempts to read the disk, clean the heads, then try to read it again. Head-cleaning kits are available. Read the disk in another drive to be sure it is a good disk. If the message "No ROM basic, system halted" appears, the system cannot find the boot device, either floppy or hard. Check CMOS SETUP, controller installation, and cables.
   4. Swap the floppy controller adapter card with a known good one. If no change, swap cables with known good ones.

5. If you have an identical drive with different problems, swap the electronics PCB(s). Keep the mechanisms and PCBs that work for possible future replacement parts. Add nonworking PCBs to nonworking mechanisms for depot repair (Appendix C).

6. If the head is not moving, check for physical binding. If the spindle is not turning, check for free spin and that cables from the drive electronics to the motor card are tight.

**Note:** Steps 7–11 are recommended for technicians.

7. Will the drive format a blank disk? Will it read its own format? If the drive can read and write its own format but not another's, one of the two is out of alignment. Confirm which. Copy any disks the misaligned system has made by reading in the bad drive and saving to a good drive, then radially align the bad drive.

8. Will the drive read its directory but give an error when trying to execute any program? Try typing a text or ASCII (.TXT, .BAT, .DOC, READ.ME, etc.) file with the DOS TYPE filename command. If the drive cannot seek and find the file to type, but it does a DIR properly, it may be out of radial alignment.

9. If the drive noisily bumps against 0 stop, check the track 0 indicator. If it reads but will not write, check the write protect indicator to be sure the disk is not write protected.

10. Is the media sensor switch working?

11. Is the write protect switch working?

## 7.5 CHAPTER REVIEW/QUIZ QUESTIONS

1. The head positioning motor moves the heads
   a. Only to "parked" position.
   b. From one track to another.
   c. Sideways, from sector to sector.
   d. From track 0 to 40 only.

2. The spindle motor
   a. Moves the heads back and forth.
   b. Moves the heads side to side.
   c. Rotates the disk.
   d. "Parks" the heads.

3. The write protect sensor
   a. Is located very close to the heads.
   b. Is a magnetic sensor.
   c. Protects your heads.
   d. Is located just inside the "door."

4. The track 0 sensor
   a. Is located just inside the "door."
   b. Is located very close to the head mechanism.
   c. Reads the disk to identify track 0.
   d. Feels the disk to identify track 0.

5. The high-density media sensor
   a. Is found only on 3.5" drives.
   b. Is found only on 5.25" drives.
   c. Is located very close to the head mechanism.
   d. Is a magnetic sensor.

6. The index sensor
   a. Is located just inside the "door."
   b. Is located very close to the head mechanism.
   c. Reads the disk to identify the index.
   d. Shines a light through a hole to identify the index.

7. The drive select jumper on ISA systems should be set at
   a. DS0.
   b. DS1.
   c. DS2.
   d. DS3.

8. The terminating resistor (if not distributed)
   a. Should be inserted on drive A:.
   b. Should be inserted on drive B:.
   c. Should never be installed anymore.
   d. Should be inserted on both drives.

9. On the MFM floppy control cable, how many lines are twisted?
   a. 3
   b. 5
   c. 7
   d. 9

10. The drive at the MFM floppy controller cable end is
    a. A:.
    b. B:.
    c. C:.
    d. D:.

11. On a 5.25" floppy, to protect a disk from being written, the write protect hole should be
    a. Open.
    b. Closed.
    c. Enlarged.
    d. Erased.

12. On a 3.5" floppy, to protect a disk from being written, the write protect hole should be
    a. Open.
    b. Closed.
    c. Enlarged.
    d. Erased.

13. What can the error "No ROM basic, system halted" on booting indicate?
    a. The wrong disk is in drive A:.
    b. The hard drive is not functioning.
    c. Neither the hard drive nor floppy drive A: is functioning.
    d. It has nothing to do with the drives.

14. What may be wrong if the drive can read the directory of a disk but cannot find files?
    a. You have a bad disk.
    b. The drive has an electronic failure.
    c. Your disk is the wrong size.
    d. The drive may be set up incorrectly in CMOS SETUP.

15. How does an AT system know what size floppy drives it has?
    a. Automatically senses them
    b. From the CMOS SETUP
    c. Assumes size
    d. DIP switches on the motherboard

16. In Windows 95 or 98, to see a pie chart of floppy capacity and usage
    a. Go to DOS and enter the "chart" command.
    b. Use the Windows floppy drive troubleshooter.
    c. Click on My Computer, Drive A:, File, Properties.
    d. There is no pie chart of usage.

17. In Windows 95 or 98, to see detailed information on byte usage by floppy drives
    a. Go to DOS "files" command.
    b. Use ScanDisk.
    c. Use Defrag.
    d. Run system setup.

18. In Windows 98, to see the DMA, IRQ, and I/O Addresses of the floppy drive
    a. Go to system SETUP.
    b. Use professional troubleshooting software.
    c. Use Control Panel, Device Manager, Floppy Drive, Resources.
    d. Use ScanDisk.

# 8

# Hard Drives

**OBJECTIVES**

After completing this chapter you should be able to

- List differences between ST506, IDE, EIDE, ATA, and SCSI drives
- Identify these five types of drives by sight
- Identify these five types of controller cards by sight
- List physical preparation to install each drive (jumpers, terminators)
- Perform physical installation of each type of drive
- Perform CMOS setup of each type of drive
- Perform software setup of each of three types (low, fdisk, high)
- Perform tests of hard drives
- Troubleshoot and correct problems

## 8.1 GENERAL DISCUSSION

The main problem with floppy disks is their limited storage capacity. To overcome this problem, a new type of drive was designed. The disks were made of metal, and many were stacked on top of each other as shown in Figure 8.1. Because the disks are made of metal, they are called **hard disks.**

Although a floppy drive has two heads, the hard disk shown in the figure has an eight-head four-platter arrangement. With four times as many heads, it can read four times the information without moving the head stack. When the heads do move, they move as a group. The platters are locked together and spin as a group or stack. While a floppy spins at 300 RPM, these stacks of platters can spin from 3600 to 10,000 RPM. This allows the information to be accessed ten to twenty times faster than on a floppy drive. Drives cost $400 for 6.4 GB with 9.5 ms access time, a 512 K cache, spinning 5400 RPM, with 33 MB/s data transfer rate.

Hard disks (sometimes called fixed disks or "Winchesters") also have many more tracks on each side of the disk or platter, so they hold much more information than a floppy. Sizes range from 40 MB to 9 GB, access times range from 16 to 8 ms, and data transfer rates range from 500K/s to 40 MB/s. This not only speeds up storing and loading information by the program, but also eliminates fumbling and swapping floppy disks. On the other hand, unless you have a special hard drive with removable medium, you cannot remove the information from the hard drive and port it to another machine or location without physically removing the hard disk or using floppies.

**FIGURE 8.1    Hard drive platter stack**

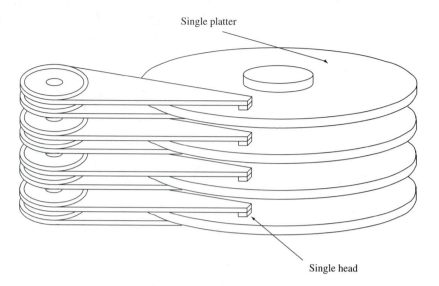

Single platter

Single head

The easiest hard drives to install are called **hardcards,** which have a very thin drive mounted right on the controller card. Amazingly, these fit in a standard expansion slot. They include software to automatically initialize, prepare, or format the drive and load DOS. They are available in sizes to 240 MB.

A new cutting-edge solution to storage is the optical hard drive, marketed by at least two companies. Using optical storage similar to read/write CD-ROMs, these have removable media in capacities of 35, 65, 105, 128, 150 MB. Access speed is 18 ms, and cost is approximately $2000. One of these manufacturers has what they call a magneto-optical drive, which can read and write 1 GB or 1000 MB on each disk of removable media. Cost for the 1 GB drive is $5,000. These drives do not have a solid standard as of yet, and may not be compatible with future drives.

## ENCRYPTION METHODS

Encryption is the method used for organizing the information written on the hard drive. Two methods of encrypting data on the magnetic medium of the hard disk are modified frequency modulation (MFM) and run length limited (RLL).

MFM is the method used by most of the early hard drives. It stores approximately 8.5K per track. It uses a 34-pin control cable and a 20-pin data cable. The control cable has five lines twisted between the two drive connectors. Typically, seventeen sectors are used.

The RLL method of encrypting data compresses data by leaving out the clock pulses, for higher density of information, giving approximately 13K per track. Typically, twenty-six sectors are used.

## INTERFACING STANDARDS

There are a number of standards for interfacing the computer and the hard drive controller. Among them are the ST506, SCSI, ESDI, IDE, EIDE, ATA, and SCSI interfaces.

### ST506

Named for the SeaGate ST506 drive, the ST506 or ST412 interface has a throughput of about 5 Mbits per second. It uses a 34-conductor ribbon cable for control and a 20-conductor ribbon cable for data. On the controller end, these may end in either edge card connectors or dual headers. They almost always have edge card connectors on the drive end. For a two-drive system, the cable will have two connectors on the end separated by five swapped lines. The drive select jumpers on both drives should be set to the second choice, drive select

2 (DS2) if labeled DS1, DS2, DS3, DS4, or DS1 if labeled DS0, DS1, DS2, or DS3. The drive at the end of the cable will be drive C: and must have a terminating resistor pack installed. The second drive is designated as drive D: and must *not* have a terminating resistor pack. MFM/ST506 interfaces typically have seventeen sectors.

Figure 8.2 shows two types of controller card connectors, and Figure 8.3 shows the drive connectors for an ST506 MFM hard drive.

**FIGURE 8.2   ST506 MFM controller card connectors**

Horizontal MFM card pins

Vertical MFM card pins

**FIGURE 8.3   ST506 MFM drive connectors**

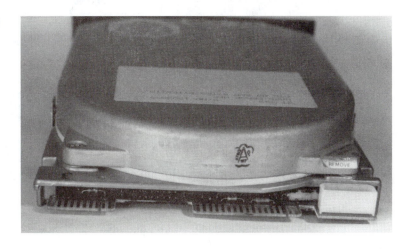

## SCSI

Small computer systems interface (SCSI) allows up to eight controllers to communicate. Typically this interface uses an MFM or RLL encryption method. It also handles devices other than disk drives. A 50-conductor ribbon cable is used with a dual-header jack on both ends. If the cable is for two drives, there will be two jacks close to one end. The cable will not have any swapped lines. Throughput can be up to 10 Mbits per second. Some drives have 64K disk caching to speed up media transfer, and 16- and 32-bit adapter cards are available.

The SCSI interface has a host bus adapter, and seven more devices can be connected, so a system is needed to identify which device is which. This is usually provided with the standard shunting (shorting) jumpers. One convention on Western Digital drives is a set of three jumpers (six pins in sets of two). The jumpers stand for a "high" or "on" or "1," and give three bit choices, for eight states and numbers from 0–7. It is this identity, and *not* the position on the cable, that determines which SCSI in a two-drive system is C: and which is D:. As in a network, the ends of the SCSI bus must be "terminated" (have a resistance to stop reflection of signals from the end of the bus). Some SCSI drives have distributed termination, in which each drive is partially terminated, so you don't have to change the termination for a single- or dual-drive system. Do *not* terminate intermediate devices.

Figure 8.4 shows the SCSI controller adapter connector on the top and the SCSI drive connector on the bottom. Signals for the SCSI port connector are given in the next chapter (Tables 9.7 and 9.8). This interface has been improved to SCSI-2 and fast SCSI-2, which may have an access time of 8 ms, rotate at 7,200 RPM, and transfer 10 MB of data per second.

## Ultra SCSI

Increases in efficiency were able to bring the SCSI speed up from 5 to 10 MB/second, and this was called Fast SCSI or SCSI II (for two). With the need for even more speed in hard drives, the data lines for SCSI were increased from fifty lines for 8 bits to sixty-eight lines for 16 bits. This speeded data transfer up to 10 MB/s for 8 bits and 20 MB/s for 16 bits, and

**FIGURE 8.4  SCSI controller and drive connectors**

SCSI controller

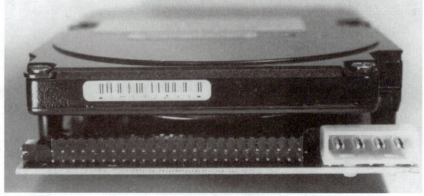

SCSI drive's 50 pins

222

was called Wide SCSI. The 8-bit, fifty-wire SCSI increased to 20 MB/s to be called Fast-20 SCSI. For 16-bit lines, this is called Ultra SCSI and gives 40 MB/s. A final improvement brought the 8-bit interface to 40 MB/s and the 16-bit to 80 MB/s; this is called Fast-40 or Ultra-2 SCSI. The AT Attachment interface (ATA later in this chapter—same thing as IDE) was designed with hard drives in mind. To allow CD-ROM drives and tape drives to use this interface, the ATA Packet Interface (ATAPI) was developed. An ATAPI-compatible CD-ROM or tape drive may be plugged into the IDE motherboard connectors. Because the CD-ROM drive is much slower than a hard drive, if they share an IDE channel the drive will be slowed down. Therefore, you may be better off connecting the CD-ROM drive to the secondary IDE channel.

## ESDI

Enhanced Small Device Interface (ESDI) is designed for high performance and high reliability. This interface usually uses the RLL encryption method. This interface allows the controller to run and communicate with the drive at a 1:1 interleave, so all information from a cylinder can be read in one revolution. ESDI has typically thirty-four to fifty-four sectors per track. It has a high throughput of about 15 Mbits per second and low access time. It is the fastest interface available. The ESDI interface uses a 34- and a 20-conductor ribbon cable.

## IDE or ATA

Integrated device electronics (IDE) is similar to SCSI, but it is newer. IDE hard drives have the controller circuits built into the circuit board of the drive. This allows just a little circuitry on the computer system motherboard to replace the controller adapter used with MFM drives, so many systems include an IDE connector on the motherboard. This is known as the ATA interface.

A standard IDE interface may have a 12-ms access time, rotate at 3600 RPM, and transfer 4.3 MB of data per second. There will typically be twenty-six to fifty-four sectors. IDE uses a forty-conductor ribbon cable to a 40-pin dual header (2 rows of 20, 0.1" apart) with no twist between the two drives. The tenth pin on the bottom row (if pin 1 is considered to be in the top row, and on the right) is missing as a key. Pin 1 is identified by a little imprinted triangle arrow on the female connector.

On a two-drive IDE system, the drives work in a master/slave or primary/secondary relationship. There may be three jumpers (6 pins in two rows of three, the top labeled 1, 3, and 5 and the bottom labeled 2, 4, and 6) on the IDE drive. Typically, no jumper installed sets the drive to be the main drive (C:) in a one-drive system. Jumping 5 and 6 set it to C: in a two-drive system. In a two-drive system, jumping 3 and 4 sets the drive to drive D:, the slave or secondary drive. It is this identity, *not* the position on the cable, that determines which IDE drive in a two-drive system is C: and which is D:. Most IDE drives have distributed termination, in which each drive is partially terminated, so you don't have to change the termination for a single- or dual-drive system.

Usually, you do not low-level format an IDE drive, because it is more complex and this is done at the factory. If you feel you must, use *only* a program written for IDE drives. Figure 8.5 shows the IDE controller adapter connector on the top and the ribbon cable connector on the bottom. Figure 8.6 shows the IDE hard drive connector. This interface has been improved to enhanced IDE (EIDE), which may have a 10-ms access time, rotate at 4500 to 7200 RPM, and transfer 13.3 MB of data per second.

## EIDE and Ultra ATA

Most modern motherboards incorporate a primary and secondary IDE interfaces, each handling up to two IDE drives, for a total of four hard disks, CD drives, and tape drives. The original ATA (IDE) standard was improved and became the Enhanced IDE (EIDE) interface. Also known as Ultra ATA and Fast ATA, this interface has a transfer rate of 16.6 MB/s. Ultra-DMA is a transfer protocol that may double the data transfer rate of IDE (ATA) drives to 33 MB/s.

**FIGURE 8.5    IDE controller and cable connectors**

IDE controller

IDE cable connector

**FIGURE 8.6    IDE hard drive connector**

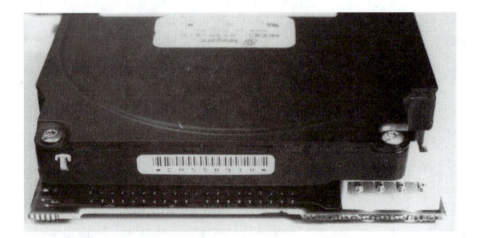

## ATA-2 and ATA-3

ATA-2 has faster Peripheral Input-Output (PIO) modes and Direct Memory Access (DMA) modes than its predecessor does. More Plug-and-Play peripherals were made compatible also. ATA-3 added a warning reporting system and security, but no speed increase.

## ATAPI

Although the ATA standard was great for hard drives, it left CD-ROM drives and tape drives behind. The ATAPI was developed to allow compatible CD drives and tape drives to be plugged directly into the motherboard IDE connectors and handled by the system. This is extremely handy. Since having two IDE (ATA) devices on one channel makes them take turns sharing the bus, the slower drives slow down the hard drive. Therefore it may be best to install your CD or tape on the secondary IDE channel.

## ZIP and JAZZ Drives

ZIP drives have removable disks like floppies that hold 100 MB of data. Eight million ZIP drives are in use, so it is the standard for removable media drives. External versions plug into your SCSI or parallel printer port and are able to be easily moved from system to system. This makes them an excellent backup system for offices.

The JAZZ drive uses a 4"-by-4" cartridge holding up to 1 GB of data. Access time is 15.5 ms and transfer rate is 6.6 MB/s, so it is a little slow. But consider the advantage of being able to pop a 1-GB drive into and out of your system and a different one in. They are sold in either internal or external versions for the SCSI, parallel, or PCMCIA ports.

Just a note, there is now a 3-GB hard drive available for the PCMCIA port that is Plug-and-Play, with a 16.6 MB/s transfer rate!

## SPECIFICATIONS AND PARAMETERS

Logical Block Addressing (LBA) is a storage system that allows the BIOS to be set up with different-sized clusters. The File Allocation Table (FAT) stores files by clusters. DOS and all Windows versions before revision 2 were 16-bit operating systems. They limited the number of bits for addresses to 16 bits, giving a maximum of 64K clusters. Dividing a 512-MB drive by 64K clusters gave an 8K cluster. A 1-GB drive would have 16K clusters. A 2-GB drive would have 32K clusters. The problem with this is that one cluster is the smallest sized file you can save. So saving a 100-byte file still uses one cluster, or 32K on a 2-GB drive. If you have a lot of small files, that is a significant loss of hard drive space.

Windows 95 revision 2 and newer versions allow a 32-bit FAT, which allows up to 4 GB of addresses. This allows the system to use more clusters of smaller size and reduce the wasted space.

## DISK CACHE

While hard disk access times have been pushed down from 25 ms to about 8 ms, they have no hope of keeping up with data reads or writes from the new fast CPUs. Thus the hard disk is a real bottleneck to the flow of information.

To help overcome this, two types of hard disk cache are now in use. First, the drives themselves have an internal RAM cache in their own electronic circuitry. The capacity can be from 32K to 1 MB. There is also external hard disk caching, in which software loads areas of the hard disk in use into RAM. With the system doing the majority of its work with the files in RAM, overall system operation speeds up tremendously.

Of course this cache needs managing, and many operating systems supply that software. DOS has the program SMARTDRV.EXE to manage the cache. Syntax is DEVICE=SMARTDRV.SYS NNN, where NNN is the size of the cache.

### Landing Zone

Heads resting or landing on the media endanger the surface. For this reason, when the drive is turned off, the heads are parked in (placed over) an area where there is no information. This is the landing zone, and it is where the heads are moved when you "park," or ship, the drive. If the heads bounce on this unused track, no data on the other tracks is destroyed. On some newer drives, parking the heads is done automatically. Programs that accomplish this are usually named PARK or SHIP. If asked in SETUP to choose a landing zone, choose one less than the number of cylinders.

### Interleave

It would make sense to read sector one on the disk, then read sector two, then three, and so on. This is done by fast drives, and the interleave is then 1:1. But it takes drive electronics time to figure out what to do in between reading sectors. If it is too slow, sector two has already spun by underneath and is gone until the next revolution.

Instead of waiting a whole revolution to read the physical sector two, the head just dips down and reads physical sector three and calls it logical sector two. So it writes or reads every other sector. This is a 2:1 interleave and, as you can see, the disk now has to spin around twice to have all sectors read, which slows data flow. Interleave optimizers are available that actually test the disk by writing and reading information and set the interleave to give maximum speed. Speed increases of 50 percent to 400 percent have been realized. One of these optimizers, Spinrite, also reformats your hard disk with the information intact. This "refreshing" of the format and data minimizes losses from weakened fields. All modern drivers now have a 1:1 interleave, so setting this is only required by older drivers.

### Precompensation

As you move in from the perimeter of the disk, from track 0 to the higher tracks, the length of the track shortens. To make up for this, the write current is raised for the shorter tracks. The track at which this occurs is referred to as "precomp." If asked in SETUP for precomp, enter 65535, which is the same as none.

### Drive Type

While it may sound like this is the type of drive you have, nothing so straightforward should be expected. The BIOS of the system in which you are installing the drive has a parameter table that lists different drive type numbers for different parameter combinations. Examples of types for different BIOS tables are:

> SeaGate ST251 820 cyl, 6 heads, 17 sect, no precomp, 820 LZone
> Award BIOS parameter table: type 40
> Phoenix BIOS parameter table: type 23
> American Megatrends Inc.: type 40

Table 8.1 is a typical type table.

If the parameters for your drive are not available, try type 47, which is usually user defined and allows you to enter your own values. Some newer hard drive controller cards require the CMOS SETUP to be No Type or Not Installed. The controller card lets the system know it is there.

Hard drive manufacturers give different specifications or parameters for their systems. Table 8.2 shows an example of specifications from a major manufacturer. Appendix E has numerous hard drive parameters.

### CAPACITY

You can figure the storage capacity of a hard disk with the same formula used for floppies. Because there is one head per side and cylinders of many tracks, we only change a few names. By far the most common number of sectors is seventeen, followed closely by twenty-six. A few systems have thirty-one, thirty-four, thirty-five, and fifty-two sectors.

$$(\text{Cylinders} \times \text{Sectors} \times 512 \text{ Bytes} \times \text{Heads}) \div 1024 = \text{Capacity}$$
$$(615 \times 17 \times 512 \times 4) \div 1024 = 20.91 \text{ MB}$$

Although we are not generally discussing upgrading, Quantum Corporation has a drive that is very simple to install, uses little power, and is fast. The Plus Hardcard™ actually is installed in a slot just like a card—no separate controller card, control cable, data cable, or power cable. It is as close as I've seen to plug and run. One model's specifications are in the last column of Table 8.2. If you are looking for an easy way out of installing or replacing a hard drive, Quantum has a 105-MB model also.

**TABLE 8.1   Drive type table**

| Type | Cylinder | Heads | Precomp | Land Zone | Sectors | Size (MB) |
|------|----------|-------|---------|-----------|---------|-----------|
| 1 | 306 | 4 | 128 | 305 | 17 | 10 |
| 2 | 615 | 4 | 300 | 615 | 17 | 20 |
| 3 | 615 | 6 | 300 | 615 | 17 | 30 |
| 4 | 940 | 8 | 512 | 940 | 17 | 62 |
| 5 | 940 | 6 | 512 | 940 | 17 | 46 |
| 6 | 615 | 4 | −1 | 615 | 17 | 20 |
| 7 | 462 | 8 | 256 | 511 | 17 | 30 |
| 8 | 733 | 5 | −1 | 733 | 17 | 30 |
| 9 | 900 | 15 | −1 | 901 | 17 | 112 |
| 10 | 820 | 3 | −1 | 820 | 17 | 20 |
| 11 | 855 | 5 | −1 | 855 | 17 | 35 |
| 12 | 855 | 7 | −1 | 855 | 17 | 49 |
| 13 | 306 | 8 | 128 | 319 | 17 | 20 |
| 14 | 733 | 7 | −1 | 733 | 17 | 42 |
| 15 | | | | | | |
| 16 | 612 | 4 | 0 | 663 | 17 | 20 |
| 17 | 977 | 5 | 300 | 977 | 17 | 40 |
| 18 | 977 | 7 | −1 | 977 | 17 | 56 |
| 19 | 1024 | 7 | 512 | 1023 | 17 | 59 |
| 20 | 733 | 5 | 300 | 732 | 17 | 30 |
| 21 | 733 | 7 | 300 | 732 | 17 | 42 |
| 22 | 733 | 5 | 300 | 733 | 17 | 30 |
| 23 | 306 | 4 | 0 | 336 | 17 | 10 |
| 24 | 977 | 5 | −1 | 976 | 17 | 40 |
| 25 | 653 | 14 | −1 | 654 | 17 | 75 |
| 26 | 1024 | 4 | 0 | 1023 | 17 | 34 |
| 27 | 965 | 10 | −1 | 967 | 17 | 80 |
| 28 | 1024 | 8 | 0 | 1023 | 17 | 68 |
| 29 | 914 | 14 | −1 | 915 | 17 | 106 |
| 30 | 1004 | 12 | −1 | 1005 | 17 | 100 |
| 31 | 989 | 5 | 0 | 989 | 17 | 41 |
| 32 | 809 | 6 | 128 | 809 | 17 | 40 |
| 33 | 1024 | 10 | −1 | 1023 | 17 | 85 |
| 34 | 1024 | 12 | −1 | 1023 | 17 | 102 |
| 35 | 1024 | 9 | −1 | 1024 | 17 | 76 |
| 36 | 1024 | 5 | 512 | 1024 | 17 | 42 |
| 37 | 830 | 10 | 0 | 830 | 17 | 68 |
| 38 | 823 | 10 | 256 | 824 | 17 | 68 |
| 39 | 615 | 4 | 128 | 664 | 17 | 20 |
| 40 | 614 | 8 | 128 | 664 | 17 | 40 |
| 41 | 917 | 15 | 0 | 918 | 17 | 114 |
| 42 | 1023 | 15 | 0 | 1024 | 17 | 128 |
| 43 | 823 | 10 | 512 | 823 | 17 | 68 |
| 44 | 820 | 6 | 0 | 820 | 17 | 40 |
| 45 | 1024 | 8 | 0 | 1024 | 17 | 68 |
| 46 | 925 | 9 | 0 | 925 | 17 | 69 |
| 47 | 699 | 7 | 256 | 700 | 17 | 40 |

TABLE 8.2  Sample
manufacturer hard drive
specifications

| | Capacity | | |
| --- | --- | --- | --- |
| | 52 MB | 80 MB | 1050 MB |
| Physical dimensions (inches) | 0.93 | 3.5 | 3.5 |
| Heads/platters | 2/1 | 4/2 | 14/7 |
| Track density (tracks per inch) | 1330 | 1330 | 2127 |
| Seek time (ms) | 17 | 15 | 11 |
| Data transfer rate (MB/s) | 4 | 4 | 10 |
| Cache size (K) | 64 | 64 | 512 |
| Mean time between failures (hours) | 250,000 | 250,000 | 250,000 |

## 8.2  GENERIC SYSTEM OPERATION

**Note:** This section is recommended for technicians.

Table 8.3 gives the signals for the MFM hard drive controller. The 20-pin is the data connector, and the 34-pin is the control connector.

As with floppy drives, the drives are both set up as the same drive (the second drive, or jumper setting). The cable has some lines twisted or reversed, or swapped. In the hard or fixed drive cable, these are lines 25 through 29. (As a memory aid in recognizing cables, you can associate the five lines with the five letters in the word "fixed.") Note that swapping these five lines swaps line 26, Drive Select 1, and line 28, Drive Select 2.

Again, as with floppies, the terminating resistor pack (either DIP or SIP) should be removed from the drive that is *not* at the end of the cable (D: in a two-hard-drive system), and installed in the drive at the end of the cable (C: in a two-drive system).

TABLE 8.3  MFM hard drive
signals

| 20-Pin Connector Data Cable | | |
| --- | --- | --- |
| Pin* | Signal | Direction |
| 8 | Index | From drive |
| 1 | Drive select | |
| 13 | +MFM write data | To drive |
| 14 | −MFM write data | To drive |
| 17 | +MFM read data | From drive |
| 18 | −MFM read data | From drive |

| 34-Pin Connector Control Cable | |
| --- | --- |
| Pin* | Signal |
| 2 | Reduced write current |
| 6 | Write gate |
| 8 | Seek complete |
| 10 | Track 0 |
| 12 | Write fault |
| 14 | Head select 0 |
| 18 | Head select 1 |
| 20 | Index |
| 22 | Ready |
| 24 | Step |
| 26 | Drive select 1 |
| 28 | Drive select 2 |
| 34 | Direction |

*Unlisted pins are reserved or grounds.

**TABLE 8.4   Hard drive
controller port addresses**

| Device | Address | Interrupt |
|---|---|---|
| Hard drive controller (8088 system) | 320H–32FH | IRQ5 |
| Hard drive controller (80286) | 1F0H–1F8H | IRQ14 |

Glancing at the signals on the 34-line connector, line 2, Reduced Write Current, should be true, or high, when writing inner tracks, giving less write current. Line 8, Seek Complete, should pulse when arriving at a track. Line 10, Track 0, signals the heads' arrival at track zero, the outermost track. Lines 14 and 18 select the head. On reading or writing a file that fills more than one track, these should change, as another track of the cylinder is read or written. Line 20, the Index, should pulse at 60 cps for a period of 16.7 ms. You can see from this that access times pushing 16 ms are as low as they can go, unless the rotation of the spindle itself is speeded up. Line 22, Ready, should be true when the drive is selected. Line 24, Step, pulses each time the head is stepped in or out. Line 34, Direction, sets whether the head steps in or out.

Table 8.4 gives the I/O port addresses for the hard disk drive controller.

## 8.3   ASSEMBLING ST506 MFM HARD DRIVES IN OLDER SYSTEMS

### HARDWARE

Installing an ST506 MFM hard drive requires a controller card, a cable, and a drive. Matching the card, the drive, and the cables is essential. Ensure that you do *not* have a floppy drive cable. Perhaps the most common mistake is using a floppy control cable, which has seven twisted lines instead of five.

The method of data encryption also must match. MFM or RLL drives need matching controllers. The controller may be integral with the system board, or it may be a dedicated hard drive controller, half of a hard/floppy controller, or part of a multi-I/O card. The following four steps for the hardware installation must be done in order.

1. Set the jumpers on the controller adapter card, connect the cable, and insert the card in an expansion slot. Appendix B has the phone numbers and Internet addresses of many hard drive manufacturers from which you can get drive type data, jumper settings, and low-level format instructions.
2. Set the jumpers on the drive and physically mount the drive in the case. On PC/XTs, this was directly screwed to side brackets. In the AT, it will slide in the case on slide rails screwed to the sides of the drives. In towers, it is back to direct bracket connection. You may need a 5.25" frame for 3.5" drives, disk drive rails, and possibly drive block-out bezels for empty slots.
3. Connect the thirty-four conductor controller cable and the 20-pin signal cable to the drive, and connect a power cable from the supply to the drive. If your supply doesn't have a free power plug, a power "Y" splitter cable may be needed to change one plug into two. An adapter card may be required to change the 34-pin dual header to an edge card connector and the mini-Molex™ power supply to the standard supply. Most newer systems have enough power connectors with the proper plugs.
4. Once the hard drive is physically installed, the PC/XT and the AT systems must be approached differently. On a PC/XT, the hard drive uses interrupt IRQ5, and it is automatically noticed by the BIOS. The AT system uses IRQ14 and must usually be informed that the hard drive is there and what type it is. The CMOS SETUP must be done, installing the cylinders, heads, and sectors. The proper capacity of the drive is then calculated automatically by the BIOS. A listing of drive parameters is essential. The Western Digital technical assistance phone number, (800) 832-4778, is a starting point for a few major drives. Appendix E gives numerous hard drive parameters.

The remaining steps in this section must now be done to make the drive functional.

## CMOS SETUP

The CMOS SETUP is entered in different ways on different systems. On some Ami BIOS systems, just press DELETE when booting. On some Award BIOS systems, just press CTRL+ALT+ESC. Old ATs had a SETUP.EXE program you could run. Find out how to enter SETUP on your system from the system documentation. Typically, you move through the fields with the arrow keys and change information with the Tab or PageUp and PageDown keys. At the hard drive field, as you change the files with those keys, the type will change and show the parameter table entries for that type. As you exit and save this information to the CMOS memory, the system reboots with the new information.

SCSI drives have a BIOS on the adapter card, so CMOS should be set up as Not Installed. Eight-bit cards may be used in an AT with CMOS set to Not Installed. IDE drives have a translator on them, allowing CMOS to be set up to an equal or smaller capacity (equal total number of sectors); the translator makes this setting physically match the drive by "translating" illegal addresses to legal ones.

Many hard drive controller cards have shunt jumpers to enable or disable their BIOS. This BIOS then overrides control from the CMOS SETUP on ATs. In general, set the CMOS type choice to equal or less than the drive's specifications. If it does not respond, try Not Installed as the CMOS SETUP, and try again.

## LOW-LEVEL FORMATTING

The low-level format totally destroys any previous data. As it prepares the hard drive to accept partitions and high-level formatting, it actually physically marks the sectors and flags bad sectors so they will not be used.

There are two ways to low-level format a hard drive. One is with a program designed to do this. Again Western Digital has a help line. This time, it is for modem calls. The numbers are (714) 753-1234 for 1200 or 2400 bps and (714) 753-1068 for 9600 bps. These programs are offered by Western Digital as a public service. (Other companies' numbers are in Appendix B.) Some handy programs currently offered are

| | |
|---|---|
| ISPFMT.EXE | low-level formatter for IDE (if you really must) |
| WDFMT.EXE | a better low-level formatter |
| EASYDIAG.EXE | a hard drive diagnostic program |

The other way to low-level format the hard drive is to use a program in the ROM on the controller. Not all controllers have this program, and the above number for Western Digital will tell you about their controllers that do. It does no harm to attempt this. Boot DOS from a floppy. Put a disk with DEBUG on it in the drive. Type DEBUG. The new prompt is a hyphen. Now, to run the program, you have to know where it starts. A number of starting locations are given for different controllers. If the address is the wrong one, your system will probably crash. Just reboot and try another address.

```
G=C800:5
G=CC00:5
G=CA00:5
G=C800:CCC
```

If the system starts asking you questions such as "Format which drive?" you have found the correct location. Answer the questions and the program does the rest.

## FDISK

The DOS FDISK program partitions the drive and writes the boot record so the drive is bootable. You can logically partition one physical drive into two logical partitions. Put a disk in the drive with the FDISK program on it, and type FDISK. Answer the questions and it will work quickly.

Be sure that the partition you want to boot from is active. Otherwise, you get a perfectly functioning hard drive that will not boot. The system will boot from floppy, and then you can read or write from or to the hard drive, but it just does not boot. Luckily, activating the partition does not destroy data.

## HIGH-LEVEL FORMATTING

This is done with the DOS FORMAT command. The syntax is FORMAT C:/S. The C: may be drive D: if it is the second hard drive. The /S transfers the DOS system files to the hard drive so it can be booted. Most technicians then create a subdirectory named DOS and copy all DOS disks to it with the following commands, or they use the DOS SETUP.EXE. DOS is self-installing.

```
MD\DOS
CD\DOS.
COPY A:*.*/V
```

Always test any device you install. (In fact, it is a good idea to test *all* devices after any change, just to be sure there are no conflicts.) For a simple test, do a DIR command, copy a text file from a disk in the A: drive with the COPY command, and then read the file with the TYPE command. An example of a simple test of drive C: follows.

Put a known good floppy of the correct capacity in drive A:. The disk must have a text file on it—we'll call it FILE.TXT.

| | |
|---|---|
| DIR C: | If this gives the files properly, fine. |
| COPY A:\FILE.TXT C:/V | V is for verify, which proves proper reading and writing. |
| TYPE C:\FILE.TXT | FILE.TXT should show on the display if it was written correctly. |

If the above works, you probably have a properly functioning hard drive. If not, see the next section, Basic Troubleshooting, or use a diagnostic program for a more thorough test.

The hard drive is now ready for installing applications, such as word processors, databases, and spreadsheets. Some programs and ideas for organizing and protecting your hard drive follow.

## DIRECTORIES AND SUBDIRECTORIES

Directories are like the drawers in a filing cabinet, and subdirectories are like the manila folders in those drawers. (See Chapter 10 to learn to move around the directory system.) When installing application software, always create a directory for that application only, then copy all disks to that directory. The root or first-level directory should be kept as clear as possible. Efforts at intelligent organization when first setting up a hard drive will pay appreciable dividends in future saved time and effort.

## OPTIMIZING THE INTERLEAVE

You can do your own interleave optimizing with only a low-level format program: Low-level format, setting the interleave at 1. Run FDISK. Run the high-level format, starting a stopwatch the instant you touch the last key and stopping it the instant FORMAT.COM asks you for a volume name. Note this time, and repeat for an interleave of 2, then 3, and so on. The interleave that allows the fastest formatting is the one to finish with. Following is example data from a 40-MB drive with 809 cylinders and six heads.

| Interleave | Time |
|---|---|
| 1 | 23 minutes and 26 seconds. (Really!) |
| 2 | 3 minutes and 1 second. |
| 3 | 4 minutes and 21 seconds. |
| 4 | 5 minutes and 42 seconds. |

Note that an interleave of 2 is almost twice as fast as an interleave of 4 (often the default). The low-level format in all cases took 6 minutes 45 seconds, so the whole procedure takes about an hour. If you are going to do this often, get a program as mentioned above.

## BACKUP

Always encourage customers to back up their data on a consistent basis. Always ask them *before* you touch their system if they have backed up their data. Always attempt to back up any hard drive you need to work on. Format enough floppies to hold the data the hard drive holds, then use the DOS BACKUP.EXE program to save this information on floppies. BACKUP C:\ A:/S backs up the entire hard disk to floppies in A:. The /S is for all subdirectories. BACKUP C:\ A:/M backs up only files modified since last back up.

The first approach is the safest. It retains all program setups. It can take a long time, as in hours. The second approach can be used then in the future (by your customer). It takes less time, since only files actually changed are saved again.

## RESTORE

If data was lost or a new drive was installed, you use the DOS RESTORE.EXE program to copy the data from the backup floppies to the hard drive. The syntax is RESTORE A: C:\/S.

## DEFRAG

If you write many small files to your hard disk, they will be placed in order, from track 1, sector 1 on up. There will be no wasted space. If you erase some of these files at random, there are now gaps in the information. Now if you write a large file, it will be broken up and pieces of it will be placed, in order, in the gaps. This file is **fragmented.** Reading fragmented files takes longer than reading nonfragmented files because the heads have to do more jumping around. This slows down disk accesses. The DOS program DEFRAG defragments a disk, speeding it up. It gives a visual representation of what it is doing that really makes its function clear. It has a number of switches, and typing HELP DEFRAG gives much information on the program.

## CHKDSK

CHKDSK is a DOS program to check the status of the disk. It gives a quick look at how the disk is used. As with most DOS programs, to run it just type CHKDSK at the prompt. A sample response follows.

```
Volume MS-DOS_6 created 07-26-1994 11:10a
Volume Serial Number is 1CFA-595B
130,912,256 bytes total disk space
147,456 bytes in 3 hidden files
253,952 bytes in 88 directories
90,421,248 bytes in 4,161 user files
40,089,600 bytes available on disk
```

```
2,048 bytes in each allocation unit
63,922 total allocation units on disk
19,575 available allocation units on disk

655,360 total bytes memory
570,864 bytes free
```

Note that CHKDSK only sees conventional memory, not expanded or extended memory. If it finds lost clusters of information, it will inform you. If you use the /F switch or parameter, it will attempt to fix the problem by repairing the FAT table and saving the lost information in a file named something like FILE0001.CHK. Some say using the /F switch is not really safe.

## SCANDISK

ScanDisk, included in DOS 6.0 up, is a newer, better version of CHKDSK. CHKDSK is still included, and it is useful for a quick report, but to really repair a damaged disk use ScanDisk. For more information, type HELP ScanDisk. Besides checking the FAT table, ScanDisk can verify the quality of the magnetic media by doing a surface scan. This is a valuable addition to DOS and to your tool kit for repairing damaged disks.

## 8.4 INSTALLING IDE, EIDE, AND ULTRA-ATA HARD DRIVES

Installing a hard drive is similar to installing a floppy drive, except for more system setup. The motherboard will have two IDE connectors, one primary and one secondary, as shown in Figure 8.7.

1. For the first hard drive, set the drive jumper as the "master" drive (C/D in the photo) as shown in Figure 8.8. This will make it the first drive on the primary IDE port. Jumper settings are usually printed on a label on the top of the drive. A second drive on the primary port would be set to "slave."
2. Now physically mount the drive in a bay of the same size (use brackets if bay is larger than drive). Two screws on each side secure the drive to the case drive bay. Figure 8.9 shows a drive being physically installed. Of course the connectors must be accessible toward the rear of the bay.

**FIGURE 8.7   Primary and Secondary IDE ports on a modern system**

**FIGURE 8.8** Setting the drive jumper for master or slave

Jumper

**FIGURE 8.9** Hard drive being inserted into bay

3. Next, the 40-pin ribbon cable (colored wire pin 1) must be plugged into the motherboard primary IDE connector. This should have a small triangle raised on the plastic at pin 1, and/or pin 1 labeled in silk-screen printing on the motherboard. Seat the plug firmly as shown in Figure 8.10.

4. This ribbon signal cable must now be plugged into the hard drive itself. Again, polarity must be observed. Seat the plug firmly in the connector as shown in Figure 8.11.

5. Finally, for physical installation, plug a power plug from the power supply into the power jack (Figure 8.12). This is keyed for polarity, but be observant, because it can be forced in upside-down.

**FIGURE 8.10   Plugging the signal cable in the primary IDE motherboard connector**

**FIGURE 8.11   Plugging the signal cable in the hard drive**

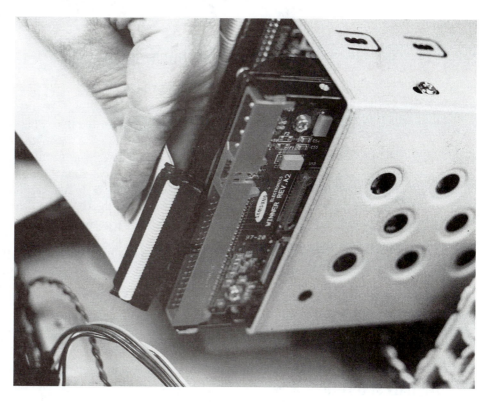

Now the system must be informed of the existence of a new hard drive, and what type it is. This is done in the BIOS SETUP program. Turn on the computer system. After the system checks itself out, it will display an instruction line on how to enter SETUP, such as "Press DEL if you want to enter SETUP." Assuming you have already set up the date, time, and floppy drive(s), select Hard Drive or Pri Master (for Primary Master). You will probably get the option of entering drive parameters yourself manually or letting the system AUTODETECT. If you enter the parameters yourself, you will need to know the number of cylinders, heads, and sectors. These are usually on the drive label or available from the manufacturer. If it asks for Precomp, enter 65335, which means none, and if it asks for landing zone, enter the number of cylinders minus one. Exit and Save from SETUP.

FIGURE 8.12   Plugging in the
power plug

## INSTALLING WINDOWS 98

To install Windows 98 on the hard drive, insert the 3.5" floppy disk labeled Startup or
Boot Disk in your floppy drive. Now turn on the computer and promptly insert the
Windows CD in the CD-ROM drive. The floppy has software to install the CD drive and
then Setup Windows 98 on the hard drive. It is automatic, just pausing a few times to ask
you questions and to have you enter your Certificate of Authenticity registration code
(product key). If your CD-ROM drive is not ATAPI compatible, Windows may not have
the driver needed to operate your CD. In that case, you may need to install an old version of DOS or Windows, install the CD-ROM driver, and then install Windows from
the CD.

For installing Windows 98 as an upgrade from Windows 3.1, click on Program
Manager, File, Run. In the dialog box, type D:\SETUP (where D: is the proper letter for
your CD-ROM drive).

For installing Windows 98 as an upgrade from Windows 95, click on Start, Run, type in
D:\SETUP, and click on OK.

## 8.5   BASIC TROUBLESHOOTING

Troubleshooting hard drive problems begins by noting any POST audio or video error codes.
Then you can proceed to noting any visual clues and taking appropriate measurements.
Details of these steps are as follows.

a. POST Audio Error Code: one short beep with good video but no boot indicates the floppies
   and hard drive are not ready
b. POST Video Error Code of 17xx
c. Measurements
   1. Call Western Digital Corporation at (800) 832-4778 for jumper settings and low-
      level format instructions for any WD controller card. WD also has drive type data
      for numerous drives. Confirm the drive parameters and jumpers, and check the
      cables' polarity. Confirm that the ST506 controller cable has *five* twisted wires,
      not seven. Check jumper settings on IDE for master/slave, and ID on SCSI drives.
      Check termination on ST506 and SCSI. See Appendix E for a list of hard drive
      parameters.

2. Check the controller card installation if one exsists. Check the cables and the drive's physical mounting. Check for pin 1 in proper place. Check CMOS SETUP for drives.

3. Do you hear hard drive "spin up" when powered up? Do you hear the head "seek" the innermost track? If so, you have a good chance of reviving the drive. If the drive is not found by the BIOS or the system goes to never-never land, try Not Installed in the CMOS to see if adapter BIOS will take over.

4. If it sounds alive but does not boot, boot from a DOS or Windows Boot floppy in drive A: and try FDISK to see if it even recognizes the drive. If it does not, run a diagnostic program from the floppy drive to test the hard drive.

5. If the hard drive does not boot, boot with a DOS floppy in drive A: and DIR C: to see if you can read the drive. If you can get a directory but the drive does not boot, look for COMMAND.COM on the root directory. If it is not there, SYS C:. If COMMAND.COM was there, FDISK from drive A: and see if hard drive partition 1 is active. If it was set inactive, it will not boot.

6. If the hard drive is unreadable, warn the customer that all data is probably gone, but if it is *extremely* important to recover it, there are places that will remove the platters and read them to recover data. This costs a small fortune. If the customer will accept loss of all data, attempt a low-level format. If successful, FDISK, and high-level format with system.

7. If the drive is accessible by diagnostic or low level or FDISK, then FDISK it and high-level format it. Use ScanDisk with surface scan to check it out thoroughly.

8. If diagnostic or low-level program cannot access drive, check documentation again, check jumpers and cables again.

9. Swap controller card with a known good compatible card.

10. Swap cables with known good cables.

11. Send drive to repair depot (Appendix C). Be sure that the repair cost is sufficiently below a new drive price to justify keeping the old drive.

12. Keep all "bad" drives to swap the circuit board and revive a drive.

13. If the drive whines excessively, the bearings may be worn. This drive must be refurbished or replaced.

14. If the drive is working enough to run, Windows 98 has some really great help screens. While it does not have a distinct "hard drive troubleshooter," it has a number of programs under System Tools. You get to them by clicking on Start, Help, Contents, Using Windows Accessories, and System Tools (Figure 8.13). This is the screen to get to the troubleshooters also.

Note many of these tools refer to the hard drive. Backup, Compression Agent, Disk Cleanup, Disk Defragmenter, Drive Converter (FAT32), DriveSpace, and ScanDisk.

This System Tools window may be reached more directly by clicking on Start, Programs, Accessories, System Tools. System Information is particularly helpful (Figure 8.14). It gives the version of Windows, the version of Internet Explorer, the type of microprocessor, the amount of system RAM, and the free space on the hard drive.

Under System Tools, System Information you will find Hardware, Components, and Software Environment. Clicking on Hardware allows you to check on Conflicts, DMAs, Forced Hardware, Input/Output, IRQs, and Memory. This is a very useful window. Note the IRQs shown in Figure 8.15.

Explore the windows shown in Figures 8.13–15 thoroughly to become familiar with the helps available to you in Windows 98.

## 8.6 CHAPTER REVIEW/QUIZ QUESTIONS

1. A 40-pin dual header identifies what kind of hard drive?
   a. SCSI
   b. IDE
   c. MFM
   d. Optical

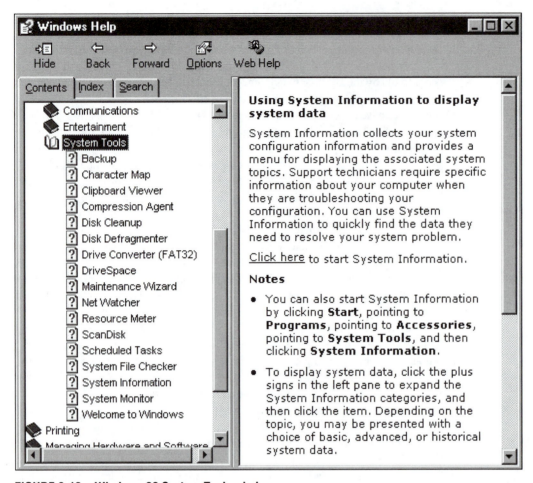

FIGURE 8.13    Windows 98 System Tools window

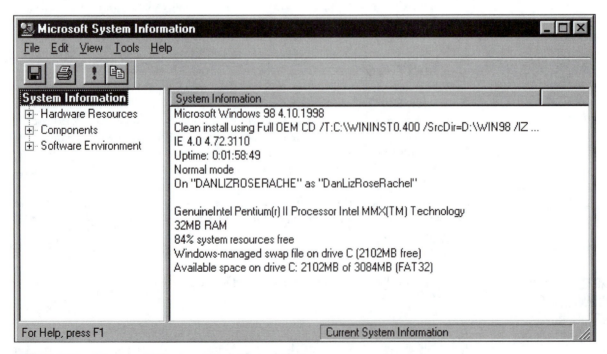

FIGURE 8.14    Windows 98 System Information window

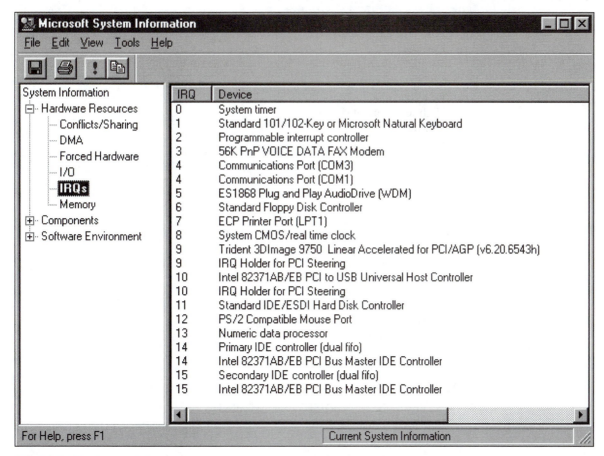

**FIGURE 8.15    System Information**

2. A 50-pin dual header identifies what kind of hard drive?
    a. SCSI
    b. IDE
    c. MFM
    d. Optical

3. A 34-conductor edge card connector identifies what kind of hard drive?
    a. SCSI
    b. IDE
    c. MFM ST506
    d. Optical

4. On MFM hard drive controller cables with two drives, how many lines are twisted?
    a. 3
    b. 5
    c. 7
    d. 9

5. Which interleave gives faster access to the hard drive?
    a. 4:1
    b. 3:1
    c. 2:1
    d. 1:1

6. What are the three steps to prepare a brand-new ST506, MFM hard drive?
   a. Power-up the system.
      Load DOS.
      Format the hard drive.
   b. Low-level format.
      Partition.
      High-level format.
   c. FDISK.
      High-level format.
      Load DOS.
   d. Format.
      Load DOS.
      Load Windows.

7. We usually low-level format only a _____ drive.
   a. SCSI
   b. IDE
   c. MFM ST506
   d. Floppy

8. On some drives, the DOS DEBUG program is used to
   a. Low-level format the drive.
   b. Partition the drive.
   c. High-level format the drive.
   d. Install DOS.

9. On all drives, the DOS FDISK program is used to
   a. Low-level format the drive.
   b. Partition the drive.
   c. High-level format the drive.
   d. Install DOS.

10. On all drives, the DOS FORMAT program is used to
    a. Low-level format the drive.
    b. Partition the drive.
    c. High-level format the drive.
    d. Install DOS.

11. Most hard drives built in the late 1990s had an interleave of
    a. 4:1.
    b. 3:1.
    c. 2:1.
    d. 1.1.

12. Which removable media drive type holds more information?
    a. JAZZ
    b. ZIP

13. For a CD drive to work from the motherboard IDE port, it must be
    a. Faster than ×2.
    b. ATAPI compatible.
    c. A SCSI drive.
    d. A writable CD-ROM drive.

14. Windows 98 has resources for hard drives:
    a. Backup, ScanDisk, Defrag, Compression, FAT32 conversion.
    b. Low-level formatting, partitioning, and high-level formatting.
    c. A "troubleshooter" program.
    d. All of the above.

# 9

# Peripherals

**OBJECTIVES**

Upon completing this chapter you should be able to

- Identify installed peripheral devices by external inspection
- Identify interrupts used for each device
- Install the peripherals covered here
- Troubleshoot and repair problems

**9.1 GENERAL DISCUSSION**

Peripherals are devices connected to the computer but not considered to be part of it. Displays are sometimes considered peripherals, but they are discussed in Chapter 6. A quick glance at the rear of a system gives a reasonably good idea what peripherals it supports. Figure 9.1a shows the rear panel jacks of an AT system.

Note the large round 5-pin female DIN for the keyboard in the lower left corner. This may also have been a 6-pin mini-DIN/PS2 jack. To the right is the 15-pin female high-density D-shell for a VGA display. Next is the 9-pin female D-shell for a mono or CGA or EGA display, and below it is a 25-pin female D-shell for a parallel printer. Next is a 9-pin male D-shell for a serial RS-232 above another 25-pin female D-shell for a parallel printer or SCSI port. Next comes the 15-pin female standard D-shell of a game adapter for a joystick (or a network card) above a male 25-pin D-shell for a serial RS-232. Finally, on the right is the 9-pin female mini-DIN of a bus mouse. This may also have been a 6-pin mini-DIN/PS2 jack. Figure 9.1b shows the rear panel jacks of an ATX system. On the left top is the PS/2 mouse port. Directly under that is the PS/2 (6-pin mini-DIN) keyboard port. Next to the right, the two rectangular jacks are Universal Serial Bus (USB) ports. Next to the right, on top, is the female DB25 jack for the parallel port. Below that, RS232 serial ports 2 and 1. All of these are built into the motherboard. To the right, in the rear panel bracket is the jack for a SVGA monitor from an AGP adapter card.

For all connectors, jacks are female connectors and plugs are male connectors. The D-shell connectors end in S for female and P for male, so a DB9P is a male 9-pin D-shell, and a DB15SHD is a high-density female 15-pin D-shell connector. Figure 9.2 shows two RJ11 phone jacks on the left. Usually two on one card, they identify an internal modem or fax card. Second from the left is a VGA or an SVGA video adapter as shown by the DB15SHD (high-density jack). Third from the left is a typical video adapter (mono, CGA, or EGA) as shown by the DB9S, with parallel printer port shown by the DB25S. Fourth from the left is a dual serial port card with a DB9P and a DB25P. Fifth from the left is an audio or sound card with three mini–phone jacks for auxiliary and microphone inputs and speaker output. The thumb wheel is

**FIGURE 9.1** (a) Rear panel
jacks of an AT system, (b) Rear
panel jacks of an ATX system

(a)

(b)

Keyboard  Mouse  USB          COM1    Parallel    COM2

for volume control, and the DB15S is a game port or MIDI music port. Sixth from the left is a bus mouse card with a DIN-9 jack. Seventh from the left is a card rear bracket with an RJ45 keystone jack (big phone jack) and a BNC bayonet connector. Either of these indicates a network card. Eighth from the left is a video capture card. It allows you to record, edit, and play full-motion video from many sources. On the top is an S-video port for a Super-VHS VCR or camera. In the middle is an RCA phono jack for composite video from the line out of a VCR. On the bottom is an RS-422 jack, which is an 8-pin DIN female. By becoming familiar with these, a glance at the jacks will give you a quick appraisal of what peripherals are installed.

Many peripherals and internal devices automatically modify your AUTOEXEC.BAT and CONFIG.SYS files on installation. It is wise to save your current AUTOEXEC.BAT and CONFIG.SYS with different filename extensions before installing any new device or software. You can use the following commands with filenames of your choice.

```
COPY CONFIG.SYS CONFIG.SAV
COPY AUTOEXEC.BAT AUTOEXEC.SAV
```

Other extensions might include .baq, .bak, .Jan, .Feb, or the like.

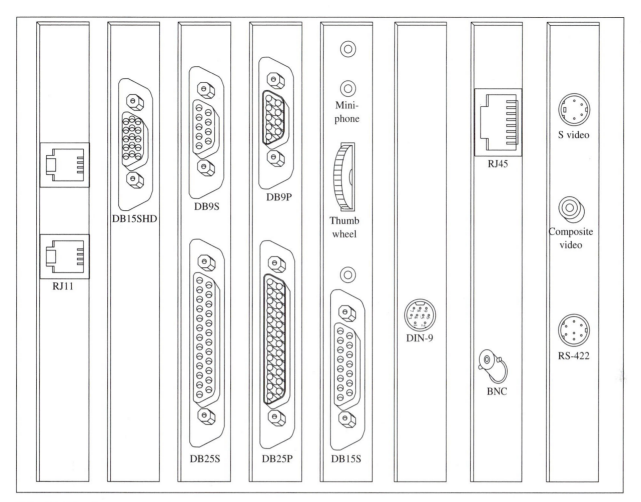

**FIGURE 9.2    Rear panel jacks**

## 9.2  THE GAME CONTROL CONNECTOR

The game control adapter connector is the 15-pin female D-shell connector shown in Figure 9.3. Pins 3, 6, 11, and 13 are the return for the potentiometers, and the current goes through a current-limiting resistor of 2.2 $\Omega$ in the adapter before connecting to the quad timer. This allows the potentiometers to be at 0 $\Omega$ without harming anything. Pins 2, 7, 10, and 14 are pulled high to +5 V by four 1 $\Omega$ resistors in the adapter.

Data from the game control adapter can be read by an OUT command to H201, followed with the IN command from address H201. Bits 0–3 are the resistive inputs, and bits 4–7 are the switch inputs.

This adapter card is not just for fun. It can be used to sense four switches and four variable resistors or voltage levels. These eight inputs can be used to sense information from locations throughout your home or business, and the parallel port can be used to control devices in your home.

### HARDWARE

Game ports can be purchased as one on a card, two on a card, or as one on a multi-I/O card that also has a serial port, a parallel port, and perhaps a floppy and hard drive controller. Examine the documentation that came with the card for switch and jumper settings that may enable or disable the game port. As always, turn the system off before inserting the card in an I/O slot. Any slot will work. Usually the device plugged into the game port is a joystick, although steering wheels are also available. A joystick schematic is also shown in Figure 9.3.

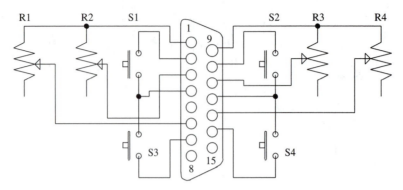

FIGURE 9.3   Joystick schematic and game control adapter jack pinout

Game adapter
(15-pin female
D-shell jack)

R1–R4 100
S1–S4 Normally open pushbutton switches
Pins 1, 8, 9, and 15 = +5 V
Pins 4, 5, and 12 = GND
Joystick has male 15-pin D-shell plug

## SOFTWARE

Game ports do not come with drivers or other software to install. They are accessed through applications, such as games. This is done automatically, so just install the game according to instructions in the documentation. The game may ask if you are using a joystick.

In Windows 98, you may calibrate your game connector by clicking on Start, Settings, Control Panel, Game Controller. This window allows you to choose a game connector type and calibrate and test it. There are General and Advanced tabs. Under General choose your type and click on properties to get Settings to calibrate your joystick, and Test to test it. This is a very handy feature of Windows 98.

Program Listing 9.1 gives a way to access this information from the game port. It allows you to build your own data-gathering system. The I/O address of the game port is 201H.

PROGRAM LISTING 9.1
Joystick test

```
10   REM Joystick example
20   REM The following print a numeric representation of the
     location of a variable resistor between pins 1 and 3 of
     the game port.
30   REM Pins 1 and 3 give stick(0).
40   KEY OFF:CLS
50   LOCATE 10,30 :PRINT "JOYSTICK TEST";:FOR I = 1 TO 5000:NEXT I
60   CLS:PRINT "Move joystick in all directions repeatedly."
70   PRINT "Resistor between pins 1 and 3"
80   FOR I = 1 TO 150
90       X= STICK(0)
100      LOCATE 5,1:PRINT X;
110  NEXT I
120  '
130  '
140  REM The following print a numeric representation of the
     location of a variable resistor between pins 1 and 6
     of the game port.
150  CLS:PRINT "Move joystick in all directions repeatedly."
160  PRINT "Resistor between pins 1 and 6"
170  FOR I = 1 TO 150
```

```
180    X= STICK(1)
190    LOCATE 5,1:PRINT X;
200 NEXT I
210 '
220 '
230 REM The following print a numeric representation of the
    location of a variable resistor between pins 9 and 11
    of the game port.
240 CLS:PRINT "Move joystick in all directions repeatedly."
250 PRINT "Resistor between pins 9 and 11"
260 FOR I = 1 TO 150
270    X= STICK(2)
280    LOCATE 5,1:PRINT X;
290 NEXT I
300 '
310 '
320 REM The following print a numeric representation of the
    location of a variable resistor between pins 9 and 13
    of the game port.
330 CLS:PRINT "Move joystick in all directions repeatedly."
340 PRINT "Resistor between pins 9 and 13"
350 FOR I = 1 TO 150
360    X= STICK(3)
370    LOCATE 5,1:PRINT X;
380 NEXT I
390 '
400 '
410 FOR I = 1 TO 1000:NEXT I
420 REM The following shows if the button from pin 2 to pin 4 is
    depressed.
430 CLS:PRINT "Button between pin 2 and 4. On = -1"
440 PRINT "Press and release button repeatedly."
450 FOR I = 1 TO 100
460    FOR J = 1 TO 250 : NEXT J
470    X = STRIG(0)
480    X$="OFF ":IF X = -1 THEN X$="on !"
490    LOCATE 5,1:PRINT X$;
500 NEXT I
510 FOR I = 1 TO 1000:NEXT I
520 REM The following shows if the button from pin 4 to pin 7 is
    depressed.
530 CLS:PRINT "Button between pin 4 and 7. On = -1"
540 PRINT "Press and release button repeatedly."
550 FOR I = 1 TO 100
560    FOR J = 1 TO 250 : NEXT J
570    X = STRIG(1)
580    X$="OFF ":IF X = -1 THEN X$="on !"
590    LOCATE 5,1:PRINT X$;
600 NEXT I
610 FOR I = 1 TO 1000:NEXT I
620 REM The following shows if the button from pin 10 to pin 12 is
    depressed.
630 CLS:PRINT "Button between pin 10 and 12. On = -1"
640 PRINT "Press and release button repeatedly."
650 FOR I = 1 TO 100
660    FOR J = 1 TO 250 : NEXT J
670    X = STRIG(2)
680    X$="OFF":IF X = -1 THEN X$="on !"
```

```
690     LOCATE 5,1:PRINT X$;
700 NEXT I
710 FOR I = 1 TO 1000:NEXT I
720 REM The following shows if the button from pin 14 to pin 12
    is depressed.
730 CLS:PRINT "Button between pin 14 and 12. On = -1"
740 PRINT "Press and release button repeatedly."
750 FOR I = 1 TO 100
760     FOR J = 1 TO 250 : NEXT J
770     X = STRIG(3)
780     X$="OFF ":IF X = -1 THEN X$="on !"
790     LOCATE 5,1:PRINT X$;
800 NEXT I
810 FOR J = 1 TO 6000:NEXT J
820 CLS:PRINT "If any of these indicators did NOT change with
    proper joystick"
830 PRINT "or button movement, do a resistance check of joystick's
    4 resistors"
840 PRINT "and 4 buttons. If these are good, suspect Game Control
    Adapter."
```

## 9.3   SERIAL CONNECTORS    SERIAL RS-232 (COM1:) CONNECTOR

The serial RS-232 connector is a 25-pin male serial connector or a 9-pin male D-shell connector on the adapter card. The 9-pin male serial connector was introduced on the AT and is the way some dual serial cards fit two connectors on the metal backplate of the adapter card. Figure 9.4 shows the 9-pin and 25-pin connectors.

Table 9.1 gives the pinouts of both serial RS-232 connectors.

**FIGURE 9.4    9- and 25-pin serial connectors with ribbon cable and dual headers**

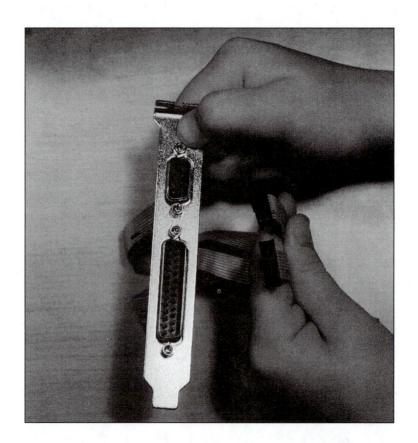

**TABLE 9.1    Serial RS-232 connector pins**

| Signal (DTE) | Name | Direction from Computer | 25-Pin | 9-Pin |
|---|---|---|---|---|
| TxD | Transmit data | Out to Device | 2 | 3 |
| RxD | Receive data | In from Device | 3 | 2 |
| RTS | Request to send | Out | 4 | 7 |
| CTS | Clear to send | In | 5 | 8 |
| DSR | Data set ready | In | 6 | 6 |
| GND | Ground | | 7 | 5 |
| CD | Carrier detect | In | 8 | 1 |
| DTR | Data terminal ready | Out | 20 | 4 |
| RI | Ring indicator | In | 22 | |

## HARDWARE

Modern motherboards are likely to have two serial and one parallel ports on board as shown in Figure 9.5a. If not on board, you must use a serial port adapter. The serial port adapter can be a single adapter on one card, two adapters on one card, or part of a multi-I/O card. In each case, switches or jumpers on each should be set according to the documentation that comes with the card. Note which motherboard 10-pin header is serial port #1 and where pin 1 is. Plug in as shown in Figure 9.5b.

The device designation for the serial ports is COM:, for communications port. Up to four ports may be designated COM1:, COM2:, COM3:, and COM4:. Table 9.2 shows that COM1: and COM3: may share interrupts but use different addresses, and the same is true of COM2: and COM4:. These may all be connected, but ports sharing interrupt requests should not be used at the same time. Set the jumpers or switches so they don't conflict with each other or with other devices in the system. Test both devices thoroughly to confirm operation.

**FIGURE 9.5    (a) Motherboard serial port dual header**

Two 10-pin serial port headers

**FIGURE 9.5** (*continued*) (b) plugging serial plug into motherboard jack

**TABLE 9.2  Serial RS-232 ports, addresses, and interrupts**

| Port | I/O Address | Interrupt |
|------|-------------|-----------|
| COM1: | 3F8H–3FFH | IRQ4 |
| COM2: | 2F8H–2FFH | IRQ3 |
| COM3: | 3EBH–3EFH | IRQ4 |
| COM4: | 2E8H–2EFH | IRQ3 |

The serial port can be tested with a loopback plug, which tricks the computer into thinking a device is connected. (Loopback pinouts are shown in Table 9.3.) A simple COPY CON: COM1: command followed by a line or two of text and a CTRL+Z will complete a test. The information you key in will be accepted by the port if all goes well.

Note that when the computer sends a request to send (RTS) on pin 4, it gives itself a clear to send (CTS) and a data set ready (DSR) on pins 5 and 6. When the computer sends out a data terminal ready (DTR) on 20, it sends itself a carrier detect and ring indicator on 8 and 22. The data it sends on 2, it also receives on 3.

The serial ports have many uses. They may connect to a plotter (Figure 9.6), a printer, a speech device, a mouse, or another serial device. Devices that communicate through a serial port are placed in one of two classifications. The computer and terminals are data terminal equipment (DTE); mice, modems, printers, plotters, and similar devices are data communication equipment (DCE). The computer, or DTE, has a male connector, and the devices, or DCE, have a female connector.

For a computer to talk to a printer, the cable is a "straight through" cable with one female end to plug into the computer and one male end to plug into the device. If one computer is going to communicate with another, they need to be connected with a different kind of cable. Since both computers are talking on line 2 and listening on line 3 (Table 9.1), those two lines must be swapped in the cable. Pin 4 is the RTS and pin 5 is CTS, so those lines must be swapped also. Line 6 is DSR and line 20 is DTR. Connecting 20 DTR out of a computer, to 6 DSR going into the other computer, the first computer is telling the other it is ready to send and receive information. Finally, line 8, carrier detect (CD) or received line signal indicator, must

**TABLE 9.3  Serial loopback plugs and null modem cable**

| Standard 25 Loopback | Alternate 25 | 9-Pin |
|---|---|---|
| Lines  2 and 3 | Lines  2 and 3 | Lines  1, 4, and 6 |
| 4 and 5 | 4 and 5 | 2 and 3 |
| 4 and 6 | 6, 8, and 20 | 7 and 8 |
| 20 and 8 | | |
| 20 and 22 | | |

*Note:* Connect these pins on one female D-shell.

**Null Modem Cable**

| Pins Plug 1 | | Pins Plug 2 |
|---|---|---|
| 2 | to | 3 |
| 3 | to | 2 |
| 4 | to | 5 |
| 5 | to | 4 |
| 6 and 8 | to | 20 |
| 20 | to | 6 and 8 |

*Note:* Connect these pins on one female D-shell to the pins on another.

**FIGURE 9.6  Serial plotters**

Desktop plotter

Floor-stand plotter

also come into the computer, so it is connected to 20 also. The null modem cable that plugs into the serial ports has a female DB-25 on both ends. The pinouts are shown in Table 9.3.

## SOFTWARE

If you attempt to use DOS or write your own software to use the serial ports, you must set up the ports with the following DOS command.

`MODE COM1:BAUD,PARITY,DATA,STOP`

Where parity = Even, Odd, or None; data bits = 7 or 8; and stop = 1 or 2, so an example is

`MODE com1:1200,n,8,1`

Programs other than your own usually do not require the MODE command. For example, a word processor should be set up according to its documentation to access the serial port, COM1: or COM2:, whichever the printer is connected to. The program may ask for baud speed and data bits, just as the MODE command requires. A drafting (drawing) program requires a similar setup for a serial plotter. Modems usually are packaged with some form of communication program that requires setting up according to instructions included. Serial mice do have driver programs, and those are covered under Section 9.6.

## TROUBLESHOOTING

Troubleshooting communications port problems is much easier with the use of a loopback connector and a breakout box. The breakout box goes between the two devices that are trying to communicate and allows you to connect the lines you want to with wire jumpers or switches. Some have LEDs to light up when the lines are active. Once you find the connections that allow the devices to communicate, you can note the pin-to-pin configuration and build a custom cable for permanent use.

Figure 9.7 shows three kinds of breakout boxes. The one on the left has open pins and requires you to insert wires to connect one side to the other. The one on the right has LEDs to light when active. The center one has LEDs, open pins for wires, and switches to connect straight across. It is obviously more expensive.

**FIGURE 9.7 Breakout boxes**

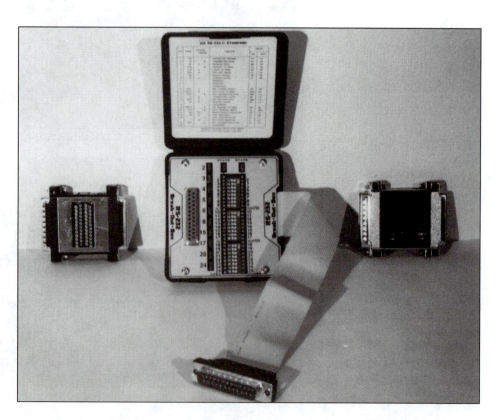

Problems with peripherals fall into one of two categories: a working serial device that fails or a newly installed device that doesn't work. Different troubleshooting approaches are used for these conditions.

## A Working Serial Device Fails

You have been using the system as is for days, or longer, and now a serial device fails to work. All the setups of hard- and software were correct and not changed, so something has failed. If the problem is with a modem, see the section on modems. Otherwise, there are four possible modes of failure:

1. Driving software has failed.
2. Adapter card has failed.
3. Cable has failed.
4. External serial device has failed.

To isolate which unit has failed, proceed as follows.

1. Confirm beyond a shadow of a doubt that no changes were made to the system: nothing new installed, nothing old removed, AUTOEXEC.BAT and CONFIG.SYS files not changed, nothing was unplugged.
2. Run MSD and note the serial port interrupts and I/O addresses. From Table 9.2 confirm that the installed devices are still recognized and still do not conflict. If a device is physically present and is recognized, but it is no longer recognized by MSD, then it is very likely the serial adapter card has failed. Replace it and try again.
3. Remove the cable and device, and insert a loopback connector. Do the simple COPY CON: COM1: or COM2: test mentioned earlier, or use a diagnostic program to send information out to the serial port in question. If this test is passed, the serial adapter card is OK, and the cable or the device or the software program being used to run the device has failed.
4. With the loopback plug connected, run the software program (application) you have been using. If it works, the cable and device are suspect. If it does not work, check the program setup for proper settings.
5. Place the loopback on the end of the cable, connect the cable to the serial adapter (or replace the cable) and test again. If the tests work, this has cleared the software, the adapter card, and the cable, and the device is suspect. If these tests fail, the cable is bad.
6. If all the above fails, the device is probably bad. You can try using the breakout box to verify that communication lines have correct signals.

## A Newly Installed Device Does Not Work

Troubleshooting a newly installed system differs from troubleshooting the failure of a working system. Six modes of failure are possible:

1. Hardware is not set up correctly.
2. Software is not set up correctly.
3. Hardware is conflicting with other hardware.
4. Software is conflicting with other software.
5. Hardware is bad (DOA).
6. Software is bad (DOA).

To isolate the problem, proceed as follows.

1. Printers, plotters, speech synthesizers, and some modems have self-test modes that totally isolate them from the computer system. These tests can be run without having the device connected to the computer. Run the self-test first to verify that the device hardware is operational.
2. See Section 9.13 on resolving conflicts to verify that the hardware and software are properly set up. The adapter card, cable, and device settings, such as printer power-on, on line, and baud rate settings, must all be checked. This can be complex, but it should eliminate the first four modes of failure.
3. If applying the information in Section 9.13 does not resolve the problem, it is likely the device or software was dead on arrival. There are inexpensive shareware programs that print on printers, drive plotters, and test mice. Try these or a diagnostic program to

drive the device. If these fail, replace the device. If these work, reinstall or replace the application software that could not make the device function.

In Windows 98, click on Start, Help, Contents, Troubleshooters, Hardware Conflicts. You may look directly at all the hardware resources by clicking on Start, Programs, Accessories, System Tools, System Information, Hardware Resources. This will give IRQs, DMAs, and I/O addresses.

## SERIAL SIGNALS

To observe the output of the computer's RS=232 port, connect an oscilloscope to pin 2 for signal and pin 7 for ground. Voltage of signal is from –12 V to +12 V, so center sweep for ground and set vertical sensitivity to 5 V/div. DC couple signal to scope. For 300 baud, the period of a bit is 3.33 ms. There will be 10 bits, so set the horizontal sweep on 2 ms/div and decalibrate until the byte received on the scope is 10 divisions long.

At the DOS prompt, type CD\DOS to change to the DOS subdirectory. Then type MODE COM1:300,N,8,1 to set the serial port up for use. Type GWBASIC or QBASIC to enter the BASIC interpreter. Copy the program in Program Listing 9.2.

**PROGRAM LISTING 9.2    Serial port signals**

```
10 OPEN "COM1:300,N,8,1,RS,CS,DS,CD" AS #1
20 A$=INKEY$
30 IF A$<> "" THEN B$=A$
40 PRINT#1,B$
50 GOTO 20
```

In GWBasic, type RUN to run the program, and press a key. In QBasic, hold the ALT key and press R, release both, then press S. The ASCII code is given in Appendix D.

The oscilloscope signal needs to be interpreted to make it match the hexadecimal code for the ASCII character sent. The first example below uses the ASCII decimal code of $49_{10}$ for the character "1." This is hex $31_{16}$ or 31H. Hex 31 is binary $0011\ 0001_2$. This is what the signal should be. Since the code is sent most significant bit (MSB) first, it will appear right-most on the scope trace, with the least significant bit (LSB) appearing on the left. To match this to our code, we must reverse the code from left to right. Now a start bit is added to one end and a stop bit to the other. Finally, although On is usually high in digital logic and Off is low, this is inverted vertically, so that we must now make lows into highs and highs into lows. Now follow the two examples that follow. The output will appear on the scope like this for the two numerals "1" and "7."

Try a few other characters until you are familiar with reading the serial output with the oscilloscope.

## USB SERIAL CONNECTOR

USB (Universal Serial Bus) is a new serial communication standard that may prove to be a real breakthrough. It is fast! It can run at two speeds, 1.5 Mbps for low speed and 12Mbps for high speed. Low-speed devices would include joysticks, keyboards, and mice. An astonishing 127 devices may be connected to the bus. Also, there are no dip switches to set; the IRO, DMA, and I/O address confusion and conflicts of the past are eliminated. The USB is Plug-and-Play and Hot Swappable. This means you can connect or disconnect any device while the system is running, and it will immediately recognize that device. The motherboard must support a USB by having a USB host controller on board, or you must install a USB Controller Card. Windows 95 does not support the USB until version OSR2.1. Windows 98 does support USB and devices that support the Windows Driver Model (WDM). This model has I/O services and compatible device drivers built in, hopefully to ensure future compatibility. USB devices are new on the market, and motherboards still include the old standard serial and parallel ports. Older devices that are not recognized as Plug-and-Play are called *Legacy* devices, which means pre–Plug-and-Play.

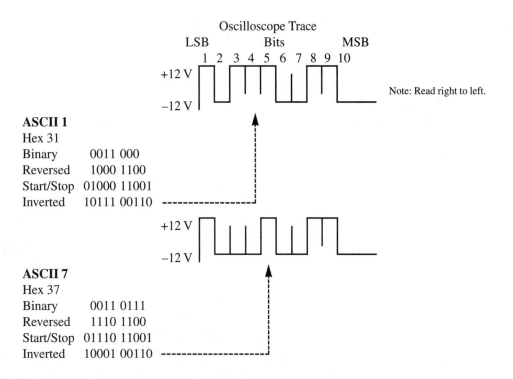

Oscilloscope Trace

LSB          Bits          MSB
1 2  3 4 5 6 7 8 9 10

+12 V

−12 V

Note: Read right to left.

**ASCII 1**
Hex 31
Binary        0011 000
Reversed      1000 1100
Start/Stop  01000 11001
Inverted    10111 00110

+12 V

−12 V

**ASCII 7**
Hex 37
Binary        0011 0111
Reversed      1110 1100
Start/Stop  01110 11001
Inverted    10001 00110

The rear panel USB jacks are four conductor connectors. On ATX boards these are directly connected to the motherboard. On some systems they are connected through a ribbon cable to a dual header on the motherboard. There seem to be three different pinouts so far for this dual header. One is an 8-pin dual header (2 × 4) pinned as follows:

    1 and 5 =    Vcc
    2 and 6 =    Negative data
    3 and 7 =    Positive data
    4 and 8 =    Ground

There are two different 10-pin dual header pinouts:
The following is the same as the 8 pin, if pins 5 and 10 are not connected.

    1 and 6 =    Vcc
    2 and 7 =    Negative data
    3 and 8 =    Positive data
    4 and 9 =    Ground
    5 and 10 =   Ground

The second posible pinout of a 10-pin dual header is different. Note that 1–5 are the same as above, but 6 –10 are reversed, as though numbered as a chip. The wrong connector on this header will damage your system. Be sure you know the pinout of the header and the connector for your system.

     1 =    Vcc
     2 =    Negative data
     3 =    Positive data
     4 =    Ground
     5 =    Ground
     6 =    Ground
     7 =    Ground
     8 =    Positive data
     9 =    Negative data
    10 =    Vcc

In Windows 98, clicking Start, Help, Search, and entering USB gives six topics about USB to review. Clicking Start, Settings, Control Panel, System, Ports you see the ports installed.

By double-clicking on Communications Port you open a window with tabs for General, Port Settings, Driver, and Resources, which lets you see the I/O addresses and IRQs.

## 9.4 PARALLEL PRINTER CONNECTOR (LPT1:)

The parallel printer connector, designated LPT1: or LPT2:, is a 25-pin female D-shell connector (Figure 9.8). Modern motherboards may have a 26-pin dual header (row 2 of 13) to connect to. If not it will be on the parallel printer card, the video card, or a multi-I/O card. Table 9.4 gives the signals. This port can also be used to send eight signals out

**FIGURE 9.8    (a) DB25F, 25-pin D-shell parallel connector with ribbon and dual-header plug; (b) plugging the 26-pin plug into the motherboard**

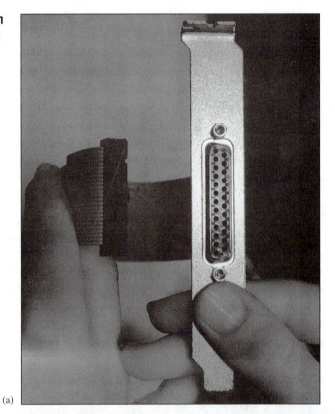

(a)

(b)

**TABLE 9.4   Parallel printer connector**

| Female DB25 Computer Connection | | Direction (from computer) | 36-Pin Centronic (printer connection) |
|---|---|---|---|
| Pin | Signal | | |
| 1 | Strobe | Out to printer | 1 |
| 2 | Data bit 0 | Out | 2 |
| 3 | DB1 | Out | 3 |
| 4 | DB2 | Out | 4 |
| 5 | DB3 | Out | 5 |
| 6 | DB4 | Out | 6 |
| 7 | DB5 | Out | 7 |
| 8 | DB6 | Out | 8 |
| 9 | DB7 | Out | 9 |
| 10 | Acknowledge | In from printer | 10 |
| 11 | Busy | In | 11 |
| 12 | Out of paper | In | 12 |
| 13 | Select | In | 13 |
| 14 | Auto feed | Out to printer | 14 |
| 15 | Error | In | 32 |
| 16 | Initialize | Out | 31 |
| 17 | Select input | Out | |
| 18–25 | Ground | | 19–30 & 33 |
| | | | 35 +5 V |
| | | | 36 select in |

to control the environment in your home or office. Two DOS programs, Interlnk and Intersvr, work together to transfer files between two computers through either the serial or the parallel ports.

## HARDWARE

Installing the hardware consists of setting the jumpers or switches on the card by the instructions that came with it, and inserting the parallel port card in an I/O slot. Be sure the settings for the interrupt and I/O address do not conflict with existing devices. It may be necessary to disable a device that is not in use. The cable from the computer to the printer has a male DB25 to mate with the computer port and a 36-pin centronics male to mate with the printer. The centronics cable connector is shown in Figure 9.9. Parallel port addresses and interrupts are given in Table 9.5.

The pinouts were given in Table 9.4. Pin 1, Strobe, is normally high, and the computer pulses it low for the printer to load a byte of data. Pins 2–9 are the 8 bits of data. Pin 10 is

**FIGURE 9.9   Centronics male connector**

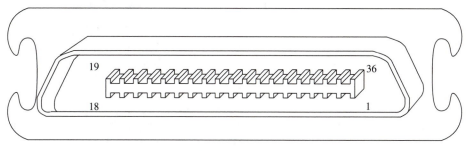

Plugs into printer

**TABLE 9.5 Parallel port addresses and interrupts**

| Port | I/O Address | Interrupt |
|------|-------------|-----------|
| LPT1: | 378H–37FH | IRQ7 |
| LPT2: | 278H–27FH | IRQ5 (On AT system—on 8088 this is hard drive.) |

the printer telling the computer it has received the byte. Pin 11, Busy, goes high for the printer to tell the computer its buffer is full.

Pin 11, Busy, pulled low tells the computer that the printer will accept characters. Pin 12, Out of Paper (or Printer Error), pulled low tells the computer that the printer has no error. Depending on your port, other lines may be needed for the loopback connector.

## SOFTWARE

DOS runs the parallel port with no setup of any kind. Pressing SHIFT+PrintScrn will print the screen to the default printer, usually LPT1:. Application software may need to be set up according to its documentation, but usually it defaults to LPT1:. If you have two kinds of printers, such as a dot matrix and a laser, connected to the system at one time, the application software must be informed which is LPT1: and which is LPT2:. Consult the application's documentation for this setup.

In Windows 98, click on Start, Settings, Control Panel, System, Device Manager. Double-click on Ports (COM and LPT) and Printer Port (LPT1). The Resource tab gives you I/O address and IRQ. Under Accessories, System Tools, System Information, Components, Printing, you can find information about your printer port also.

Direct setting and testing of the printer may be done by clicking Start, Settings, Printers, the printer of your choice, File, Properties. There are tabs and buttons to perform such tasks as print a test page; change color, paper, orientation; and check on what may be in the print buffer.

In this printer window you may also add or delete different printers and set which will be the default printer.

## TROUBLESHOOTING

There are two failure modes for printers: hardware or software. Since the parallel port is well standardized and reliable, suspect the application software first. Proceed as follows.

1. Almost all printers have self-test modes that totally isolate them from the computer system. These tests can be run without connecting the printer to the computer. Run the self-test first to verify that the printer is operational. This clears the printer only 90%, because the parallel receiving circuitry has not been tested.
2. DOS alone can test the parallel port with a parallel loopback connector by attempting to print out that port. In fact, it is usually best to try DOS first rather than some complex application software that needs to be set up. SHIFT+PrintScrn will print whatever text is on the screen out the default port. The DOS PRINT command can also be used, as follows.

```
PRINT /D:LPT1 AUTOEXEC.BAT
```

This command tells DOS to print the AUTOEXEC.BAT file on LPT1:. No application software gets in the way, and there is no setup. If this fails with the loopback plug installed, then there is a hardware problem, either in setup or due to conflicts or in the adapter.

3. If the loopback test passed from DOS or a diagnostic program, then it may be time to look at signals (Table 9.6).
4. There is a "Print" troubleshooter in Windows 98. Click on Start, Help, Contents, Troubleshooting, Windows) Troubleshooters, Print. It will lead you through a question list to isolate your problem.

| Standard Pin–Pin | Alternate Pin–Pin |
|---|---|
| 1–13 | 11–Ground |
| 2–15 | 12–Ground |
| 10–16 | |
| 11–17 | |
| 12–14 | |

## PRINTERS

The three major types of printers in use today are the dot matrix, the ink-jet, and the laser. All printers have some mechanism for moving the paper through, but the real difference is how they apply color to the paper. The sections below address this difference.

### Dot Matrix Printers

Dot matrix printers have a head that travels across the paper, printing characters or dots for graphic pictures. The head has "wires" or "pins" in a vertical column. These pins strike out from the head, hit a ribbon holding ink, and hit the ribbon against the paper, which is resting on a cylindrical rubber platen (looks like a rubber kitchen rolling pin). If all pins fire, they make a vertical line like the number 1. The head then moves horizontally just a little, and different pins fire. Modern printers usually use 9 or 24 pins. The 24-pin printer can place dots closer together, making higher quality characters and graphics. The best resolution is about 300 dots vertically per inch and 300 horizontally. At this resolution, you can tell that a picture is made up of dots. Anything less than that resolution looks bad. Figure 9.10 shows a dot matrix printer and the print head.

### Ink-Jet Printers

Ink-jet printers are like dot matrix printers because they put dots on the paper to form characters and graphics. However, instead of using pins to push an inked ribbon against the paper, the

**FIGURE 9.10  Dot matrix printer and print head**

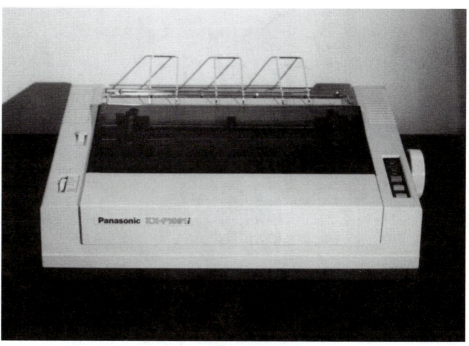

Printer

FIGURE 9.10 *(continued)*

Print head

ink jet actually squirts a tiny drop of ink at the paper. The "head" is a cartridge of liquid ink in one piece with the squirting mechanism. Resolution in ink-jet printers ranges from $300 \times 300$ for an inexpensive printer ($200) to $600 \times 300$, which looks good. A more expensive printer ($600) using special paper can now reach $720 \times 720$ dpi (dots per inch). This gives excellent graphics. Figure 9.11 shows an ink-jet printer and two ink cartridges. On the cartridge on the left, the bright dots are electrical contacts; on the cartridge on the right, the small shiny rectangle on the front near the top is the face with the tiny holes the ink squirts out of.

## Laser Printers

Laser printers are much more complex and, of course, more expensive. While an ink-jet–printed paper will spot and run when wet or damp, a laser-printed paper is more water resistant. Laser resolution starts at $300 \times 300$ dpi for less expensive models, and can go to $1200 \times 1200$ for a price. Color laser printers are available, and they cost more. Figure 9.12a shows a laser printer. The ink of a laser printer is really a plastic and iron dust or powder, called toner, that is melted onto the paper. Instead of replacing ribbons or ink cartridges, you replace a toner cartridge, shown in Figure 9.12b.

**FIGURE 9.11    Ink-jet printer and cartridges**

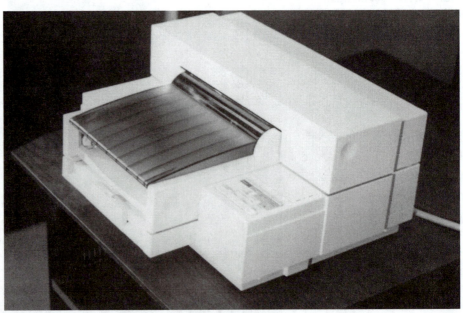

Ink-jet printer

**FIGURE 9.11**

Cartridges

**FIGURE 9.12 (a) Laser printer, (b) laser toner cartridge**

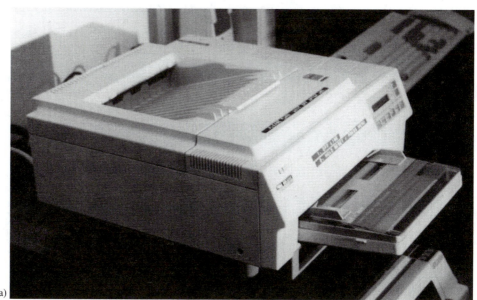

(a)

Laser printer

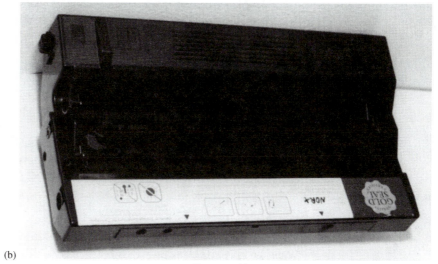

(b)

Toner cartridge

**FIGURE 9.12** *(continued)*
**(c) the laser printing process**

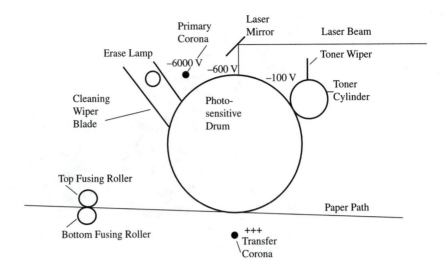

Hewlett-Packard is a company well known for a myriad of high-quality, high-tech equipment, including laser printers. Their laser service manual defines six stages in the laser printing cycle.

1. First the photosensitive drum is cleaned. This drum can be electrically charged, but light hitting a spot will lower that charge. This first step, *cleaning*, is a two-part process. First, a cleaning wiper blade wipes off any toner clinging to the drum. Then an erase lamp discharges any remaining charge, so the drum is clean and discharged (Figure 9.12c).

2. The second step is *conditioning* the drum. The primary corona wire is charged to about –6000 V. This creates negatively charged ions, which pass through a grid to charge the drum. The grid controls the charge and holds the final drum charge to about –600 V, much less than the –6000-V primary corona wire.

3. The third step is *writing* the information on the drum. A microprocessor takes the data from the computer and breaks it down into horizontal rows of dots much like a dot matrix printer does, one row of dots at a time. A laser is turned on for a dot and off for none, as a rotating octagonal scanning mirror sweeps the beam horizontally, creating a line of dots of light where the dots of "ink" should be. A diagonal mirror redirects this beam down to the drum. The laser-light hitting the drum discharges the –600 V to about –100 V. As the drum turns (clockwise in Figure 9.12c) after a row is written, it is ready for another row of dots. So –100 V at a point means an "ink" dot will be there, and a –600-V point means no "ink."

4. In the *developing* step, the toner cylinder rotates (counterclockwise in Figure 9.12c), carrying a surface dusting of toner powder, which is kept to a thin layer by the toner wiper blade. The toner is charged to an intermediate negative value of perhaps –300 V, depending on the setting for the density of print. As the toner on the toner cylinder comes close to the drum, toner is attracted to the more positive –100-V dots. Thus an image of dust now is on the drum.

5. *Transferring* the powder image from the drum to the paper is done by the positive transfer corona. This wire positively charges the paper, attracting the negatively charged toner dots from the drum. At this point the image is just powder held to paper by electrostatic charges and can be easily smeared or smudged.

6. Finally, *fusing* of the image to the paper is done by pressing and melting the toner on the paper. The heat can reach more than 300 degrees Fahrenheit, which is why the paper comes out warm. This is also why the laser-printed page is not as susceptible to damage from moisture as an ink-jet printed page.

## PARALLEL SIGNALS

With an oscilloscope looking at the printer end of the cable, which is connected to the printer, you can use the following program to send out chosen characters and check the actual signal received by the printer. Pins 2–9 are the actual ASCII code, pin 10 is Acknowledge, and pin 11 is Busy. Note in Figure 9.13 how the binary pattern is followed for the numbers chosen.

FIGURE 9.13 Parallel port
signals with corresponding
decimal numbers

| Char | Pin | | | | | | | | | |
|------|-----|---|---|---|---|---|---|---|----|----|
|      | 2 | 3 | 4 | 5 | 6 | 7 | 8 | 9 | 10 | 11 |
| None | L | H | H | H | H | H | H | H | H | L |
| 0    | L | L | L | L | L | L | L | L | H | L |
| 1    | H | L | L | L | L | L | L | L | H | L |
| 2    | L | H | L | L | L | L | L | L | H | L |
| 4    | L | L | H | L | L | L | L | L | H | L |
| 8    | L | L | L | H | L | L | L | L | H | L |
| 16   | L | L | L | L | H | L | L | L | H | L |
| 32   | L | L | L | L | L | H | L | L | H | L |
| 64   | L | L | L | L | L | L | H | L | H | L |
| 128  | L | L | L | L | L | L | L | H | H | L |

*Notes:*
1. Parallel signals on pins.
2. Number in CHR $ (XX).
3. H = 5 V; L = 0 V

Program Listing 9.3 is a simple program you can use to assist in testing parallel port signals with an oscilloscope.

**PROGRAM LISTING 9.3**
**Parallel port signals**

```
REM testing parallel port
REM with oscilloscope
LPRINT "Running"
forever = 1
WHILE forever
 REM Change the 1 in the next
 REM line to other numerals 2,
 REM 4,8,16,32,64,128
 REM 1 will raise line 2 high
 REM (+5 V), 128 will raise line
 REM 9 high.
 LPRINT CHR$(1);
 PRINT CHR$(254); ".";
 FOR j = 1 TO 400: NEXT j
END
```

## 9.5 THE SCSI PORT

The SCSI (Small Computer Systems Interface) port allows up to eight controllers to communicate. This interface handles hard drives, CD-ROM drives, scanners, tape drives, and other peripherals. Tables 9.7 and 9.8 give the signals for a single-ended SCSI port and a differential SCSI port, respectively. Figure 9.14a shows the 50-pin SCSI dual header in use on hard drives. Figure 9.14b shows a SCSI bus adapter card with a female 25-pin D-shell connector. This is the same as the parallel printer port, so care must be taken when connecting peripherals. Note the bracket is labeled to aid in identifying this as a SCSI port.

The standard was improved from the 8-bit 50-conductor port to a 16-bit 68-conductor port known as "Wide SCSI." When the data transfer speed reached 20 MB/s, it was dubbed "Ultra-SCSI." Improvements have brought the 40 MB/s Ultra-2 SCSI.

**TABLE 9.7   Single-ended SCSI port signals**

| Pin | Signal |
|-----|--------|
| 2 | –Data bit 0 |
| 4 | –Data bit 1 |
| 6 | –Data bit 2 |
| 8 | –Data bit 3 |
| 10 | –Data bit 4 |
| 12 | –Data bit 5 |
| 14 | –Data bit 6 |
| 16 | –Data bit 7 |
| 18 | –Data bit P |
| 20 | Ground |
| 22 | Ground |
| 24 | Ground |
| 26 | Terminator power |
| 28 | Ground |
| 30 | Ground |
| 32 | –Attention |
| 34 | Ground |
| 36 | –Busy |
| 38 | –Acknowledge |
| 40 | –Reset |
| 42 | –Message |
| 44 | –Select |
| 46 | –C/D |
| 48 | –Request |
| 50 | –Input/Output |

*Note:* All odd pins ground.

**FIGURE 9.14   (a) SCSI port**

(a)

SCSI

1 ▢▢ 2
49 ▢▢ 50

**TABLE 9.8   Differential SCSI port signals**

| Pin | Signal |
|-----|--------|
| 1 | Shield |
| 2 | Ground |
| 3/4 | ± Data bit 0 |
| 5/6 | ± Data bit 1 |
| 7/8 | ± Data bit 2 |
| 9/10 | ± Data bit 3 |
| 11/12 | ± Data bit 4 |
| 13/14 | ± Data bit 5 |
| 15/16 | ± Data bit 6 |
| 17/18 | ± Data bit 7 |
| 19/20 | ± Data bit P |
| 21 | Differential sense |
| 22/23 | Ground |
| 24 | Ground |
| 25/26 | Terminator power |
| 27/28 | Ground |
| 29/30 | ± Attention |
| 31/32 | Ground |
| 33/34 | ± Busy |
| 35/36 | ± Acknowledge |
| 37/38 | ± Reset |
| 39/40 | ± Message |
| 41/42 | ± Select |
| 43/44 | ± C/D |
| 45/46 | ± Request |
| 47/48 | ± Input/Output |
| 49/50 | Ground |

**FIGURE 9.14   (b) a SCSI adapter card with a female DB25 connector**

SCSI lines must be terminated properly, just like hard drive signal lines, and each SCSI peripheral will have some means of selecting a SCSI address. See how a SCSI optical scanner is installed later in this chapter.

In Windows 98 you may find information on your SCSI port by clicking on Start, Settings, Control Panel, System, Device Manager, double-click on SCSI controllers. Click on the controller that appears, and then on Properties. Tabs are General, Settings, Driver, Resources. Resources gives the I/O address being used. There is no direct "troubleshooter" for the SCSI port.

## 9.6 MOUSE OR TRACKBALL

Mice and trackballs fall under the general definition of pointing devices (Figure 9.15). A trackball is basically an upside-down mouse. Installation of mice and trackballs is the same. In general, they come with one of three types of connectors, bus, PS/2, or serial. The serial versions connect to one of the serial RS-232 ports and cost about $12. The bus version has an adapter card to fit in an expansion slot, so it costs $70. Each needs software to interface with the system. The PS/2 with a 6-pin mini-DIN connects to ATX system boards and a few other modern motherboards. See information about this jack in Chapter 5.

The type of mouse or trackball can be identified by the connector. A serial mouse has a female 9-pin D-shell plug. A bus mouse has a round 9-pin mini-DIN connector. The PS/2 is similar to the bus mouse but has only 6 pins. Following are steps for serial hardware installation, for bus hardware installation, and for software installation for either one.

### SERIAL MOUSE INSTALLATION

1. Identify COM1: and COM2: on your system. Determine which, if either, is being used. Watch for modems, which also use a COM: port. If your motherboard does not have serial ports on board, you may need to install a dual serial adapter card. Many systems have a multi-I/O card with two parallel ports, two serial ports, and a game port. Find out what you have and write it down for future documentation.
2. Plug your mouse into the serial port chosen. If your system has 25-pin D-shell connectors, you can buy adapters to change the output from 25-pin to 9-pin.

### BUS MOUSE INSTALLATION

1. Set the interrupt setting on the bus board to an unused interrupt IRQ 2–5. Interrupts for 8088 systems are found in Table 3.7. Interrupts for 286 and above systems are found in Table 3.9. On a PC/XT 8088 system board, 5, 6, and 7 are hard drive, floppy drive, and

**FIGURE 9.15   Trackball and mouse**

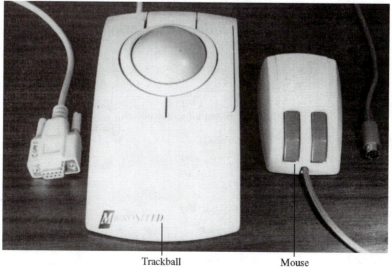

Trackball          Mouse

LPT1: parallel printer port, respectively. That leaves 2, 3, and 4. IRQ4 is for COM1:, so if you have a serial port, that's gone. IRQ3 is for COM2:, which may be open, but IRQ2 is perhaps your most likely choice. On an AT (286 or above), IRQ14 is used for the hard drive and IRQ5 is used for a second parallel port, LPT2:. If you don't have a second parallel port, then IRQ5 is open. So IRQ2, 5, and 3 are the most likely choices.

2. Install the bus board in an expansion slot and plug in the mouse. Avoid using expansion slot 8 on an XT.

## PS/2 MOUSE INSTALLATION

The PS/2 mouse is the simplest of all. Just plug into the correct port (looks just like the keyboard port, so be careful), and Windows 98 automatically identifies the mouse. You can get more information and adjust your mouse by clicking on Start, Settings, Control Panel, Mouse. Three tabs in the window give choices of selecting left- or right-handed buttons, double-click Speed, Type of pointer icon, Speed Trail. There is no "troubleshooter" for the mouse.

## SOFTWARE INSTALLATION FOR BUS AND SERIAL MICE

Most peripherals are well documented and come with software that automatically installs itself. Just read the manual, follow the installation instructions, and everything will go nice and easy. Just in case that doesn't work, I offer the following.

Since a mouse is the first peripheral we are covering that needs its own software to drive it, two sets of instructions are given showing you how to install the drivers. You should refer to these instructions for most of the following devices also. The first set of instructions is for automatic installation. This may or may not be included with your software. The second set is for manual installation.

### Automatic Device Driver Installation

1. With driver disk in drive A:, change the active drive to A:

```
A:
```

2. Find the name of the program.

```
DIR *.EXE
DIR *.COM
DIR *.SYS
```

One of these should give a program with a recognizable name, such as INSTALL.EXE or SETUP.EXE.

3. Run that automatic installation program.

```
INSTALL (or SETUP)
```

### Manual Device Driver Installation

1. With driver disk in drive A:, create a subdirectory with an appropriate name.

```
C:
MD\MOUSEDRV
```

2. Change to that subdirectory.

```
CD\MOUSEDRV
```

3. Copy all the files from the floppy to the hard drive.

```
COPY A:*.* C: /V
```

4. Identify the driver.

```
DIR *.EXE   (MOUSE.COM or MOUSE.SYS or similar)
DIR *.COM
DIR *.SYS
```

5. MOUSE.COM is run in the AUTOEXEC.BAT file. MOUSE.SYS is loaded in the CONFIG.SYS file. Edit each file as required. If you have no instructions from the documentation, for the serial mouse try a line similar to

```
MOUSE N
```
where N is 1 or 2 for COM1: or COM2:.

Usually the mouse package has detailed documentation and other programs for calibrating and testing the mouse.

## REPAIRING MICE AND TRACKBALLS

The main two failures in mice are mechanical failure (often due to dust and dirt) and cable failure. Check continuity of all lines in the cable. Clean the ball and the rollers it rests on. The ball is usually removed by sliding or twisting a retainer ring loose, allowing the ball to be removed.

Figure 9.16 shows an inside view of a mechanical mouse that requires mechanical cleaning and an optical mouse. An optical mouse senses movement by changes in the reflection from a pair of LEDs. It is usually used with a grid-pattern mouse pad. Dust must not block the optical path. Figure 9.17 shows an inside view of a trackball. Again, the ball and rollers must be clean.

Figure 9.18 shows the bus mouse pinout of the 9-pin mini-DIN.

**FIGURE 9.16   Mechanical (left), optical (right) mice, inside view**

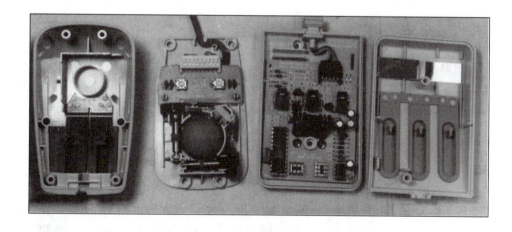

**FIGURE 9.17   Trackball, inside view**

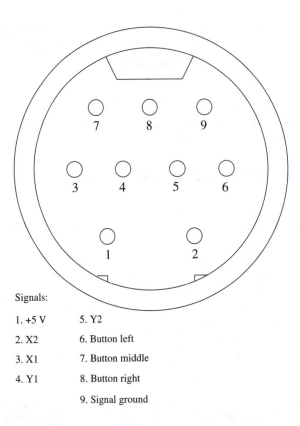

**FIGURE 9.18   Bus mouse 9-pin mini-DIN**

Signals:

1. +5 V      5. Y2

2. X2      6. Button left

3. X1      7. Button middle

4. Y1      8. Button right

           9. Signal ground

## 9.7   CD-ROM DRIVES

The incessant demand for higher-quality programs with higher graphics resolution and incorporated audio has made even the largest hard disks seem to shrink. CD-ROM drives are the solution for now. So similar to the audio compact disk that (with the right software) it can play your music CDs, the CD-ROM can store 1 GB of information.

The speed of retrieving this amount of data is a very important consideration. Two quantities are required to evaluate a CD-ROM. First is access time, measured in milliseconds, which is the time required to find the data you are looking for on the CD. Next is the data transfer rate, measured in kilobytes per second, which is how many bytes of information it can read in a second. Table 9.9 gives some typical values for these parameters.

Because of the rising popularity of multimedia, multimedia kits are being offered. These have a CD-ROM drive, a sound card, an external speaker, and five to ten CDs. The kits are almost always less expensive than the individual components. In Figure 9.19a, the top device in the drive bays is an internal CD-ROM drive. Installation is similar to that of a hard drive, installing control card, drive, and software. The controller may be an IDE interface on the motherboard, a sound card, or a stand-alone SCSI controller.

### INSTALLING AN IDE CD-ROM CONTROLLER

Figure 9.19b shows s CD-ROM drive being inserted in a bay. As with an IDE hard drive, the connections are easy. The 4-pin DC power plug needs to be plugged into the CD-ROM

**TABLE 9.9   CD-ROM transfer rates and access times**

| Data Transfer Rate | Average Access Time | Cost |
|---|---|---|
| 150K/s (parallel port) | 350 ms | $500 |
| 300K/s | 300 ms | $100 (dual/double speed) |
| 450K/s | 250 ms | $200 (triple speed) |

**FIGURE 9.19** (a) Internal CD-ROM drive, (b) set jumper, then insert CD-ROM drive in a bay, (c) plugging the IDE connector into the CD-ROM drive

(a)

Installed drive

(b)

(c)

**FIGURE 9.19** *(continued)*
**(d) installing the 4-pin audio cable into the CD-ROM drive**

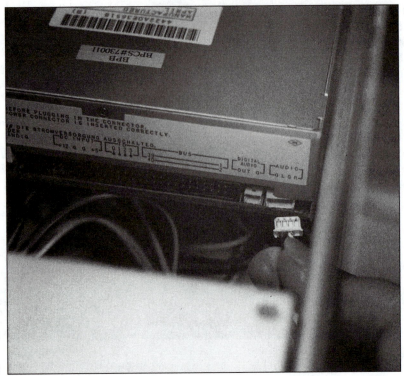

(d)

drive for the +5 and +12 V. The 40-pin IDE connector should have one end labeled with pin 1 or 2 and the other end with 39 or 40. Insert in drive as shown in Figure 9.19c. The 40-conductor ribbon cable should have a red or blue stripe on the side for pin 1. This plugs into a sound card IDE port or a motherboard IDE port. Use BIOS SETUP to select CD for this channel. A 4-pin audio connector needs to be connected from the drive, as shown in Figure 9.19d, to the audio card or a motherboard audio port. There may be a single thick wire to connect the metal case of the drive to the computer case for grounding. As typical with IDE, a drive select or identity (ID) jumper must match the ID on the IDE controller on the sound card or motherboard SETUP.

## INSTALLING A SCSI CD-ROM CONTROLLER

As with a SCSI hard drive, the SCSI CD-ROM controller can be identified by the male 50-pin dual header on the rear of the drive. It will also have a 3- or 4-pin audio output connector, a 4-pin DC power plug, and some shunt jumpers for SCSI identification. If it is the only SCSI device, it may need to be terminated with resistor packs or, sometimes, a jumper. If it is not the last SCSI device, do *not* terminate it.

## INSTALLING THE SOFTWARE FOR THE CD-ROM DRIVE

DOS cannot handle the CD-ROM drive without being "extended." It includes a program for this, called MSCDEX. You must add this to your AUTOEXEC.BAT file. HELP MSCDEX will get you into the DOS help file on this command. Do read this if trying to install a CD-ROM drive. A typical line is as follows.

```
C:\DOS\MSCDEX.EXE /D:driver /L:E
```

where driver is the name of the driver program in your CONFIG.SYS file and E: is the letter designation of the CD-ROM drive. Now you must add the driver that came with the drive to your CONFIG.SYS file. A typical line is as follows.

```
DEVICE=C:\DEVICEDRIVERS\ROMDRIVER.SYS /D:driver
```

where driver is the name given in the MSCDEX command in the AUTOEXEC.BAT file.

The system must be rebooted. The drive should then be able to be accessed just like another hard drive.

## 9.8  SOUND CARDS

The internal speaker in the typical personal computer system is digitally driven by minimal circuitry, which gives very poor quality sound. It can beep and make various sounds vaguely reminiscent of music, but there is no direct amplitude control. For quality music or speech, a sound card is needed. A cheap card is available for $50, while a quality card could be $180. It is important that the card you buy be Windows compatible.

### HARDWARE

The typical sound card has DIP switches or shunt jumpers for changing the interrupt request to IRQ2, 3, 5, or 7. Be aware that IRQ7 is usually used for the first parallel printer port, LPT1:. The card will have at least a microphone in, speaker out, and volume control on the rear bracket. Table 9.10 gives the card sound/port addresses and interrupts.

To install a sound card, power-down, insert the card in matching expansion slot (8- or 16-bit), and plug in the mike and speakers.

### SOFTWARE

Installing the software is as simple as most applications nowadays. Place the first disk in the floppy drive and type INSTALL. This will add a line to your AUTOEXEC.BAT file something like the following, where A is followed by one of the above addresses, D is a low DMA channel, I is the interrupt, and H is the high DMA channel.

```
SET SOUND = AXXX DX IX HX225
```

Now some sound cards have a game port (for a joystick) used for a MIDI (Musical Instrument Digital Interface) connection. This may conflict with an existing game port, and it may have a shunt jumper to enable or disable it.

The top photo in Figure 9.20 shows a typical sound card, with SCSI controller for a CD-ROM drive. On the rear bracket, from the top, are three mini–phone jacks: first an auxiliary input for audio, second a microphone input, and finally a stereo speaker output. Below them is a female DB15 D-shell connector that may double as a game port and a MIDI music plug. The bottom photo shows a typical set of speakers with the plugs that go into the sound card rear jacks.

**TABLE 9.10  Sound card/port addresses and interrupts**

| Port | I/O Address | Interrupt | DMA L | DMA H |
|------|-------------|-----------|-------|-------|
| Sound card | 220H–233H | IRQ5  default | 1 | 5 |
| | 240H–253H | IRQ2  alternate | 0 | 6 |
| | 260H–273H | IRQ7  alternate | 3 | 7 |
| | 280H–293H 16-bit | IRQ10 alternate | | |
| Note: The above IRQs, DMAs, and addresses can be mixed. | | | | |

FIGURE 9.20 **Sound card and speakers**

Sound card

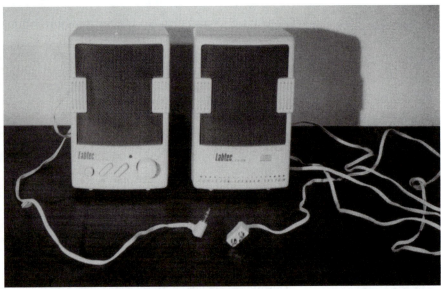

Speakers

## 9.9   MODEMS AND FAX

### INSTALLING AN EXTERNAL MODEM

#### Hardware

Modems come in two main types: internal and external. The internal modem is a printed circuit board or card that is inserted in the expansion slots on your motherboard. The external modem is in its own case, runs on its own power supply, and must connect to your system with a cable to the serial RS-232 port. The computer, which is considered DTE, has a male DB25 connector. The modem, which is considered DCE, has a female DB25 connector. The proper cable is a male-to-female straight-through cable. The software that came with your modem must be set up to know if the modem was plugged into port COM1: or COM2:.

The speed at which your modem sends information is the baud speed, which is basically how many bits per second it communicates. The standard baud speed is 56K; it costs about $60. The baud speed must be set up to be compatible with the system you are calling. Most bulletin boards today automatically determine the speed of the calling modem (autobaud) and adjust to that speed.

Two new technologies are competing for the fast modem market: K56 Flex and X2. Both claim 56K baud transfer rate. Regretfully, they are not compatible. The system you are calling must have matching technology to achieve that speed. *Home Office Computing* magazine did a test, and found neither one will achieve 56K speed unless conditions were absolutely perfect, and, even then, for data download only.

## Software

Once the modem hardware is installed, the software or programs to run the hardware must also be installed. These days, most software automatically installs itself when you type SETUP or INSTALL from the A: prompt, with the disk holding the program in drive A:.

## INSTALLING AN INTERNAL MODEM

Internal modems consist of a printed circuit board or card that is plugged into one of the expansion connectors on the motherboard (Figure 9.21a). A set of DIP switches or shunt jumpers on the board or the rear bracket allow you to set the interrupt level for COM1: or COM2: (jumper and silkscreen shown in Figure 9.21b).

Both internal and external modems hook to the phone line through an RJ-11 or similar phone jack labeled LINE. The phone plugs into the rear of the modem card in a connector

**FIGURE 9.21 (a) Installing an internal modem, (b) modem shunt jumper settings for COM port**

(a)

(b)      Dual header for jumper

labeled PHONE. This allows you to use the phone for voice when the modem is not in use. In a system with more than one phone, this handy feature also gives rise to the possibility that someone will pick up another phone while the modem is connected and passing or receiving information. The information will almost certainly be corrupted, and you may lose your connection. If you do much modem work, or if you set your modem to answer calls, then it is best to have a dedicated phone line.

In Windows 98, you may find information about the modem by clicking Start, Settings, Control Panel, Modems. The General tab shows your type of modem, and clicking the Properties button shows much more information. The diagnostic tab shows how your serial ports are allocated, and clicking on More Info gives the IRQ and I/O address of the port selected. When installing a new modem, you may use Add New Hardware in the Control Panel window. There is a modem "troubleshooter." Click on Start, Help, Contents, Windows Troubleshooters, Modem, and it leads you through isolating the problem with your modem.

Two Web URLs that may be of interest are:

| | | |
|---|---|---|
| www.hayes.com | phone 1-770-840-9200 | Hayes Microcomputer |
| www.usr.com | phone 1-800-342-5877 | U.S. Robotics |

## FAX

Fax (facsimile) machines can send graphics images over the phone lines. The lowest cost fax machine is a fax/modem combination adapter card that takes its input from a file on the computer and outputs to the phone line. Any computer can become a fax machine for about $150. Adding a scanner to the system allows you to copy and send anything to anyone who has a fax machine.

## 9.10  OPTICAL SCANNERS  GENERAL DISCUSSION

Optical scanners scan printed material onto a CCD (Charge Coupled Device) to read the image into bits. The bits are stored in a bitmapped (BMP) file for use by the computer. This is how you get pictures from a magazine or photo into your computer.

Scanners come in black-and-white or color. A 4-inch-wide black-and-white hand scanner costs $100 and has a resolution of 300 or 400 dpi. A color hand scanner costs $100, and a full-page (flatbed) color scanner costs $150 or so. Hand scanners usually include software used to join side-by-side scans into a seamless one-page document. A flatbed color scanner can range from $600 \times 600$ dpi at $1000 to $2400 \times 1200$ dpi at $3000.

If your system has a printer and an optical scanner, it is in fact a photocopier, able to scan anything printed or drawn and print it out on the printer. This process can be in full color. Figure 9.22a shows a hand scanner, and Figure 9.22b shows a flatbed scanner.

If the image is of a page of writing, Optical Character Recognition (OCR) software that comes with the scanner can "read" the file and change the bitmap file to actual characters you can then load into your word processor. While this is supposed to be easy, be aware that if the OCR software is 90 percent correct, the product is useless, because correcting it takes longer than typing it in from scratch. Also, a given style or type print may be much harder than another to "read." If you really plan to use OCR, have a very detailed demonstration done with your own material.

If the image is graphic, the bitmap file can be loaded into paint programs and modified and printed, because a painting program saves its files as bitmapped files. Drawing and drafting (CAD) programs save their files as vectored drawing files, so they cannot be used to manipulate scanned images.

## SCANNER INSTALLATION

Optical scanners have an adapter card to install in an expansion slot. As always, power is off when plugging or unplugging anything. A set of shunt jumpers on the card must be set

**FIGURE 9.22** (a) Hand scanner, (b) full-page flatbed optical scanner, (c) set SCSI scanner addresses

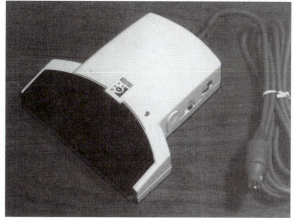

(a)

Hand scanner

(b)

Full-page flatbed scanner

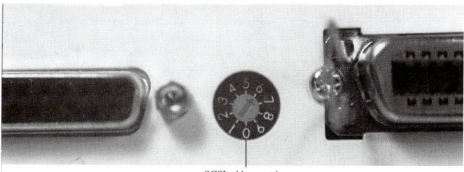

(c)

SCSI address setting

**FIGURE 9.22** *(continued)*
**(d) the SCSI scanner cable,**
**(e) the SCSI terminator, (f) the**
**SCSI scanner jack**

(d)

DB25M to SCSI scanner cable

(e)

In-Line terminator

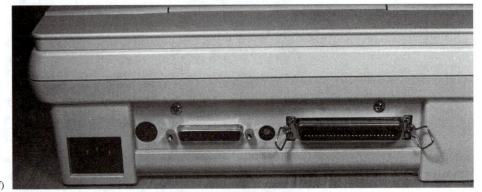

(f)

DB25 and SCSI connectors for choice

per documentation for the correct interrupt IRQ, direct memory access DMA channel, and I/O address (Table 9.11). As your system grows, finding an open interrupt can be more and more of a chore. Some diagnostic programs, such as MSD and System Information under Accessories, System Tools, examine interrupts and report those in use.

Once the board is installed, the bus scanner usually plugs into the board through a 9-pin mini-DIN, like the bus mouse does. A SCSI optical scanner needs to have its address set as shown in Figure 9.22c, where it is set to 6. The cable has a 50-pin SCSI plug on the scanner end and a male 25-pin D-shell (DB25M) for the computer side. The computer jack looks just like the parallel printer jack (see SCSI earlier in this chapter). See Figure 9.22d for both ends of the cable. Like the hard drive cable, SCSI signals need to be terminated, so a terminator (Figure 9.22e) must be installed on the cable. Finally, the cable is plugged into the rear of the scanner in the jack on the right in Figure 9.22f. Then the software that came with the scanner must be installed and set up. In DOS, as usual, this is done by putting the disk that came with the device into drive A: and entering A:, then typing DIR *.EXE. This should show a program with a recognizable name, such as INSTALL.EXE, SETUP.EXE,

TABLE 9.11  Optical scanner
port addresses and interrupts

| Port | I/O Address | Interrupt | DMA Channel |
|---|---|---|---|
| Optical scanner | 280H | 7 | 1 (default) |
| | 2A0H | 2 | 3 (on AT) |
| | 330H | 3 | |
| | 340H | 4 | |
| | | 9,11,12(AT) | |

or HANDSCAN.EXE. If you see a name similar to one of the first two of these names, just type the name and answer the questions. It should automatically copy the software to the proper subdirectory and edit AUTOEXEC.BAT or CONFIG.SYS. You should then be able to reboot the system and test the scanner.

If there is no program on the disk like the first two, but there is one like the third, use it. If none of those are recognizable, type DIR *.SYS. This may find a file with a name similar to HANDSCAN.SYS. The manual installation steps are the same as with most peripherals. If you find a file similar to HANDSCAN.SYS, read the documentation and follow the directions. Typical usage is to add a line similar to the following to your CONFIG.SYS.

```
DEVICE=C:\SCAN\HANDSCAN.SYS /A=XXX /I=X /D=X
```

A is for I/O address and XXX is one of the suggested addresses from Table 9.11. I is for IRQ, and the X is for one of the IRQs suggested. D is for DMA channel, and X is for the channel chosen from the table.

Of course, the system has to be rebooted now to load this driver. Test out the scanner with the program you intend to use with it. If there is any problem, consult your manual or carefully examine your three choices of IRQ, DMA, and I/O address again.

In Windows 98, you may install the scanner by clicking on Start, Run, and typing in your CD-ROM letter (if your scanner drivers are on CD) or A: if the drivers are on floppy disk. You may browse either drive for a program named SETUP. You may also click on Start, Programs, Accessories, Entertainment, CD Player to play music CDs on your computer if you have a sound card. Or choose interactive CD Sampler. To have your system automatically start the program on a CD when it is inserted, go to Control Panel, System, Device Manager. Double-click on the CD-ROM line, then single-click to select the line that appears beneath it. Then click on Properties and put a check in Auto Insert Notification. There is no "troubleshooter" for the CD, although System Information may be helpful.

## 9.11  NETWORKS

A **network** is a system to connect computers to transfer or share information. A LAN (Local Area Network) is typically one small office building or a floor or department of a large office. A MAN (Metropolitan Area Network) may cover an area ten miles or so in diameter. A WAN (Wide Area Network) may be worldwide.

A typical network installation involves installing an adapter card, setting the interrupt level, and setting the address of that station or node. Each computer must have a unique address on the network so other computers can indicate where they are sending their data. Finally, a cable connects the systems together. Some use coaxial cable, others use phone line cable. A network managing program must be run on each system connected, allowing it to communicate with other systems. While the basic system is not difficult to install, numerous irritating difficulties can arise, especially as the number of stations grows and as the distance between them increases. If you are installing a large network, it may be wise to bring in a professional with a lot of experience.

**FIGURE 9.23    Network cards**

DB9 connectors

BNC connectors      BNC and RJ45 connectors

The photos in Figure 9.23 show cards for networks. Note the top uses a DB9 connector with adapters to change to phone line, the bottom left uses a BNC (bayonet) connector for coaxial cable, and the bottom right uses both the BNC and the RJ45 connectors.

## NETWORK TOPOLOGY

The pattern in which networked computers are connected is the topology of the network. The three most common patterns are linear, ring, and star.

A linear network uses one cable from one computer to another, and from that one out to another, and so on. It is like a line of kids holding hands playing Red Rover. This one-series wire is called a bus, or backbone, and the computers are considered to be daisy-chained together. Because of signals reflecting off the end of a cable if it is not terminated, each end of the cable must have a terminator. This is just a resistor to stop reflection.

A ring network is a linear network with the ends connected together instead of terminated. It is like a ring of kids holding hands playing Ring Around the Rosy.

A star network has a concentrator, a central point, and one cable going out to each computer connected, like spokes going out from the hub of a wagon wheel.

As noted, a network is any connection of two or more computers together to share resources such as hard drives, printers, programs, and data. Each computer on the network must have a Network Interface Card (NIC) to communicate with the others. They must also communicate with the same protocol, like speaking the same language. The most common is the Transmission Control Protocol/Internet Protocol (TCP/IP). Besides or instead of this "direct cable connection" to a network, you may have a "dial-up network" connection. This requires a modem to connect to the phone lines. If your office network is connected to a modem also, that is all you need to communicate with your office network. To communicate with the Internet or the World Wide Web, you must have an Internet Service Provider (ISP). America Online (AOL) and the Microsoft Network (MSN) are two major ISPs. To identify your computer with as a unique system, the ISP assigns you an Internet Protocol (IP) Address. They do this through a Domain Name Server (DNS), which keeps track of and assigns IP addresses. Your ISP will let you choose a "screenname," or Internet address, such as JDOE@aol.com, or JDOE@worldnet.att.net if you use AT&T as an ISP. Servers of companies such as Intel, AOL, and Cyrix have Universal Resource Locators (URL) such as http://www.intel.com, where the http stands for Hyper Text Transfer Protocol and the www stands for the World Wide Web.

Windows 98 makes it easier to connect to the Internet. With a few clicks you can set up the Microsoft network. Of course you pay a fee to your ISP for the connection. In Windows 98, you may find information about your network in the Control Panel under Networks. Tabs are Configuration, which shows the type you have; Identification, which shows your workgroup name; and Access Control, in which you may require a password. In Windows Explorer (right-click on Start button), scroll to the bottom and you can see your Network Neighborhood. There is a network "troubleshooter" in Help, Contents, Windows Troubleshooters, Networking. A typical network has one or more "servers" that have the information, programs, or hardware to share. The computers that connect to the servers to share their resources are called "clients."

## CABLES AND CONNECTORS

Certain cables are required for different technical reasons, such as shielding from noise and allowing for the distance that signals must travel. Different jacks are used for the different cable requirements.

10base5 is "thick" EtherNet cable. It is used for long ranges, up to 1600 feet. It uses a 15-pin female D-shell connector. An Attachment Unit Interface (AUI) is required to attach to the thicknet, which may link smaller LANs together.

10base2 is "thin" EtherNet cable with a range of about 1000 feet. It uses a BNC connector with RG-58 coaxial cable with 50 $\Omega$ impedance.

10baseT is "twisted pair" EtherNet cable with a range up to 500 feet. It uses an RJ45 8-conductor (four twisted pairs) keystone (phone-like) connector. These use unshielded twisted pair (UTP) and shielded twisted pair (STP) cables.

10baseF is fiber optic EtherNet cable with a range of 60 miles. Fiber optics shoots a beam of light down the inside a solid glass fiber. Because of the angle, it doesn't come out the sides but reflects back and forth along the inside until it reaches the other end. It is an expensive technology, but it has many advantages over copper wire, such as amount of information carried and speed.

The network card may use the addresses and interrupts given in Table 9.12 as long as they don't conflict with another card. The addresses and interrupts shown in each line do not have to go together but may be mixed.

**TABLE 9.12   Network card port addresses and interrupts**

| Port | I/O Address | Interrupt |
|------|-------------|-----------|
| Network card | 300H–317H | 15 |
| | 340H–357H | 10 |
| | 360H–377H | 11 |

## 9.12   TAPE DRIVES

There are three common ways to protect yourself against loss of data on your hard drive. The first is the DOS BACKUP or MSBACKUP command. The only problem is that it takes 130 1.44 MB disks to fully back up a 200-MB drive. You need to sit there and feed them in. The second way is to have another hard drive used just to copy information from the main hard drive. This hard drive holds only the information on one similar size hard drive. That is about $150 of hard drive to hold the backup of each hard drive in your office. The third, and best way for large drives, is the magnetic tape drive. The photo in Figure 9.24 shows an external drive on the left, which connects to the parallel printer port. In the center is a 3.5" drive, and on the right is a 3.5" drive in a 5.25" mounting bracket. The tape in front is a standard quarter-inch cartridge (QIC).

The magnetic tape drive is very similar to your audio or video cassette tape recorder. The cost for the drive is approximately the same as for a hard drive, but you use a different tape for each hard drive in the office. That's only $20 for each hard drive backup, and it is much quicker than floppies.

A simple tape drive that can be moved from office to office or system to system is an external unit that plugs into the parallel printer port. Cost is about $300. Some internal units use an SCSI interface, which may require its own controller card and may cost $1000. For these two inconveniences you get more speed and higher capacity, up to 4 GB or more. Most inexpensive drives use the existing floppy controller, saving money with a cost of about $150. For this savings you lose speed and capacity, with about 250 MB per tape. A discussion on installing these more common drives follows.

### HARDWARE

Installing a tape drive is similar to installing a floppy drive. Most will fit in either a 5.25" or a 3.5" bay and may require mounting brackets. Once the drive is mounted, connect the power cable and a standard 34-pin floppy drive cable from the floppy drive controller card. If your existing floppy cable does not have a spare connector, you will need to replace the cable or add a "Y" connector. Watch for proper pinout, matching pin 1 of the cable with pin 1 on the drive.

**FIGURE 9.24   Tape drives and tape**

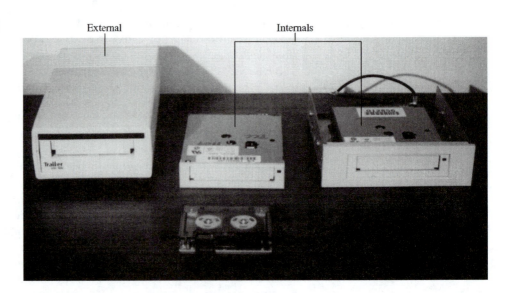

Just as the floppies have a drive select jumper, so does the drive. Set the jumper according to the documentation that comes with the drive. If in doubt, let it alone and try the drive. The default (factory) setting may work.

## SOFTWARE

Software to access the tape drive will be included with it. Install this according to the documentation. If you have no documentation, generally follow the instructions under mouse software installation. They may be command or menu driven, and will include choices to FORMAT the tape, similar to formatting a floppy, and backing up the hard drive.

## 9.13 BASIC TROUBLESHOOTING: RESOLVING CONFLICTS

If a previously functioning serial device stops working, see the troubleshooting guide in the section on that device. The following is general information for installing new devices.

One of the most frustrating problems in installing yet another peripheral to your system is avoiding conflicts with existing ones. Interrupt requests, direct memory access channels, and I/O addresses should be unique for each peripheral device. (See Tables 3.7, 3.8, 3.9, and 3.10.) It is good to keep a data sheet on your system with the used IRQs, DMAs, and addresses and what they are used for. Noting if they can be changed on each device also is wise. This may save you much time and trouble in the future.

Windows 98 "Troubleshooters" can be a great help. The improved Plug-and-Play does save a lot of headaches, although Legacy cards and devices can still be big trouble.

### INSTALLING A NEW DEVICE

1. Find out the IRQs, DMAs, and addresses of the devices already installed, and note these so you never have to do this again. You can get this information
   a. From the documentation that came with them.
   b. From a program such as MSD, Windows System Information, or a diagnostic program.
   c. By physical inspection of each card.
2. Set the new device's IRQs, DMAs, and addresses to unused ones.
   a. Leave these values at the defaults if possible, and note the default on your data sheet.
   b. Change to unused ones from information below.
3. Install software and test by operating the device.
   a. Follow instructions that come with device.
   b. Use software installation steps in section on device.
4. If the new device fails to operate properly,
   a. Run MSD, Windows System Information, or the diagnostic software to examine IRQs, DMAs, and addresses for conflicts.
   b. Change to other unused, and try again.
5. Windows, Control Panel, Add New Hardware can be a great blessing. With newer devices, it often works great.

### RESOLVING TROUBLE OR CLEARING UP A MESS

1. Remove all cards except the video, hard, and floppy controllers.
2. Do MSD or diagnostic program, and write down (PrintScrn) current settings of these three.
3. Arrange cards to be installed in order of worst to best by the following criteria.
   a. No documentation, no adjustable settings.
   b. No documentation, few adjustable settings.

c. Poor documentation, few adjustable settings.

d. Poor documentation, many adjustable settings.

e. Good documentation, many adjustable settings.

4. Install the worst card first.

a. Use the new device steps above.

b. Use MSD or a diagnostic program to add these settings to current data sheet.

5. Repeat step 4, moving from the worst through the best card. If you cannot find a free IRQ, you can attempt to share one with a device that will not be used at the same time as the new device. Test both thoroughly!

6. Finally, test the entire system thoroughly.

## CHOOSING AN IRQ

The following IRQs are generally *not* available for use by peripherals.

| | |
|---|---|
| IRQ 0 | is always used for the system timer |
| IRQ 1 | is always used for the keyboard |
| IRQ 6 | is always used for the floppy drive |
| IRQ 8 | isn't on an XT, and is the real-time clock on an AT |
| IRQ 13 | isn't on an XT, and is the math coprocessor on an AT |
| IRQ 14 | isn't on an XT, and is the hard disk on an AT |

IRQs blacked out in Figure 9.25 are also not available if you have the item shown in the left column.

## CHOOSING A DMA CHANNEL

The following DMAs are generally *not* available for use by peripherals.

| | |
|---|---|
| DMA channel 0 | on XT is used for RAM refresh |
| DMA channel 2 | is always used for the floppy drive |
| DMA channel 3 | on XT is used for the hard drive |
| DMA channel 4 | isn't on XT, on AT is cascaded from 0–3 |
| DMA channels 5, 6, 7 | aren't on XTs; on ATs, they are available to 16-bit only |

DMAs blacked out in Figure 9.26 are also not available if you have the item shown in the left column.

**FIGURE 9.25  Choosing an IRQ**

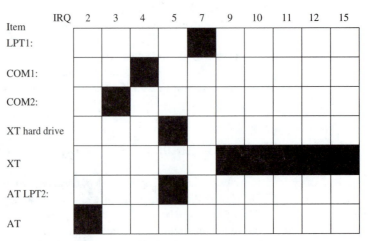

*Note:* IRQs blacked out are used by item.

FIGURE 9.26  Choosing a DMA

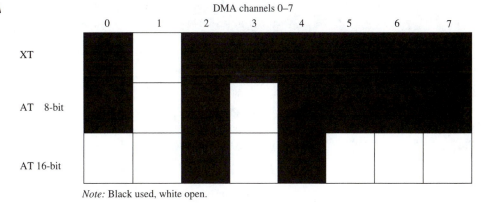

Note: Black used, white open.

## CHOOSING AN I/O ADDRESS

These are "typical" addresses for the devices shown.

| Port/Card | I/O Address | |
|---|---|---|
| Game | 200H to 207H | |
| Sound | 220H to 233H | |
| Sound | 240H to 253H | |
| Sound | 260H to 273H | |
| LPT2: | 278H to 27FH | |
| Sound | 280H to 293H | |
| Scanner | 280H to 267H | |
| Scanner | 2A0H to 2FFH | (This conflicts with COM2: and COM4:) |
| COM4: | 2E8H to 2EFH | |
| COM2: | 2F8H to 2FFH | |
| Scanner | 330H to 367H | |
| Scanner | 340H to 377H | |
| LPT1: | 378H to 37FH | |
| COM3: | 3EBH to 3EFH | |
| COM1: | 3F8H to 3FFH | |

## FITTING IT ALL TOGETHER

Each peripheral listed in this chapter had its IRQs, DMAs, and I/O addresses listed if needed. By referring to those, then to Tables 3.7, 3.8, 3.9, and 3.10, you should see how it all starts fitting together.

## 9.14  CHAPTER REVIEW/QUIZ QUESTIONS

1. A female DB25 connector indicates a
   a. Serial RS-232 port.
   b. Parallel printer port.
   c. Bus mouse port.
   d. Game port.
   e. Video port.

2. A male DB25 connector indicates a
   a. Serial RS-232 port.
   b. Parallel printer port.
   c. Bus mouse port.
   d. Game port.
   e. Video port.

3. A female DB9 connector indicates a
   a. Serial RS-232 port.
   b. Parallel printer port.
   c. Bus mouse port.
   d. Game port.
   e. Video port.

4. A male DB9 connector indicates a
   a. Serial RS-232 port.
   b. Parallel printer port.
   c. Bus mouse port.
   d. Game port.
   e. Video port.

5. A 9-pin mini-DIN connector indicates a
   a. Serial RS-232 port.
   b. Parallel printer port.
   c. Bus mouse port.
   d. Game port.
   e. Video port.

6. A female DB15 connector indicates a
   a. Serial RS-232 port.
   b. Parallel printer port.
   c. Bus mouse port.
   d. Game port.
   e. Video port.

7. Serial and parallel ports can be tested without electronic equipment attached by inserting a(n)
   a. Meter.
   b. LED.
   c. Serial or parallel loopback plug.
   d. Cable with the other end unconnected.

8. The game port can be used for
   a. A joystick only.
   b. Games only.
   c. Various analog input also.
   d. Video input.

9. Does the game port usually come with software drivers to install?
   a. Yes
   b. No
   c. Always

10. A serial port can be used for
    a. Printers.
    b. Plotters.
    c. Mice.
    d. Modems.
    e. All of the above.

11. Up to how many serial ports can be installed?
    a. 1
    b. 2
    c. 3
    d. 4
    e. 5

12. A parallel port can be used for
    a. Printers.
    b. Tape backup systems.
    c. File transfers between two computers.
    d. Output signals to control equipment.
    e. All of the above.

13. If the head has solenoids with wire pins that push out, hitting a ribbon against the paper, against the platen, it is a(n) _____ printer.
    a. Thermal
    b. Daisy wheel
    c. Dot matrix
    d. Laser
    e. Ink-jet

14. If the head has no moving parts, but shoots "sparks" or arcs through the paper to a metal flat platen, it is a(n) _____ printer.
    a. Thermal
    b. Daisy wheel
    c. Dot matrix
    d. Laser
    e. Ink-jet

15. If the head has fully formed letters and a hammer (or the full head) strikes the ribbon against the paper, it is a(n) _____ printer.
    a. Thermal
    b. Daisy wheel
    c. Dot matrix
    d. Laser
    e. Ink-jet

16. If there is no ribbon, but the head "squirts" ink at the paper, it is a(n) _____ printer.
    a. Thermal
    b. Daisy wheel
    c. Dot matrix
    d. Laser
    e. Ink-jet

17. If there is no ribbon, but charged powder is attracted to a charged cylinder and then attracted to the paper and heated to "melt" onto the paper, it is a(n) _____ printer.
    a. Thermal
    b. Daisy wheel
    c. Dot matrix
    d. Laser
    e. Ink-jet

18. The printer that usually requires very special paper is the _____ printer.
    a. Thermal
    b. Daisy wheel
    c. Dot matrix
    d. Laser
    e. Ink-jet

19. For instant access of information, hard drives are superior to tape drives because
    a. They spin at a higher speed.
    b. The head travels faster.
    c. They access information randomly.
    d. They read more than one track at a time.

20. Internal tape drives are superior to hard drives for archiving because
    a. They access information randomly.
    b. They have a removable medium.
    c. They cost less.
    d. They spin at a higher speed.

21. External tape drives may be more cost efficient than internal tape drives in an environment with many PCs because
    a. They cost less.
    b. They are easier to install.
    c. They use less expensive tapes.
    d. They can be moved from system to system.

22. Any tape drive can reformat a tape previously formatted in another brand and type of tape drive.
    a. True
    b. False

23. There is no software to install for tape drives.
    a. True
    b. False

24. Installing an internal tape drive requires
    a. Jumper setup, physical installation, software installation, software configuration.
    b. Physical installation only.
    c. Physical installation and software installation.
    d. Patience and determination.

25. Installing an external tape drive requires
    a. Installing the software on each system to be used and plugging it into the parallel port.
    b. Plugging it into the parallel port and keying in BACKUP.
    c. More expertise than a nontechnical user usually has.
    d. Less patience and determination than installing an internal drive.

26. CD-ROM stands for
    a. Compact Disk Read-Only Memory.
    b. Compact Digital Random-Ordered Music.
    c. Certified Driver Ready On Multimedia.
    d. Compact Disk Ready On Multimedia.

27. CD-ROM drives have two typical ways of inserting the CD, namely
    a. Label up and label down.
    b. Clockwise and counterclockwise.
    c. Extending tray and door with caddy.
    d. Analog and digital.

28. The principle the CD-ROM works on is
    a. Magnetic.
    b. Analog.
    c. Optical.
    d. Thermal.

29. The tracks on a CD-ROM are about _____ apart.
    a. 0.1 inch
    b. 0.01 inch
    c. 1 mm
    d. 1 μm

30. In operation, the reflective surface of the CD-ROM reflects
    a. Random light.
    b. White light.
    c. A laser beam.
    d. A magnetic field.

31. The information is stored on the CD-ROM as
    a. Horizontal zigzags.
    b. Vertical zigzags.
    c. A series of pits in a flat surface.
    d. A series of magnetic domains.

32. A typical CD-ROM can store
    a. 10 MB.
    b. 100 MB.
    c. 680 MB.
    d. 6800 MB.

33. The massive storage capacity of a CD-ROM is required for
    a. Business data.
    b. Movies, still video, audio, and large files.
    c. A big program.
    d. General computer work.

34. CD-ROM recorders are
    a. Commonly available.
    b. Very expensive.
    c. Not intended for home use.
    d. Not available at any price.

# 10

# DOS Through Windows 98

**OBJECTIVES**

After completing this chapter you should be able to

- Install DOS on a hard drive system
- Enter the common DOS commands
- Give a short description of each common command
- Create your own batch files
- Create and modify your own AUTOEXEC.BAT file
- Create and modify your own CONFIG.SYS file
- Run DEBUG, minimally
- Use the EDIT word editor
- Run EDLIN, minimally
- Troubleshoot and repair DOS software problems
- Install Windows on a hard drive system
- Demonstrate basic knowledge in use of Windows

## 10.1 DOS

### INSTALLING DOS

DOS is installed on your computer's hard drive with the SETUP command. Insert your first DOS disk in your boot drive (usually A:) and type A:, press ENTER, type SETUP. DOS will install itself if you just answer the questions it asks you. You will need two formatted but empty disks for use with this program.

With the release of DOS 6.0, numerous operations became much easier, and the system gives much support and help along the way. DOS 6.0 was improved to 6.20 and then to 6.22. While this chapter is not meant to replace the DOS manual, it should give you a firm grip on using DOS to simplify your work.

### Getting Help

Typing FASTHELP followed by a command name gives a one-screen description of the command. Typing COMMAND /? gives the same full screen for most commands. Typing HELP followed by the command name gives much more information. This takes you into an editor and you can use the Page Up and Page Down keys or the up and down arrow keys to move up and down through many pages of information. If there are angle brackets <> around any phrase, you can arrow to the first letter and press ENTER to go to that topic.

The help file on a command has three main sections. You enter the SYNTAX section. The TAB key moves you from NOTES to EXAMPLES, to the <bracketed> subtopics. Press ENTER at any of these to go to that section of help. The Esc key backs you up to the section you just came from. Together these sections contain many pages of help on the command you asked for.

You can print the help out to your printer with Alt+F (for File), and pressing P for print. Typing HELP alone at the DOS prompt gives you an index, or list of all the topics available. Browsing this can be quite interesting. Function key F3, or Alt+F+X will get you out when you want to quit.

DOS 6.0 has MEMMAKER to manage your memory more automatically, DBLSPACE in 6.0 and DRVSPACE in 6.22 to give you twice the storage on your hard disk, and MSAV and VSAFE to protect against viruses. A great help to troubleshooters is that pressing F5 while the "Starting DOS" message is displayed will bypass the AUTOEXEC.BAT and CONFIG.SYS, eliminating the need to rename them.

Many DOS commands and other programs are menu driven. The word "menu" in this context means a list of commands is given, from which you choose the one you want. This is done by pushing the first letter of the menu item or by moving the cursor to the first letter of the item using the arrow keys or the Tab key, then pressing ENTER.

## ENTERING A COMMAND

You give DOS a command by typing the command and pressing the ENTER key. Enter tells the computer you are done typing. If you make a mistake in the command and notice it before you press ENTER, you can use the Backspace key to delete characters to the left. Then you can finish the command properly and follow it with an ENTER. If the line is so messed up you want to just forget it, you can press ESC. That shows a backslash, moves you down one line, and totally ignores what has been typed. No other means of editing is available to correct a line until you press ENTER. Since commands often interact with files, it is important to know how to name files.

### Naming Files

A filename can have up to eight characters followed by a period and three more extension characters. An asterisk is a wild card for any number of characters and a question mark for exactly one character. It is conventional to use the extension in a meaningful way. Some extensions are already standardized for certain types of files (Table 10.1).

### Directories and Subdirectories

The hard drive has a large capacity, and can hold perhaps 1 GB of information. If you had one giant bucket to dump information into, wouldn't it be hard to find one small piece of information? In the same way, information must be placed on the hard drive intelligently, or the task of finding it will be much more difficult.

The hard drive is broken up into **directories** and **subdirectories,** much as a file cabinet is divided into drawers. In the drawers are further divisions of green hanging dividers, and in those, even more divisions of manila folders. It is wise to place related information in its own directory to keep it separate from unrelated information. DOS includes commands for managing directories.

PROMPT $P$G shows the directory you are currently in. This command only has to be given once, and it remains in effect until it is changed or the system is shut down. This is usually placed in the AUTOEXEC.BAT file.

TREE shows the directories in a graphical layout, similar to a family tree. TREE/F shows every filename in every branch of the tree.

PATH sets the path DOS will use when it looks for files. If no path has been set, DOS will search only the current directory.

**TABLE 10.1  Typical standardized file extensions**

| Extension | Type of File |
| --- | --- |
| ARC | Archived (compressed) file |
| ASC | ASCII format file |
| ASM | Assembly program |
| BAS | BASIC program file |
| BAT | Batch DOS program file |
| BMP | Bitmapped graphics file |
| CFG | Configuration file |
| CHK | CHKDSK, RECOVER, or SCANDISK recovered file |
| CLP | Windows clipboard file |
| CO_ | If the last letter of a filename extension is the underscore character, the file is compressed. Use the EXPAND.EXE program to decompress it. |
| COM | Command program file |
| DAT | A data file |
| DOC | An ASCII text documentation file |
| DRV | Device driver file |
| EXE | Executable program file |
| FLC | PCPaintBrush graphics files (PCX) with $640 \times 480 \times 256$ colors |
| FLI | PCPaintBrush graphics files (PCX) with $320 \times 200 \times 256$ colors |
| FNT | Font (see FON and TTF) |
| FON | Fonts the printer has in it |
| GIF | Graphics Interchange Format picture file |
| GRP | A Windows group file |
| HLP | Help text file |
| INI | Initialization file |
| PCX | PCPaintBrush graphics files |
| PIF | Windows program information file |
| PLT | Plotter picture data file |
| PRO | Autodesk Animator Program (movie) files |
| REC | Windows macro recorder file |
| SCR | Script of an assembly program or a Windows screen saver |
| SYS | System information file |
| TIF | Graphics format |
| TMP | Temporary file |
| TTF | True Type Font (can be any size on screen and on printer) |
| TXT | An ASCII text file |
| VOC | Voice sound file |
| WAV | Waveform audio sound file |
| WRI | Windows Write document file |
| ZIP | A Zipped (compressed) file |

MD stands for "make directory," and it makes a new directory, which is, of course, empty. CD stands for "change directory"; it changes your location from the current directory to the specified one. RD stands for "remove directory"; it erases the directory so that it no longer exists.

Here's how you might use these commands to set up and move to a new directory.

You just bought a new game and want to put it on your hard drive. Do you install it in the "root" directory (the beginning one, the big bucket)? No! The root directory would soon become so full of files it would be unmanageable. You should create a new directory to hold the game. Let's say it is a chess game called Chess Master. You try the commands:

| | |
|---|---|
| `PROMPT $P$G` | This shows the full path so you know where you are. |
| `MD\CHESSMAS.TER` | This creates the new directory. |
| `CD\CHESSMAS.TER` | This changes the current directory to the new one. |
| `COPY A:*.*` | This copies all files from the disk to the hard drive. The instructions may say to type INSTALL or SETUP. |
| `CD\` | Since there is no name, this takes us to the root. |

Now you try to run the game by typing CM, as the instructions with the game told you. A File Not Found error is displayed. You type DIR to see a directory of what is on the hard drive. You see CHESSMAS.TER <DIR>. The <DIR> shows that CHESSMAS.TER is not a file but a directory. There is no CM.EXE or CM.COM to run!

Because DOS cannot find CM.EXE in the current directory and does not know where to look, you must do one of two things. Either move to that directory with the following:

| | |
|---|---|
| `CD\CHESSMAS.TER` | This changes to that directory. |
| `CM` | This runs the game. |

Or you set a path so DOS knows where to find CM.EXE. That is done by typing

```
PATH=C:\DOS;C:\CHESSMAS.TER
CM
```

The first command makes DOS look in its own directory first (always a wise inclusion) and then in the game directory. The second command runs the game.

To move up and down the directories, you can use the CD (Change Directory) command followed by one of three characters. One is the backslash (\), one is the blank, and the last is two periods (..). The backslash is an absolute command meaning "Go to this subdirectory no matter where it is in the directory tree." The space means "Go one directory down the current branch of the tree only." The double dot means "Go back up the tree one level from where we are." Windows is a program in a directory named WINDOWS, and it has a subdirectory named SYSTEM. The following shows the above commands in use. The comments set in regular type in the right column show the effect of each command.

| | |
|---|---|
| `C:\>` | |
| `C:\cd\windows\system` | Go to the system subdirectory in the Windows directory. |
| `C:\WINDOWS\SYSTEM>` | Here we are. |
| `C:\WINDOWS\SYSTEM>cd..` | Back up one directory. |
| `C:\WINDOWS>` | Here we are, one up. |
| `C:\WINDOWS>cd system` | Go back down one. |
| `C:\WINDOWS\SYSTEM>` | And here we are. |
| `C:\WINDOWS\SYSTEM>cd\` | Go to root. |
| `C:\>` | And here we are. |
| `C:\>cd system` | Try to go down two directories. |
| `Invalid directory` | Cannot go down two. |
| `C:\>cd windows` | Try to go down one. |
| `C:\WINDOWS\>` | OK. |
| `C:\WINDOWS\>cd system` | Try to go down one more. |
| `C:\WINDOWS\SYSTEM>` | OK. |
| `C:\WINDOWS\SYSTEM>cd..` | Back up one. |
| `C:\WINDOWS\>` | Here we are. |
| `C:\WINDOWS\>cd..` | Back up one more. |
| `C:\>` | OK, we are back at the root. |

Practice moving through the directories. These commands work the same for Make Directory (MD) and Remove Directory (RD).

## EDITING A COMMAND

Once you have typed a command and pressed ENTER, you may see that you want to modify the command. The function keys are set up by DOS for editing the command line. The F3 key repeats the last command you entered by writing it to the screen. All you have to do is press F3 and ENTER, and the last command is repeated. This is handy if the end of the command should be changed. The F1 key repeats one character of the last command given each time it is pressed. This allows you to modify one or more characters quickly. Pressing F2 and a character repeats the last command up to but not including that character. This is a little quicker than F1 in some cases.

An example follows with the DIR command. Try typing these commands now to get familiar with them. You cannot hurt the computer by anything you type on the keyboard. The FORMAT, DEL, DEBUG, and ERASE commands may destroy data or programs, so use them with caution.

| | |
|---|---|
| DIR | Lists your files. |
| F3 | Gives you the same list of files. |
| F1 | Gives you a D (don't press Enter yet). |
| F1 | Gives you an I (don't press Enter yet). |
| F1 | Finally gives you R for DIR. Now press ENTER. |
| DIR *.COM | Gives you all COM files. |
| F2C | Gives you DIR*. all up to the C in COM. |
| EXE | Press ENTER for DIR *.EXE, which is all EXE files. |
| F7 | Lets you see the last ten commands. |

## COMMON COMMANDS

DOS commands come in two types: internal (resident) and external (transient). They can be identified by the fact that the transient commands are actually separate programs and will be seen in a directory of the disk as COM or EXE files.

The major commands you should be familiar with are listed in Table 10.2. Table 10.3 gives the syntax for the most commonly used commands. Remember that more information about the commands can be found by using the HELP or FASTHELP commands in DOS or by typing the command name followed by the /? switch, as in DIR /?.

## GETTING HARDCOPY

The Print Screen Key (or PrintScrn on some keyboards) will print to the printer the twenty-five lines of the screen usually with the command CTRL+PrintScrn. CTRL+PrintScrn causes everything written to the screen afterwards to also go to the printer. A second CTRL+PrintScrn toggles or cancels the first. Combined with the TYPE command, this allows the printing of files more than twenty-four lines long.

As an example, if GRANDMA.LTR is a text file holding a three-page letter to Grandma, the following commands would print it out.

CTRL+PrintScrn

```
TYPE GRANDMA.LTR
```

Note that the TYPE command only types ASCII text files, not all word processor files. Those have control codes for fonts and formats that TYPE cannot interpret.

The device names LPT1: and PRN: are for the first parallel printer. COM1: is for the first serial printer (or communications port). You can print a file using redirection by typing TYPE JUNK > PRN:, although the PRINT command above is usually the better choice.

**TABLE 10.2 Major DOS commands**

| | |
|---|---|
| ASSIGN | Lets you switch physical and logical devices. |
| ATTRIB | Lets you change the attribute of a file. |
| BACKUP | Makes backup copy of files (from hard) to a floppy. |
| CD | Changes to an existing directory. |
| CHKDSK | Checks disk and reports status of disk and RAM. |
| CLS | Clears the video display. |
| COMP | Compares two files to see if they are identical. |
| COPY | Makes duplicate of a file. |
| DATE | Gives date and asks for correction. |
| DBLSPACE | Compresses the data on the hard drive, giving more capacity. |
| DEBUG | A program for advanced programmers only. |
| DEFRAG | Defragmentizes a drive, saving room and speeding access. |
| DEL | Delete is the same as erase. |
| DELTREE | Deletes the files in the directory and all lower subdirectories. |
| DEVICE | Loads device drivers in conventional memory. |
| DEVICEHIGH | Runs device drivers from upper memory. |
| DIR | Shows files on disk with size and date modified. |
| DISKCOMP | Compares two disks to see if identical. |
| DISKCOPY | Makes exact duplicate of entire disk, will format. |
| DOSKEY | Keeps a history of DOS commands given for quick reuse. |
| DOSSHELL | A file handling utility for those not familiar with DOS. |
| DRIVER | Similar to DRIVPARM, but for floppies only. |
| DRIVPARM | A program that lets you tell the system the physical parameters of a floppy or hard drive. |
| DRVSPACE | A new form of DBLSPACE, compressing a drive to give more capacity. |
| EDIT | A screen-oriented editor, a great improvement over EDLIN. |
| EDLIN | A line-oriented editor. The pits, but better than nothing. |
| EMM386 | Memory manager for 386/486. |
| ERASE | Erases a file from the disk. |
| EXPAND | Expands a compressed file in DOS 6.0. |
| FASTHELP | Shows all commands with short description. |
| FASTOPEN | Allows files to be opened faster by tracking high-use files. |
| FC | File compare. |
| FDISK | Partitions the hard drive. |
| FIND | Looks through file for a string of characters. |
| FORMAT | Formats a disk to prepare for use. /S makes disk bootable. |
| GRAPHICS | Puts printer in graphics mode. SHIFT+PrintScrn will now draw graphics on the printer. |
| GWBASIC | Runs the BASIC interpreter. |
| HELP | DOS 5.0 or newer gives a description of commands. |
| LOADHIGH | Loads TSR programs or drivers or even DOS above 640K. |
| MD | Makes a new directory. |
| MEM | Gives the status of RAM memory (DOS 5.0 or higher). |
| MEMMAKER | Attempts to free as much conventional memory as possible. |
| MIRROR | Saves information about files to aid UNFORMAT. |
| MODE | Sets up video screen, printer, or communications port. |

TABLE 10.2 *(continued)*

| | |
|---|---|
| MORE | Stops screen output at twenty-four lines and asks if you want more. |
| MSAV | Searches for and removes viruses when run. |
| MSBACKUP | DOS 6.0 backup to save information to floppies. |
| MSD | Shows information about your system, including interrupts. |
| MWAV | A Windows antivirus program that checks and cleans when run. |
| MWAVTSR | A Terminate and Stay Resident antivirus program for Windows. |
| PATH | Shows current path or sets new one. |
| PRINT | Spools a file so it can be printed while doing other work. |
| PROMPT | Allows you to change prompt or video function keys. |
| QBASIC | Runs the QBasic interpreter. |
| RAMDRIVE | Creates a drive in RAM. |
| RD | Removes an empty directory. |
| RECOVER | Recovers files lost due to corrupted FAT table. |
| REN | Renames a file. |
| RENAME | Renames a file. |
| RESTORE | Restores backed-up files to the hard drive. |
| SCANDISK | An advanced CHKDSK, finds and flags bad sectors, repairs FAT table, recovers files. |
| SETVER | Sets the DOS version so older programs can run under newer DOS. |
| SMARTDRV | A hard disk caching program to speed up hard drive access. |
| SORT | Alphabetizes a file; many options. |
| SUBST | Allows you to substitute physical and logical devices. |
| SYS | Puts the DOS system on a disk, if tracks 0 and 1 have room. |
| TIME | Gives the time, asks for correction. |
| TREE | Shows the tree structure of the subdirectories. |
| TYPE | Types a text file on the screen. |
| UNDELETE | Attempts to rebuild an erased or deleted file. |
| UNFORMAT | Attempts to rebuild data lost when a disk is formatted. |
| VER | Tells what version of DOS was booted. |
| VERIFY | Verifies data in disk writing commands. |
| VOL | Tells the name of the disk, if one was given. |
| VSAFE | Continually checks for virus activity. |
| XCOPY | Copies files, including subdirectories. |

**TABLE 10.3  Syntax of the more common commands**

**A:** How is that for a simple command? The computer may have two floppies and a hard drive. Which one will it usually use? It uses the default drive. That is the drive it boots on, until you tell it differently. The command C: causes the prompt to change from an A> to a C>. Now a DIR will be of C: drive because you have made it the default drive. Of course, B: and C: are valid commands also. In future commands, if you do not designate the drive, the default is assumed.

**ASSIGN** A=B will cause the computer to look at B when it's told to look at A. ASSIGN B=A causes a DIR B: to run drive A:. ASSIGN alone cancels any assignments.

**ATTRIB** sets the attributes of a file. Switches are preceded by a + or – to set or clear. Switches include R for read only, A for archive, S for system, H for hidden, and /S to process all subdirectories. An example is ATTRIB –R A:*.* to clear the Read Only attribute from files on a backup disk, so they can now be erased. If unable to erase a file, it may be read only. You can display or change a file's attributes with the ATTRIB command.

**TABLE 10.3** *(continued)*

**BACKUP** and its opposite, RESTORE, are used so you have extra copies of all your data in case of a failure of your working data disk or the hard drive. The general form is BACKUP FROMDRIVE TODRIVE:/PARAMETER. Usually you are backing up a hard (or fixed) drive, C: or D:, to a floppy, A: or B:. The /(slash) always precedes a parameter, and you may continue adding slashes and parameters. Possible parameters are /S for all subdirectories; /M for only files modified since last backup; /D:date for only files modified since a certain date (date format is MM-DD-YYYY); /T:time for only files modified since time (format is HH:MM). The /S is critical, for it will cause all the files everywhere on the disk in all subdirectories to be backed up for safety. The /M is extremely useful for after your initial overall backup using ONLY the /S parameter (which uses many backup disks and takes ten to thirty minutes). Future backups can be with the /S and /M parameters (or switches) such as BACKUP C: A:/S/M, which will back up the files anywhere on the disk that have been modified since last backup, saving much time and trouble. Use a new formatted disk for this. In fact, it is always good to have a few formatted disks lying around.

**CD** is Change Directory. The general form is CD\DIRNAME, where DIRNAME is the full name of the directory, such as CD\WP51\WP51DOCS. This moves you "from manila folder to manila folder." CD\ gets you back to the root, or base directory.

**CHKDSK** checks the disk for clusters of lost information that the File Allocation Table (FAT) has no pointer for. These can be repaired by the /F switch, which fixes the problem by setting a pointer to the location and naming it FILE0001.CHK or some similar name. You can then look at the file with a word processor. If it appears as garbage, it may have been part of a program that will crash the next time it is run. If you can read it, you may be able to identify it as part of a data file and use it to repair the file. Either way, at least the FAT is now repaired. The /V switch gives the full path of every file on the disk. The general form is CHKDSK DRIVE:/SWITCH, and it may be very wise for you to place the line CHKDSK C:/F in your AUTOEXEC.BAT file.

**CLS** is Clear Screen, and it does just that.

**COPY** does just that, makes a duplicate of a file. The general form is COPY FROMDRIVE:FILE TODRIVE:NEWNAME/V. COPY A:MYFILE.DAT B: copies the data file named MYFILE.DAT from drive A: to drive B: with the same name. COPY A:MYFILE.DAT B:MINETOO.DAT does the same, except with a new name of MINETOO.DAT on drive B:. COPY C:*.BAS A: copies all your BASIC programs you have written from drive C: to drive A:. The /V switch tells DOS to verify that the copy was perfect.

**DATE** gives the date the computer thinks it is, and it gives an opportunity to correct it if needed. If not needed, an ENTER is required. Format for the date is MM-DD-YYYY. This is often used in conjunction with the TIME command discussed later.

**DBLSPACE** is a data compression program that almost doubles the storage capacity of your disk. This is DOS 6.0 only, and, as with most DOS 6.0 commands, it is menu driven and has help available with the HELP command or the /? switch after the command. DOS 6.2 and higher have replaced this with DRVSPACE because of problems.

**DEBUG** is a program for changing a program on the machine or assembly level. This is for advanced programmers only. Usage is DEBUG PROGRAMNAME. The prompt changes to a hyphen. A few commands follow.

| | |
|---|---|
| ? | Help screen. |
| A | Assemble the following assembly program. |
| D | Dump or display lines of program in machine code in hexadecimal. |
| E | For entering new data in an address. |
| R | Displays contents of all registers. |
| RAX | Shows that register and accepts new data. |
| RBX | Shows that register and accepts new data. |
| RCX | Shows that register and accepts new data. |
| RDX | Shows that register and accepts new data. |
| U | Unassemble the program in memory. |
| W | Writes program with changes to disk. |
| Q | Quits program. |

It is best to not use DEBUG unless following explicit directions from some book or article or if you're just practicing. It can change and destroy the loaded program, so keep a backup.

**DEFRAG** is used when numerous files have been erased and others added over time. This leaves gaps between information, wasting space and slowing access. A drive in this condition is said to be fragmented. Use DEFRAG alone to pack the information together with fewer and smaller spaces and to speed disk access. Pressing X exits the program.

**DEL** deletes files. It has a /P switch for prompting before erasing, and it uses wildcards. This makes it dangerous. DEL *.* will delete all files in the current directory. DEL *.EXE will erase all executable files in the current directory.

**TABLE 10.3** *(continued)*

**DELTREE** is even more dangerous but also more useful. It erases all files in the current or specified directory and all subdirectories below it! The /Y switch means YES, do it without asking. Don't use the /Y prompt unless you are *positive* you want to erase all those files and subdirectories. DELTREE C:\JUNK will erase all files in the junk directory and all subdirectories under it.

**DIR** is often used just like that. However, the Wide or Paginate parameters or switches are often used, as in DIR/W or DIR/P. Also, the wildcards are often used as in DIR PROGRAM?.BAS, which shows PROGRAM1.BAS, PROGRAM2.BAS, PROGRAM3.BAS, and so on, or DIR *.EXE, which shows all the executable (program) files, or DIR *.COM, which shows another type of program (command) files. DIR *.BAS shows all your BASIC program files. Note that in DOS 5.0 or higher the /O switch gives a sorted or alphabetized directory. The /A switch shows files with the given attribute (A,D,H,R,S) when the A is followed by a colon and the attribute. DIR C:/A:H shows hidden files on drive C:, while the command DIR C:/A:+R shows read-only files. DIR C:/A-R shows read and write files. A is for archived, S is for system, and D is for directories.

**DISKCOPY** makes a duplicate of a disk. The original and the copy must be the same size disk. For reasons of fragmentation, the COPY *.* command may be superior in many cases.

**DOSKEY** is an extremely useful command, but only in DOS 5.0 or above. It remembers the past DOS commands you gave, and keeps a list or history. You can use the arrow keys to up arrow through the list or down arrow back to the most recent command. The F7 command gives a numbered list or history of commands. F9 allows you to select one of the past commands by number. ALT+F7 erases the history for a fresh start. This is a good one to put in your AUTOEXEC.BAT file.

**DOSSHELL** runs a file manager program that some find easier than using the DOS commands. The Tab key moves you from window to window, and the cursor arrow keys move between fields. Holding Alt and pressing the highlighted character of a command on the command menu line will activate that command. It functions much the way QBasic does.

**DRIVER.SYS** is very similar to the DRIVPARM command, except it cannot be used for hard drives. It creates a logical drive and assigns it to a physical drive.

**DRIVPARM** sets the specifications for a physical drive. Important switches are: /D:X, where X value 0 is drive A:, 1 is B:, 2 is C: /F:X, where X value 0 is 360K, 1 is 1.2 MB 5.25", 2 is 720K 3.5", 5 is hard disk, 7 is 1.44 MB 3.5", 9 is 2.88 MB 3.5". If using 5 for F:X, use the following also: /H:X, where X is heads; /S:X, where X is sectors; /T:X, where X is tracks.

**DRVSPACE** is a new form of DBLSPACE. It compresses the information on the drive and gives you almost twice the storage capacity. It is almost automatic, and the /? switch and the DOS HELP command will give you much information on this command.

**EDIT** is a very handy screen editor available in DOS 5.0 and above.

**EDLIN** is a not-so-handy line editor available before DOS 5.0.

**ERASE** is a dangerous command, as you could well guess. ERASE JUNK.DAT will remove that file from the disk. Not so bad; in fact, often needed. But ERASE *.* will erase *all* the files from the disk. ERASE *.DAT will erase all the files you gave the extension .DAT.

**EXPAND** is used to decompress files compressed on the setup disk. Syntax is EXPAND A:\SMARTDRV.EX_ C:\DOS\SMARTDRV.EXE. This copies the compressed file ending in "_" to the hard drive in decompressed or normal form, shown by the "E" last character.

**FASTHELP** is a quick glance at all commands with a short description of each.

**FASTOPEN** is run from the CONFIG.SYS file with INSTALL = C:\DOS\FASTOPEN.EXE /X. The /X switch makes the cache for tracking heavily used files in expanded memory EMS.

**FC** is the File Compare command that compares two files to see if they are the same. If different, it shows you those differences. An important switch is /A for abbreviate, which shows only the first and last of different lines, instead of all. Syntax is FC FILENAME FILENAME.

**FIND** hunts for a string in the file you name. Syntax is FIND /PARAMETER "THIS" MYFILE, which will hunt through the file named MYFILE looking for lines containing THIS. The parameter is optional: the quotes are required. Parameters or switches are /V for variant, or the lines that don't have the string; /C for just the count of lines that do have it; /N for the line numbers to be shown with the lines; /I for ignoring the case (capitalized or lower case).

**GRAPHICS** will prepare the printer to print the graphics screens you design in BASIC programs. You must run GRAPHICS before running BASIC. Without this command, trying to print a graphics screen will print garbage.

**GWBASIC** runs the BASIC (Beginners All-purpose Symbolic Instruction Code) interpreter. This allows you to program (once you learn BASIC). Newer DOS versions have dropped GWBASIC and use QBasic (QuickBASIC) instead.

**HELP** HELP COMMAND gives complete information about that command. This is included in DOS 5.0 or higher. With separate pages on syntax, notes, and examples and related topics shown with angle brackets (< >), this is a fount of information.

**TABLE 10.3** *(continued)*

**LOADHIGH** FILENAME will put DOS or Terminate and Stay Resident (TSR) programs or drivers above 640K in high memory, leaving more room for programs that run in the base or conventional memory range. HIMEM.SYS and EMM386 must be installed before LOADHIGH.

**MD** is Make Directory. MD\BUSINESS makes a directory named BUSINESS, in which to keep your business files. MD\BUSINESS\OOS makes a subdirectory under BUSINESS directory called OOS for your Out Of State business. Any names that do not conflict with DOS commands can be used, following the same rules as filenames.

**MEM** shows what is in memory at the current time. Three switches are /P for programs; /D for DEBUG, which also shows drivers; and /C for classify.

**MEMMAKER** frees up as much conventional memory as possible. Express setup is more automatic than Custom setup, therefore easier. Custom may squeeze out a little more memory. This will change your AUTOEXEC.BAT and CONFIG.SYS files, so always keep a copy with the BAK or BK1 extension.

**MIRROR** keeps track of information on the hard drive to help UNFORMAT rebuild a disk that is accidentally formatted. The /T switch loads a TSR file tracking program. Syntax is MIRROR C: /TC.

**MODE** reconfigures ports. For the parallel port, you can choose 80 or 120 characters per inch and 6 or 8 lines per inch vertically. Example is MODE LPT1:132,6. For the serial port, you can set baud to 12, 24, 48, or 96 hundred, or 19 for 19.2K baud. Parity can be set to None, Even, or Odd. The number of data and stop bits may be set at 5–8 for data and 2 or 1. 2 is only used for 110 baud, so is not used any more. An example is MODE COM1:24,N,8,1.

**MORE** will stop a file being displayed on the screen at line 24, print MORE, and pause. Pressing the Enter key allows another 24 lines to be typed, then it will stop again. This allows pagination of files being written on the screen.

**MSAV** is the Microsoft antivirus in DOS 6.0. Two switches are /P for automatically scanning on bootup and /I for local drives only, not network drives. It works when it is run, then stops.

**MSBACKUP** is the new DOS 6.0 backup routine. It is menu driven, has both help modes, and includes RESTORE.

**MSD** is a very useful program that gives about fifteen pages of information about your system. When first checking out a system, the command MSD/F C:\FILENAME lets you enter the name of the customer. Then it prints that and the info in a file you named in FILENAME. You may then print this out as a record of that customer's unit. Just typing MSD brings a menu to the screen, and you can choose by typing the highlighted letter of the entry. This goes to an information screen. The PageUp and PageDown keys or arrow keys let you move to more screens of information. Enter brings you back to the main menu. Function key F3 allows you to quit or exit the program.

**MWAV** is a Windows antivirus program that checks and cleans when run.

**MWAVTSR** is a Terminate and Stay Resident antivirus program for Windows that stays running all the time once started. Do not exit the program, just minimize it. You must run VSAFE in DOS before starting Windows to run MWAVTSR in Windows.

**PATH** tells you what DOS thinks is the default path it will follow when looking for files. With PATH=C:\WP;\DOS you tell DOS which path to follow.

**PRINT** is extremely useful. It allows you to print a file on the printer. EXE and COM files cannot be printed, and if attempted, will cause the printer to go crazy. The general form is PRINT FILENAME, and if used more than once it adds files to a pile or queue. /T for terminate stops the printing, and /C for cancel empties the queue.

**QBASIC** runs the QBasic interpreter, if you have a newer DOS version.

**RD** is for Remove Directory. You must first ERASE *.* all files in that directory. Be careful with the ERASE *.* command! If you are in the root, it can erase your entire disk.

**RECOVER** recovers files lost due to corrupted FAT table. Syntax is RECOVER [filename]. This is recovered in a file named FILE0001.REC without the bad sectors. You must reenter that lost data yourself.

**RENAME** or **REN** will change the name of a file. RENAME OLDNAME NEWNAME is the syntax, and you may put a path in front of OLDNAME only.

**RESTORE** is one you hope you will never use. It is used after a hard disk failure or crash to retrieve the backed up files from the backup disks and back onto a new or repaired hard drive. The general form is RESTORE FROMDRIVE TODRIVE:/PARAMETER. Since it is usually from the floppy to the hard drive, the typical command would be RESTORE A: C:/S/M to do the opposite of the command given in backup above.

**SCANDISK** is great for disk repair. Type SCANDISK alone to check the directory and file structures. You will be prompted if it finds errors in the FAT table and asked whether to proceed with fixing it. Then you will be asked if you want a surface scan. This is to find areas of the disk that are physically failing; it will flag them as bad and not use them anymore. This can take up to an hour, and you are given an estimate of the time required. The parameter /AUTOFIX stops the prompting and /SURFACE automatically runs the surface scan. For a fast, automatic run, try SCANDISK C:/AUTOFIX/SURFACE.

**TABLE 10.3** *(continued)*

**SETVER** reports to older programs the version of DOS they need to run. This allows a newer DOS to still run older programs.

**SORT** does just that, sorting a file alphabetically, line by line. SORT < MYFILE will sort a file named MYFILE by lines and replace it under the same filename. SORT < MYFILE > YOURFILE will leave MYFILE intact and put the sorted file under YOURFILE. The /R switch reverses the alphabetization, Z-A, as in SORT/R < MYFILE. /+N sorts on the Nth column. Using column 10 to sort for last names is SORT /+10 < NAMES > SORTNAMES.

**SUBST** A: B:\ (the backslash is important) turns drive B: into drive A: . SUBST alone shows any substitutions you have made. SUBST A:/D (that's a slash, not a backslash) cancels the substitution.

**TIME** is the command that gives you the time the computer thinks it is and gives you the opportunity to correct that if needed. If not needed, an Enter is required. Format for time is HH:MM.

**TREE** gives a graphic presentation of the subdirectory structure of the directory you are in. Of course, being in the root gives the most complex structure. The /F switch shows all files in the lower branches. On a large disk, this can be a lengthy graph.

**TYPE** is a useful command that will write a file on the screen. EXE, COM, and BAS files print garbage. Word processor files will have garbage here and there. ASCII and DOS text files may be typed or BASIC files saved with the ASCII switch or parameter, as in SAVE "MYFILE",A.

**UNDELETE** recovers deleted files. A useful switch is /LIST, which shows deleted filenames for possible recovery but does not recover. Syntax is UNDELETE FILENAME. It uses the MIRROR command file.

**UNFORMAT** tries to undo the damage done by accidently formatting a disk. Switches /J or /TEST show what may happen, but they do not do anything. Syntax is UNFORMAT DRIVE.

**VER** gives the version number of DOS.

**VERIFY** tells DOS to check files it copied to ensure that it copied accurately. Syntax is VERIFY ON, which does just that; VERIFY OFF, which does just that; and VERIFY, which is asking DOS.

**VOL** gives the volume number or what you called the disk when you formatted it (usually nothing).

**VSAFE** is a TSR program that continuously checks for viruses.

**XCOPY** is similar to the COPY command, but it includes all subdirectories. This can be a real time saver. Syntax is XCOPY DRIVE1:FILE1 DRIVE2:FILE2. Important switches are /S for subdirectories, /V for verify, and /M. When used with the ATTRIB command, this switch is very helpful in copying many files to two or more floppies. First, in the subdirectory you want to copy to floppies, use the command ATTRIB +A *.*/S to set the archive bit on all files in the directory and any subdirectory under it. Now use the command XCOPY *.*/M/S. The S gives all subdirectories again. The M copies files and turns off the archive bit. So it will fill the first floppy and then give an error of a full disk. You insert a blank formatted disk and type the exact command again. XCOPY ignores those files already copied and starts where it left off. You may repeat this again and again.

## PIPING

Piping means to route a command through another command. It uses the ¦ symbol. For example, TYPE SORT displays the file named JUNK on the screen as is, but TYPE JUNK ¦ SORT pipes it through SORT instead, so the outcome is an alphabetized screen display of JUNK. JUNK itself is *not* sorted; all you changed was its display format. If JUNK is more than twenty-four lines, TYPE JUNK ¦ MORE will cause the screen display to stop at line twenty-four and display the word MORE at the bottom of the screen.

## REDIRECTION

### Redirection of Drives

If your drives have different capacities, sometimes a program will try to look at drive A: even though its disk only fits in B:. There are two ways to make the computer think of B: as A:, and vice versa. ASSIGN A=B looks at B: when it's told to look at A:. ASSIGN B=A causes a DIR B: to run drive A:. ASSIGN alone cancels any assignments. In newer DOS versions, the SUBSTitute command does the same thing. SUBST A: B:\ (the backslash is important) turns drive B: into drive A:. SUBST alone shows any substitutions you have made. SUBST A:/D (that's a slash, not a backslash) cancels the substitution.

### Redirection of Input and Output

Redirection of input and output is similar to piping. It causes the input or output of a file to come from or go to a different direction than normal. TYPE JUNK types the file named JUNK to the screen. TYPE JUNK > JUNK2 causes nothing to be written to the screen, but a copy of JUNK is written to a new file named JUNK2. If a file named JUNK2 had already existed, it would have been written over. The APPEND redirection sign is two greater than signs, >>. The command TYPE JUNK >> JUNK2 adds (appends) JUNK to the end of JUNK2.

SORT receives its input by redirection or piping only. Using input and output redirection, you can type SORT < JUNK > SORTED, and the contents of the file named JUNK are sorted and written into a file named SORTED.

Examples of piping and redirection are shown in the section on batch files that follows.

## 10.2 CREATING AND RUNNING BATCH FILES

## CREATING BATCH FILES

A batch file is a file that you can write and run as your own program. A batch file consists of DOS commands that are typed in an ASCII or DOS text file (just plain old characters with no word processor control codes in it). You run the file by typing the name of the file.

You can create a batch file in a word processor if you save the file as a DOS or text file. When you exit your word processor, do *not* save the file as usual, because that may overwrite the text file with one that contains control codes. You have already saved it as text, so just choose NO, don't save, and exit. Following are some examples of batch files.

### Using the COPY Command to Create Files

To create a file without a word processor, you use the DOS command COPY so that what you type on the keyboard is placed into a file named JUNK.BAT on drive A:.

```
COPY CON: A:JUNK.BAT
```

You must press ENTER after each line to make it take effect, and after entering the lines desired, you end with CTRL+Z (that's holding down control while pressing Z). So let's create one. Type the following exactly! (^ means Control.)

```
COPY CON: JUNK.BAT
DIR/w
^Z
```

It should have responded with "1 file copied." Doing a DIR after the above should show the file JUNK.BAT. This way of creating a file is the pits, because once you press the Enter behind a line, that line is terminated, and no errors can be repaired. To repair a line, you must re-create the whole file again. (Using the same name erases the old version; this is good if the old version had an error.) But it's quick and easy. You *can* edit a line while you're still in it with the Delete key.

While the EDLIN editor is only slightly better than the COPY command creating files, the EDIT program with DOS 5.0 is very handy. It is worth learning to use it to create and modify files. As a bonus, it is the same editor used for creating QBasic programs.

### Using the EDLIN Program to Create Files

The EDLIN program is a line-oriented text editor that came with DOS in the past. A line editor is just slightly better than the COPY command. You start it by typing EDLIN FILENAME, where FILENAME is whatever you choose; JUNK is a good practice run name. The screen will have an asterisk prompt. The following is a short list of the commands.

| Command | Function | Example(s) | |
|---------|----------|-----------|---|
| I | Insert text | i | Starts off at line 1. |
| | | 5i | Insert new lines before 5. |
| ^Z | Stops insert | | Hold Ctrl, press Z. |
| L | Lists lines | L | Lists twenty-four lines. |
| | | 12L | Lists twenty-four lines from line 12. |
| | | 2,4L | Lists lines 2 through 4. |
| D | Deletes line | D | Deletes current line. |
| | | 6D | Deletes line 6. |
| | | 3,5D | Deletes lines 3 through 5. |
| E | Exits, saves | E | Stops program, saves file. |
| Q | Quits | Q | Stops program, does not save. |

## Using the EDIT Program to Create Files

With DOS version 5.0 or greater, the EDIT program is a screen-oriented editor. It is a great improvement over EDLIN. If you have it, try it. Just type EDIT to start it. The following is a brief overview of commands to get you started. With the F1 help key, you can soon become proficient.

The main menu is listed along the top of the screen. The Alt key activates the menu so that you can press one of the highlighted letters on the keyboard to give a pop-up menu. Then you can either press one of the new highlighted letters or scroll with the down arrow to a selection and press Enter to choose it. The main menu with all the pop-up menus popped up follows.

| Files | Edit | Search | Options |
|-------|------|--------|---------|
| New | CuT | Find | Display |
| Open | Copy | Repeat | Help Path |
| Save | Paste | Change | |
| Save As | ClEar | | |
| Print | | | |
| EXit | | | |

Only one of these is visible at a time. There is an instruction line at the bottom of the screen you should pay attention to. Some keystroke commands follow

| | |
|---|---|
| Home | Moves cursor to beginning of line. |
| End | Moves cursor to end of line. |
| Insert | Toggles between "typeover" and "insert" modes. |
| Delete | Deletes character or selected text. |
| Tab | Moves forward through the "help" fields. |
| Shift+Tab | Moves backward through the "help" fields. |
| Arrow | Moves one character forward or backward or one line up or down. |
| Shift+Arrow | Blocks or selects characters for editing. |
| Ctrl+Arrow | Moves one "word" forward or backward. |
| Esc | Cancels or backs you down the command structure. |

## RUNNING A BATCH FILE

You run your file by typing its name (it's like a new command). Just typing JUNK should now give you a wide directory. To stop a batch file that is running, CTRL+C causes the system to ask if you want to exit the batch file.

## Special Commands to Put in Your Batch File

You may use almost any DOS command in your batch files. However, DOS includes the following batch commands to do the processing. The DOS help program gives more information on these commands.

| | |
|---|---|
| APPLY | Uses a list of data one line at a time in another batch file. |
| CALL | Calls another batch file as a subroutine. |
| CHOICE | This allows the input of a choice of actions. |
| ECHO | With ECHO OFF in effect, this prints the text that follows ECHO on the screen. |
| ECHO OFF | Keeps DOS from printing the commands on the screen. It neatens your program. The @ symbol in the first line "hides" that command. |
| ECHO ON | DOS will print commands on the screen. |
| FOR | Performs a command on a set of files. |
| GOTO label | Label must start with a colon. |
| IF cndtn cmd | Does the command if condition is true. |
| PAUSE prompt | Stops and prints the prompt that follows and displays "Strike any key to continue." |
| REM | Causes DOS to ignore the following text as a remark for your help only. |
| SHIFT | Makes %1=%2 moves all up one in line. |

You can also add DOS commands to test for certain conditions.

| | |
|---|---|
| ERRORLEVEL | An error level is set when an error occurs. You can test for this with a command such as the IF command: IF ERRORLEVEL 2 ECHO. (Problem encountered.) |
| EXIST | Searches for a file and does a command if found: IF EXIST AUTOEXEC.BAT ECHO. (There is already an AUTOEXEC file.) |
| == | Checks if the sides are the same: IF word1 == word2 ECHO. (These are the same.) |
| NOT | Used with another condition to show the opposite: IF NOT EXIST file ECHO. (There is no such file.) |

## Replaceable Parameters

If you type a word after the batch filename, that word goes into a variable name called %1. The second word goes into %2, and so on, up to 10. These allow input into your batch file. An example follows. This is ADD.BAT. If you type ADD and the name of two text files that exist, and the name of a third to hold the results, they will be added together in the third.

```
COPY %1+%2 %3
```

With DOS 6.0, another type of replaceable parameter is allowed. If you use the block header (see CONFIG.SYS in this chapter) name as a label name, you can GOTO that label with the GOTO %config% command. The example under A Menu for CONFIG.SYS in this chapter uses [ADAM] and [EVE] as block headers. In the AUTOEXEC.BAT file used with that CONFIG.SYS, you can use the GOTO %config% command to select your AUTOEXEC.BAT commands to be run. The REM lines tell you what actions are being performed. They are ignored by the system. A simple example follows.

```
@ECHO OFF
CLS
REM  The above two lines clear the screen and stop unwanted
REM  comments from being displayed.
VERIFY ON
REM  The following line will go to whichever label the
REM  config file selected.
GOTO %config%
:ADAM
DATE
TIME
```

```
REM  The following line jumps over EVE.
GOTO END
:EVE
PROMPT $p$G
REM  If there were a third choice, you would need a
REM  GOTO END here to jump over it.
:END
```

## 10.3  AUTOEXEC.BAT

AUTOEXEC.BAT is just another batch file, but DOS automatically looks for the CONFIG.SYS file first, then the AUTOEXEC.BAT file when it boots up and runs them, if they exist in that order. This gives you the ability to make your system start off with the configuration you want, doing the commands you want. Pressing F8 as DOS loads allows you to single step through the CONFIG.SYS, then the AUTOEXEC.BAT file. On each command, it asks if you want to perform that operation. This is handy for troubleshooting. Pressing F5 as DOS loads is also handy and totally bypasses both CONFIG.SYS and AUTOEXEC.BAT.

Before you create a file named AUTOEXEC.BAT, check to see if one already exists. If so, copy it to an AUTO.BAK backup file for safety, then either create a new one or add to the existing one.

Be aware that sometimes important setup is done in the AUTOEXEC.BAT file, and your monitor or printer may not work properly without it. This is also the file that initially loads many TSR programs. TSR (Terminate and Stay Resident) programs are those that load, attach to DOS and become part of it, then terminate. They therefore remain resident and lie in wait for some trigger such as a set of "hot" keys to activate them. Some older systems do not have a battery-backed-up date or time. When these come up, typing the command date or time gives you 1/1/1980 and midnight 00:00:00 or some such thing. People with these systems often add an AUTOEXEC file (or line to the file) with the date and time commands. This can also produce a main menu. Examples of both follow.

| | |
|---|---|
| `DATE < CR` | CR is a file with one ENTER you created. |
| `TIME < CR` | |
| `PROMPT $t` | Gives you the time at every prompt. |
| `C:\MOUSE\MOUSE` | Looks in the directory with the mouse software and runs it. |
| `PATH=C:\DOS;C:\WP51` | Tells DOS where to look for files. |

This example shows a typical AUTOEXEC.BAT file.

| | |
|---|---|
| `ECHO OFF` | Do not display commands on the screen. |
| `CLS` | Starts with empty screen. |
| `DATE < CR` | Displays date on screen. |
| `TIME < CR` | Displays time on screen. |
| `DOSKEY` | Keeps a queue of the last few commands for easy recall. |
| `COMMAND /E:1024` | Enlarges the environment for exotic prompt commands. |
| `PROMPT $p$g` | Print path and greater than sign for prompt. |

The last line above shows which subdirectory you are in by printing the path. If you are always worried about time, try PROMPT $t.

## 10.4  EXAMPLES OF BATCH FILES

### PRINTING DATE AND TIME

The following prints the date and time when the system boots. First, you must create a file named CR holding one carriage return. Do this with COPY CON: CR. Once the cursor has fallen one line, press Enter once then Ctrl+Z and enter again. The PATH command is for the program to locate FIND.EXE. The PROMPT command is to help you see where you are.

```
PROMPT $p$g
PATH = c:\dos
DATE < CR > DT
TIME < CR >> DT
TYPE DT ¦ FIND "Current"
```

## USING THE CHOICE COMMAND

The following lines are a batch file example of the CHOICE command. Again, the REM lines explains the purpose of the line above.

```
ECHO OFF
CLS    REM  Demo of the choice command.
REM  Demo of the choice command.
CHOICE /C:YNC Yes, No or Continue /t:n,10
REM  YNC are to receive pressed key. ERRORLEVEL is set to
REM  1,2,3 etc y=1, n=2, c=3 etc. Yes...Continue is prompt
REM  /t is timeout with default set to n for No
REM  after 10 seconds.
REM  put ERRORLEVELS below in descending order.
IF ERRORLEVEL 3 goto cont
IF ERRORLEVEL 2 goto no
IF ERRORLEVEL 1 goto yes
GOTO END
:CONT
ECHO This was the choice to Continue!
GOTO END
:NO
ECHO This was the choice of NO!
GOTO END
:YES
ECHO This was the choice of YES!
:END
```

## CLEARING THE SCREEN

The following lines are a batch file that just clears the screen and displays one line.

```
ECHO OFF
CLS                    CLS clears off the garbage that's already there.
ECHO Hi ya Buck!       ECHO types the greeting on the screen.
```

## USING THE FOR COMMAND

The FOR command allows you to execute another command for large numbers of files automatically. You can use this command for many repetitive tasks. The only things you change are the *.RPT to your filenames and the PRINT command to any DOS command. %%F is a parameter. The following example will print three files. Use an editor to create the three files with extensions of .RPT. They must be printable from DOS, i.e., text or ASCII files. Also create the CR file with one carriage return in it. This will print all three on the printer.

```
PRINT < CR
FOR %%V IN (*.RPT) DO PRINT %%V
```

## SORTING A DIRECTORY LISTING AND ERASING FILES

In DOS 5.0 or up, DIR/O will show files on the disk alphabetically, piping the directory into the sort program before showing output. If you have an older version, you can use this batch file. If you have more than twenty-four files, the MORE pauses the screen.

```
ECHO OFF
CLS
ECHO
DIR | SORT | MORE
```

The following prints a sorted directory of drive B: on the printer.

```
DIR B: | SORT > PRN:
```

The following makes a backup file, runs the assembler, and types out results.

```
COPY %1.ASM %2.ASM
MASM %2,%2,%2;
TYPE %2.PRN
```

The following is like the ERASE command, but it accepts a list of files to erase.

```
ECHO OFF
CLS
:LOOP                        Just a label.
IF "%1" == "" GOTO DONE      Compares a replaceable parameter for a decision.
ERASE %1                     Erases the filename typed after the batch name.
SHIFT                        Moves %2 to %1, %3 to %2, and so on.
GOTO LOOP                    Unconditional GOTO the label above.
:DONE                        Label to come to if next parameter blank.
ECHO FILE(S) ERASED          Message back to operator.
```

## SEARCHING FOR FILES

The following I call DIRNOT.BAT because it looks for files *without* the name-part you follow it with. For example, DIRNOT.BAS will show all the files that are *not* BASIC.

```
ECHO OFF
CLS
DIR/O | FIND/I/V "%1" | MORE
```

The following searches for a file in the directory.

```
ECHO OFF
IF EXIST %1 GOTO X                    Decision on parameter.
ECHO SORRY, THAT FILE NOT FOUND
GOTO END
:X
ECHO THAT FILE IS ON DISK
:END
```

## COPYING FLOPPY DISKS

The following is like the DISKCOPY command, but it is for one-floppy systems with a hard drive. If you want the disk created to be bootable, it has to have been formatted with the /S switch so it has the system on it.

```
MD\KOPYTEMP
CD\KOPYTEMP
PAUSE PUT DISK TO BE COPIED IN DRIVE A:
```

```
COPY A:*.*
PAUSE PUT A NEW, FORMATTED DISK IN DRIVE A:
COPY *.* A:
DEL *.*
CD\
RD\KOPYTEMP
```

## SAVING THE PATH TO YOUR FILE

The following saves the path that exists before your batch file runs by redirecting output to a file. The single > creates a file; the >> appends to a file. Now you run your batch file, changing the path as needed. When the command to exit your batch file menu system is given, make it also run the pathback batch file, to restore the system as it was. The third line should set your desired path.

```
PATH > PATHBACK.BAT
ECHO CD\ >> PATHBACK.BAT
REM  Here change path to whatever you need.
REM  Here insert lines to run your commands.
REM  When done, the following restores the old path.
PATHBACK
```

## CREATING YOUR OWN HELP FILE

You can create a handy help file of your own. Create the following batch file named HELP!.BAT and the text files that follow. We are using the extension .HEP instead of .HLP to avoid conflicts with existing help files. If you already have a HELP!.BAT, name this one H.BAT or ANYTHING-YOU-LIKE.BAT

```
ECHO OFF
CLS
REM  This file is HELP!.BAT.
REM  This file is a help utility that types files.
REM  named TOPIC.HEP
REM  when entered as HELP! topic. When entered as HELP!, a list
REM  of topics for which files exist is displayed. If you add a
REM  help topic, you must append it to the HELP!.HEP file.
REM  The DOS program MORE.COM must exist in the directory.
IF EXIST %1.HEP GOTO DOES
IF "%1"=="HELP" GOTO HELP
IF "%1"=="" GOTO HELP
ECHO Help unavailable for %1
GOTO END
:DOES
TYPE %1.HEP | MORE
GOTO END
:HELP
TYPE HELP.HEP | MORE
:END
```

Now, under the name of BASIC.HEP create a file that says

```
This is a BASIC help file.
You invoke BASIC by one of the two following commands.
        BASICA   on an IBM system
        GWBASIC  on a clone system
```

Under the name of HELP.HEP create a file that says

```
To receive help on a topic, type HELP! Topic.
Help is available on the following topics ...
BASIC
```

Now create other helpful notes in files with the extension .HEP and add their filenames (without extension) to the HELP.HEP file. The file below helps do this.

```
ECHO OFF
CLS
REM  This is ADD.BAT, which adds a line to a file then sorts it.
REM  To run, type ADD filename
TYPE %1
ECHO  Type data in identical format to above, then press Ctrl+Z.
COPY CON: TEMP
TYPE TEMP >> %1
COPY %1 TEMP
SORT < TEMP ¦ FIND/V "-----" > %1
TYPE %1¦MORE
```

## CREATING MENUS

You can create your own menus using batch files.

```
REM This is MENU.BAT
ECHO OFF
CLS
TYPE MENU.TXT
PROMPT Please enter choice
```

Now, under the name MENU.TXT you create a text file with a number and program name on each line, such as the following.

```
1. Word Processing
2. Basic
```

You can make this much fancier, with entries for all the programs you expect to run often. For this example, you must make batch files named 1.BAT and 2.BAT with the proper commands in them, as follows.

```
REM  This is 1.BAT.
ECHO OFF
CLS
CD\WORDPERF
WP
CD\
MENU
```

This batch file will clear the screen, change to a subdirectory named WORDPERF, run WordPerfect, and, when done, return to the root directory and run the menu again. It is like a closed loop. Some programs will break the loop. 2.BAT would be the same except for the subdirectory it changes to, and the program (GWBASIC or QBASIC) it runs. The following is a possibility.

```
REM  This is 2.BAT.
ECHO OFF
CLS
CD\BASIC
```

```
GWBASIC
CD\
MENU
```

If you experiment with the above, you can make a professional-looking menu system to run your programs.

## 10.5 PROMPT COMMANDS

### STANDARD PROMPTS

The PROMPT command changes the prompt according to the following list of parameters. They can be used in combination, each preceded by a $ (dollar sign). Maximum length after the command PROMPT is 120 characters. You must include the line DEVICE = ANSI.SYS in your CONFIG.SYS file to use the following commands.

| | |
|---|---|
| $ | Gives dollar sign. |
| t | Gives time. |
| d | Gives date. |
| p | Gives path. |
| v | Gives DOS version. |
| n | Gives drive. |
| g | Gives greater than >. |
| l | Gives less than <. |
| b | Gives the piping character ¦. |
| _ | Underline gives the CR/LF (go to new line). |
| s | Gives space (leading only). |
| h | Gives backspace, erasing previous character. |
| e | Gives escape (1B hex). |

A very common prompt is

```
PROMPT $p$g
```

This gives the path and then a greater than sign, letting you always know what directory you are in, such as C:\WORD>.

### KEY REASSIGNMENT PROMPT COMMANDS

You can also change the character that appears on the screen when a key is pressed. The syntax is PROMPT $e[old#;new#p. Entering the ASCII code (or keyscan code) of the characters reassigns that key (see Appendix D). Some keys that might be available for reassignment are the reverse apostrophe (`), the tilde (~), and the braces ({}). Some characters you might like to use from the extended ASCII code are alpha (α), beta (ß), omega (Ω), and pi (π). You may also assign a string of characters.

The following are some key reassignment examples.

```
PROMPT $e[96;224p     Changes the reverse apostrophe to alpha.
PROMPT $e[126;234p    Changes the tilde to omega
```

The following reassigns the key and returns the standard prompt.

```
PROMPT $e[96;224p$p$g
```

The following replaces the tilde with a DIR command. The 13 is the code for an ENTER, to complete the DOS command.

```
PROMPT $e[126;"DIR";13p
```

The following replaces the tilde with a DIR command, but it does not have an ENTER (13) after it, so it waits for completion. For example, you might add the parameter "/O" then press ENTER, to get a directory in alphabetical order.

```
PROMPT $e[126;"DIR"p
```

In the following, 125 is the right brace. The word "DATE" will appear, that is, the DOS command DATE. The first 13 is an ENTER. The DATE command asks to confirm or change the date, so a second 13 for ENTER is needed. The reassignment PROMPT command is ended with a "p." Finally, "$p$g" is added for the standard prompt

```
PROMPT $e[125;"DATE";13;13p$p$g
```

## SCREEN POSITION PROMPT COMMANDS

Screen position PROMPT commands are used to position the cursor on the screen.

| | |
|---|---|
| PROMPT $e[#;#H | Moves cursor to position = row #, column #. |
| e[#A | Moves cursor up # rows. |
| e[#B | Moves cursor down # rows. |
| e[#C | Moves cursor right # columns. |
| e[#D | Moves cursor left # columns. |
| e[s | Saves cursor position. |
| e[u | Restores cursor position. |
| e[2J | Clears display, moves cursor to the home position. |
| e[K | Erases from cursor to end of line. |

The following are two examples of screen position PROMPT commands.

```
REM  The following moves the cursor to row 20, column 40.
PROMPT $e[20,40H
REM  The following gives date and time at the left
PROMPT $e[s$e[1;1H$e[K$e[7m$d $t $e[0m$e[u$p$g
```

When used with a text file of a menu, the following is a nice menu display. It has the date and time centered above the menu and the words "Enter Command!" below the menu.

```
ECHO OFF
CLS
TYPE MENU
PROMPT
$e[1;1H$e[K$e[1;28H$d$t$e[1;48H$e[K$e[23;1H$e[K$e[22;1H$e[K$e[22;2
8H$ Enter Command!
```

A sample menu text file might be as follows:

```
        Word processing

        Spreadsheet

        Database

        Games
```

## SCREEN COLOR PROMPT COMMANDS

You can control the color of the text on the display screen by using the following commands. You may use one, two, or three numbers, and the order does not matter. End it with a small "m."

PROMPT $e[#;#;#m

**Attributes**

| | |
|---|---|
| 0 | Black on white |
| 1 | Bold white |
| 4 | Underline on text display only |
| 5 | Blink |
| 7 | Reverse video |

**Foregrounds**

| | |
|---|---|
| 30 | Black |
| 31 | Red |
| 32 | Green |
| 33 | Yellow |
| 34 | Blue |
| 35 | Magenta |
| 36 | Cyan |
| 37 | White |

**Backgrounds**

| | |
|---|---|
| 40 | Black |
| 41 | Red |
| 42 | Green |
| 43 | Yellow |
| 44 | Blue |
| 45 | Magenta |
| 46 | Cyan |
| 47 | White |

The following are examples of screen color PROMPT commands, which set foreground and background colors. The ECHO ON/OFF command at the end is needed to activate the prompt if it is used in a batch file. If not, just use the prompt line.

The following allow you to pick foreground (in the 30s) and background (in the 40s) colors.

```
REM  Red foreground.
PROMPT $e[31m
REM  Green background.
PROMPT $e[42m
```

The following example sets the foreground to dark blue letters on a cyan (light aqua) background.

```
ECHO OFF
CLS
PROMPT $e[34;46m
ECHO ON
ECHO OFF
```

The following gives blinking green letters on red background.

```
PROMPT $e[32;41;5m
```

The following gives a foreground of dark blue letters on a cyan (light aqua) background, bolded, with the time above the prompt.

```
PROMPT $e[44;36;1m$t$_$p$g
```

To return your screen to normal, use the following.

```
PROMPT $e[0m
```

## BATCH FILE SCREEN COLOR CONTROL

The following batch file prints in color on the screen when you run it by typing the name. Copy the lines below with any processor or editor that will save an ASCII file. The spaces between the word "echo" and what follows are important, because they will later hold control codes. Save this as a batch file, such as COLOR.BAT.

```
ECHO OFF
CLS
ECHO [36mDAN'S COLORING BOOK
ECHO [5;35mThank You! [0m
```

Now, to insert control codes that your word processor would not accept, run DEBUG FILENAME or DEBUG COLOR.BAT. At the hyphen prompt, type D for dump, and you will see the file.

```
13BD:0100 65 63 68 6F 20 6F 66 66-0D 0A 63 6C 73 0D 0A 65 echo off..cls..e
13BD:0110 63 68 6F 20 20 20 20 5B-33 36 6D 44 41 4E 27 53 cho       mDAN'S
13BD:0120 20 43 4F 4C 4F 52 49 4E-47 20 42 4F 4F 4B 0D 0A  COLORING BOOK..
13BD:0130 45 43 48 4F 20 20 20 20-5B 35 3B 33 35 6D 20 20 ECHO    [5;35m
13BD:0140 20 54 68 61 6E 6B 20 59-6F 75 21 20 5B 30 6D 20  Thank You!.[0m
13BD:0150 20 0D 0A 0A 1A EF 5E 5B-72 06 E8 8A E3 E9 2D FF  .....^[r.....-.
```

The hex code for a blank is 20 and for the square bracket ([) is 5B. Find the three blanks (20s) before the three 5Bs and change them to ESCapes (1B). To do this type E 0100. You will see a number and a period. Press the space bar to accept each character as is until you come to the 20 you want to change, then type 1B. When you finish all three 20s before the 5Bs, you can repeat the D 0100 command to view your results. They should appear as follows.

```
13BD:0100 65 63 68 6F 20 6F 66 66-0D 0A 63 6C 73 0D 0A 65 echo off..cls..e
13BD:0110 63 68 6F 20 20 20 1B 5B-33 36 6D 44 41 4E 27 53 cho  .[36mDAN'S
13BD:0120 20 43 4F 4C 4F 52 49 4E-47 20 42 4F 4F 4B 0D 0A  COLORING BOOK..
13BD:0130 45 43 48 4F 20 20 20 1B-5B 35 3B 33 35 6D 20 20 ECHO   .[5;35m
13BD:0140 20 54 68 61 6E 6B 20 59-6F 75 21 1B 5B 30 6D 20  Thank You!.[0m
13BD:0150 20 0D 0A 0A 1A EF 5E 5B-72 06 E8 8A E3 E9 2D FF  .....^[r.....-.
```

When it is correct, type W to write the file to disk and then Q to quit DEBUG. You can run the file as you do any other batch file, by typing its name.

You may create the above batch file in EDLIN. On line 3 type ECHO ^V[[36m YOURWORDS, where the ^V is a real CTRL+V. Remember that in EDLIN 1,20L lists lines 1–20, 1I puts you in insert mode for line 1, ^Z (CTRL+Z) gets you out of insert mode, and e exits and writes the file to disk. You may also create or modify the file in DEBUG. You may insert an ESC in a text file by pressing CTRL+P, then holding down the Alt key and pressing 027 on the keypad.

To run the file, just type the filename. You should get a colored printout. If you get a really weird screen display you can't read during this experimentation, just type the command PROMPT $e[0m$p$g to get the screen back to normal.

The following is an example of a cute file. When you type it using the TYPE FILENAME command, it prints colored letters on the screen, but you see no commands. However, when you run the DEBUG program and display it, you see its commands in hex code. The point is, it does not have to be a batch file but just a file that gets typed on the screen. The technique can be used to print out a colorful menu from a batch file that just types the menu text file. The file can be created as the batch file above was, and DEBUG can be used to insert the 1Bs. The file should *not* have the .BAT extension. Just key in TYPE FILE-NAME to use it.

Create the file by using EDLIN or EDIT. Type a space, a left bracket, a number from 31 to 36, then the first letter of your word. Repeat for each letter, using a different number for a different color. Save the text file when done. Then use DEBUG as described below to change the spaces (hex 20) to escapes (hex 1B).

```
14CD:0100  1B 5B 33 31 6D 44 1B 5B-33 32 6D 61 1B 5B 33 33  .[31mD.[32ma.[33
14CD:0110  6D 6E 1B 5B 33 34 6D 27-1B 5B 33 35 6D 73 0D 0A  mn.[34m'.[35ms..
14CD:0120  1B 5B 33 36 6D 43 1B 5B-33 37 6D 6F 1B 5B 31 3B  .[36mC.[37mo.[1;
14CD:0130  33 31 6D 6C 1B 5B 31 3B-33 32 6D 6F 1B 5B 31 3B  31ml.[1;32mo.[1;
14CD:0140  33 33 6D 72 1B 5B 31 3B-33 34 6D 69 1B 5B 31 3B  33mr.[1;34mi.[1;
14CD:0150  33 35 6D 6E 1B 5B 31 3B-33 35 6D 67 20 1B 5B 31  35mn.[1;35mg .[1
14CD:0160  3B 33 36 6D 42 1B 5B 31-3B 33 37 6D 6F 1B 5B 33  ;36mB.[1;37mo.[3
14CD:0170  31 6D 6F 1B 5B 33 32 6D-6B 1B 5B 33 37 6D 0D 0A  1mo.[32mk.[37m..
```

You can use the E DEBUG command with the exact address to change the 20s to 1Bs. For example, you type E 0100 and DEBUG gives 20. You type 1B and press ENTER. Then you type E0106 and press ENTER. DEBUG gives 20 again, and you type 1B ENTER. Do this to fill all spaces before the 5Bs with 1Bs. Type W, then Q to write to disk and quit. This file is not a batch file, and it must be printed to the screen with the TYPE command to be used.

## 10.6 USING DEBUG TO CREATE COM FILES

This is the help file from DEBUG.COM. The ? command displays it.

```
assemble         A [address]
compare          C range address
dump             D [range]
enter            E address [list]
fill             F range list
go               G [=address] [addresses]
hex              H value1 value2
input            I port
load             L [address] [drive] [firstsector] [number]
move             M range address
name             N [pathname] [arglist]
output           O port byte
proceed          P [=address] [number]
quit             Q
register         R [register]
search           S range list
trace            T [=address] [value]
unassemble       U [range]
write            W [address] [drive] [firstsector] [number]
allocate expanded memory    XA    [#pages]
deallocate expanded memory XD    [handle]
map expanded memory pages   XM    [Lpage] [Ppage] [handle]
display expanded memory status  XS
D 0100 0105   dumps 6 2 digit hex #s in addresses 100 through 105
                    for display.
```

Before writing a file to the disk, you must inform DEBUG how many bytes to write and what to name the file.

| | |
|---|---|
| N FILENAME | Sets filename. |
| RCX | Displays contents of CX register. |
| 99 | Types hex number of bytes to write. |
| W | Writes to disk. |
| DEBUG FILENAME > LOG | You can no longer see output, but it is going into file named LOG. You must type in commands blind. U,U,U,U,Q will |

list sixteen lines of unassembled code for each U and then quit. This captures the assembly for a COM file.

E 130                         Shows contents of 0130 and waits for new data. A space bar shows next address and waits for new data. This is how you enter data.

If you remove the numbers in front of the assembly commands on each line, add an A as a first line for "assemble," and add the five lines starting with RCX; you can use DEBUG to assemble a program. You save as an ASCII text file and then type DEBUG < FILENAME.SCR (the SCR is for "script"). This runs DEBUG automatically.

The following script file (saved under CLRSCRN.SCR) creates CLRSCRN.COM if given as DEBUG < CLRSCRN.SCR. It clears the screen. Do not type in the semicolon or anything thereafter. They are remarks.

```
A
MOV   AH,0F        ; function code 15 decimal, 0F hex returns video state
INT   10           ; video interrupt vector 10 hex, 16 decimal
OR    AL,02        ; mode 2 is 80 × 25 characters, put in AL register
MOV   AH,00        ; A high is loaded with zero value
INT   10           ; video interrupt again
POP   BP           ; where were we?
INT   20           ; return to DOS
RCX                ; prepare the CX register
E                  ; this is hex for 14 bytes
N CLRSCRN.COM      ; name it
W                  ; write it to disk
Q                  ; quit DEBUG
```

Now you can run it in DEBUG by the command G=0100, which is always the starting address. It tells if the program terminated normally.

Note that to move a value into a register, the SCR value has reversed the first and second pair of digits when assembled into the COM file. See the example that follows.

```
24F9:0140 B86735 MOV AX,3567
```

The following lines should be copied into a text file. This can be done using the EDIT program of DOS or any word processor that saves an ASCII text file. Then at the DOS prompt, type the command line DEBUG < ABCDE.SCR. You should see the action on the screen. Then type ABCDE to have it print "ABCDE" on the screen. Do not type in the semicolons, or the comments that follow them.

```
A                  ; debug command to assemble what follows
STI                ; enable interrupts
LEA DX, [010B]     ; DEBUG loads at 0100. 010B is down below at DB 00
MOV AH,09          ; video string display function
INT 21             ; video interrupt
INT 20             ; terminate and return to DOS interrupt
DB 00
DB 41              ; hex ASCII for A
DB 42              ; hex ASCII for B
DB 43
DB 44
DB 45
DB 0D              ; hex ASCII for return
DB 0A              ; hex ASCII for line feed
DB 24              ; a dollar sign to end the string
N ABCDE.COM        ; name this file
RCX                ; look at CX register
```

```
16                    ; put 16 hex in CX as the byte length of program to save
W                     ; write it to default disk
Q                     ; DEBUG command to quit DEBUG
```

With a little practice, you can create your own COM files.

## 10.7   CONFIG.SYS

Although CONFIG.SYS is not a batch file, it is a very important file that works in conjunction with the AUTOEXEC.BAT file to configure your system. At bootup, DOS examines the CONFIG.SYS file, then the AUTOEXEC.BAT file. The CONFIG.SYS file is used by your system automatically to set up the environment under which it will be running programs. Pressing F8 as DOS loads allows you to single step through the CONFIG.SYS, then through the AUTOEXEC.BAT file. On each command, it asks if you want to perform that operation. This is handy for troubleshooting. Pressing F5 as DOS loads is also handy; it bypasses both CONFIG.SYS and AUTOEXEC.BAT.

CONFIG.SYS file commands are given in the form of device drivers, which are programs that help DOS control a device such as the keyboard, display monitor, drives. Drivers are installed with the DEVICE or DEVICEHIGH command. Different "blocks" or choices of configurations may be chosen from a menu.

### COMMANDS FOR CONFIG.SYS

The following are the typical commands used in a CONFIG.SYS file.

| | |
|---|---|
| BREAK = ON, | or BREAK = OFF tells DOS whether to check for the control C or control break keys. |
| BUFFERS = 20 | tells DOS to save twenty buffers in memory to help speed the transfer of data between memory and disks. |
| DEVICE = DRIVER | tells DOS to load the installable device driver to aid in controlling that device. |
| DEVICEHIGH = DRIVER | tells DOS to load that driver into upper memory (640K–1 MB), saving conventional memory. |
| DOS = HIGH | or = HIGH,UMB loads DOS in high memory, saving conventional memory. The UMB switch allows DOS to use the upper memory block. |
| DRIVPARM | The DRIVPARM command sets the specifications for a physical drive. Important switches follow: /D:X, where X value 0 is drive A:, 1 is B:, 2 is C:, and so on. /F:X, where X value 0 is 360K, 1 is 1.2 MB 5.25", 2 is 720K 3.5", 5 is hard disk, 7 is 1.44 MB 3.5", 9 is 2.88 MB 3.5". If using 5 for F:X, use the following also: /H:X, where X is heads; /S:X, where X is sectors; /T:X, where X is tracks. |
| DRIVER.SYS | Very similar to the DRIVPARM command except it cannot be used for hard drives. It creates a logical drive and assigns it to a physical drive. |
| FCBS | Sets number of file control buffers. The default is four. |
| FILES = 30 | tells DOS to keep track of up to thirty files at one time. While this uses some memory to keep track, it speeds file access. The default is 8. The range is 8 to 255. |
| INCLUDE | Includes other configuration blocks in the current one. |
| INSTALL | Loads a TSR program into memory. |
| MENUCOLOR | Sets foreground and background colors on bootup. Syntax is MENUCOLOR=FOREGROUND, BACKGROUND, where colors are the same as in BASIC.<br>0=black<br>1=blue<br>2=green |

|                | 3=cyan |
|                | 4=red |
|                | 5=magenta |
|                | 6=brown |
|                | 7=white |
|                | 8=gray |
|                | 9=light blue |
|                | 10=light green |
|                | 11=light cyan |
|                | 12=light red |
|                | 13=light magenta |
|                | 14=light brown (or yellow) |
|                | 15=light white (or bright white) |
|                | 16 Added to the number gives a blinking character of that color |
| MENUDEFAULT    | Tells DOS what configuration to use if none is selected. Syntax is MENUDEFAULT=BLOCKHEADER, WAITTIME, where the system waits for WAITTIME before automatically choosing the block named in BLOCKHEADER. |
| MENUITEM       | Allows you to define an item or choice on the menu shown on bootup. |
| NUMLOCK        | Allows you to choose whether the NumLock key should be on or off. |
| REM            | Tells DOS to ignore this remark line. |
| SET            | Alone shows system environment variables; SET VARIABLE = STRING sets that variable; for example, SET SOUND= D:\SBPRO. |
| SHELL          | Chooses which command interpreter DOS will use, normally COMMAND.COM. One handy use is to increase the bytes of memory used for environment variables with the following syntax: SHELL = COMMAND.COM/E:1024. |
| STACKS         | Allows you to select the number and size of stacks of memory to use to handle hardware interrupts. Syntax is: STACKS NUMBER,SIZE. The number can be from 8 to 64 and the size from 32 to 512 bytes. |
| SUBMENU        | Is used in a menu block to designate a part of a menu. Syntax is SUBMENU=BLOCKNAME,TEXT. |
| SWITCHES       | Gives options on loading DOS. /F means fastload DOS, and skips the two-second wait when loading. /K makes an enhanced keyboard act nonenhanced. /N stops F5 and F8 from bypassing CONFIG.SYS and AUTOEXEC.BAT. |

More on the above commands can be found by using the HELP command in DOS.

## DRIVERS FOR CONFIG.SYS

Some installable device drivers that can be used in the CONFIG.SYS file follow.

| ANSI.SYS      | Allows you to use some graphic commands for terminals. |
| DBLSPACE.SYS  | Moves DBLSPACE.BIN in memory to free conventional memory. |
| DISPLAY.SYS   | For displays. |
| DRIVER.SYS    | Creates a logical drive name to use with a physical drive. |
| EMM386.EXE    | A program of DOS 5.0 and 6.0 (XMAEM.SYS and XMA2EMS.SYS in DOS 4.0) that enables the use of expanded memory (EMS) created from extended memory (XMS). *Note:* Do not use DEVICEHIGH to load! |

| | |
|---|---|
| INTERLNK.EXE | Allows one computer to use another's drives through the serial or parallel port. |
| HIMEM.SYS | Manages extended memory. *Note:* Do not use DEVICEHIGH to load! |
| RAMDRIVE.SYS | Creates electronic disk, same as virtual or RAMDISK. |
| SETVER.SYS | Helps DOS run programs written for older DOS versions. |
| SMARTDRV.EXE | Caches the hard drives (was SMARTDRV.SYS in DOS 5.0). *Note:* Do not use DEVICEHIGH to load! |

The HELP command in DOS or the command /? switch will give more information on these programs.

## IGNORING AUTOEXEC.BAT AND CONFIG.SYS

On bootup, DOS 6.0 displays the message "Starting MS-DOS" for a moment. While this display is on the screen, you can bypass the CONFIG.SYS and AUTOEXEC.BAT files by holding down the Shift key or by pressing and releasing the F5 function key.

## A MENU FOR CONFIG.SYS

DOS 6.0 allows choosing from a menu of configuration options upon bootup. Instead of one CONFIG.SYS file in which all commands are run, you can now make a menu and configuration blocks (similar to subroutines) from which to choose. The configuration blocks are identified by square brackets ([ ]) around a name. Together, the brackets and name are called a block header. This is similar to the :LABELS in batch files.

Two predefined block header names are [COMMON] and [MENU]. The [COMMON] block header makes a block that will be run no matter what menu choice is made, just the way the old CONFIG.SYS file was run. The [MENU] block header sets up a menu that is displayed on bootup, from which you make a selection. The other new MENU commands may be used in this block. An example follows.

| | |
|---|---|
| `[MENU]` | |
| `MENUITEM=ADAM` | |
| `MENUITEM=EVE` | |
| `MENUCOLOR=7,9` | This line makes the system use white characters on light blue. |
| `MENUDEFAULT=ADAM, 10` | This line makes the system choose ADAM if no choice is made in 10 seconds. |
| `[COMMON]` | This line would be commands for both ADAM and EVE. |
| `[ADAM]` | This line commands for ADAM ONLY. |
| `[EVE]` | This line commands for EVE ONLY. |
| `INCLUDE=ADAM` | This line includes ADAM commands in EVE. |

When the system boots, it will give choice 1 as ADAM and choice 2 as EVE, and it will prompt you for a choice. At the end of 10 seconds, it will choose ADAM automatically.

## EXAMPLE CONFIG.SYS FILES

Examples of possible CONFIG.SYS command lines follow.

HIMEM.SYS is used to handle extended memory. Setting HIMEM.SYS in your CONFIG.SYS file (and rebooting to activate it), as in the following lines, prepares your system to use extended memory. The line after it loads DOS into the extended area, saving conventional memory. HIMEM.SYS must be installed *before* EMM386.EXE. The first line allows access to extended (XMS) memory. It must come before other memory lines.

The second line allows using extended XMS as expanded EMS; 256K is the default. HIMEM.SYS must be installed before EMM386.EXE.

To increase the amount of EMS memory, just add the size desired as shown in the following lines, which load DOS into high memory in order to save conventional memory.

```
DEVICE=C:\DOS\HIMEM.SYS
DEVICE=C:\DOS\EMM386.EXE
DEVICE=C:\DOS\HIMEM.SYS
DOS=HIGH
DEVICE=C:\DOS\EMM386.EXE 1184
```

The following line modifies the screen and keyboard to use the extended ASCII character set graphic characters.

```
DEVICE=C:\DOS\ANSI.SYS
```

In DOS 5.0 and 6.0, the RAMDRIVE.SYS file (VDISK.SYS in DOS 3.0 and 4.0) allows you to designate your own area of RAM to use as an electronic disk drive. This is called a RAMDRIVE, EDISK (electronic disk), VDISK (virtual disk), or RAMDISK. Because it has no mechanical parts, it is far faster than a floppy or hard disk. Because it is DRAM, it loses all information when the system is powered-down. The syntax is: DEVICE = DRIVE:PATH RAMDRIVE.SYS BYTES SECTORSIZE NUMBEROFDIRECTORIES /E/A. The /E parameter places the RAMDRIVE in extended memory, and the /A parameter places the RAMDRIVE in expanded memory. Of course, you must have activated the extended memory. The following two lines for the CONFIG.SYS file are typical. Remember to reboot to implement the new CONFIG.SYS file.

```
DEVICE=C:\DOS\HIMEM.SYS
DEVICE=C:\DOS\RAMDRIVE.SYS 960/E
```

These two lines create an electronic drive of 960K in extended XMS memory. An /A switch would use expanded EMS instead. The following are three more examples

```
DEVICE=SMARTDRV.EXE 512
```

This line creates cache in extended XMS memory that gives a 512K cache to speed hard drive access. An /A switch would create it in expanded EMS.

```
DEVICE=C:\MOUSE\MOUSE.SYS
```

This line loads the mouse driver in conventional memory.

```
DEVICEHIGH=C:\MOUSE\MOUSE.SYS
```

This line loads the mouse driver in upper memory.

```
DEVICE=C:\MOUSE\MOUSE.COM SER 1
```

This line loads the serial mouse driver on COM1:.

The following lines create thirty 512-byte buffers to speed disk access and thirty files to be open at once.

```
BUFFERS=30
FILES=30
INSTALL=C:\DOS\FASTOPEN.EXE /X
```

The last line in this sequence opens files faster and works in EMS.

## 10.8 BASIC TROUBLESHOOTING: DOS

DOS is usually installed in its own directory named DOS. If some DOS commands are not working, check to make sure the path is set by typing PATH. This should give PATH=C:\DOS or a more complex line with C:\DOS in it. If not, set the path by typing PATH=C:\DOS. Without the proper path, resident or internal commands will work but transient or external commands won't.

Another possible problem is a mixing of DOS versions. The VER command will give the version of DOS the system is booted in and running. Programs from different versions may state "Incorrect DOS version" and refuse to run. Running the SETVER program will tell the offending program that an older version of DOS is running, thus allowing the offending program to operate.

If you are having a problem with a particular DOS command, try the HELP command to confirm the proper usage of the command. If all else fails, you may have a corrupted program and may need to reinstall DOS.

## 10.9   WINDOWS 3.XX

Windows is an entire system of programs that is intended to give a simple user interface. It gives the ability to leave one program running in a "window" and start another program in another window. You can then switch back and forth, carrying information with you if you please. Windows has its own

- Word processor
- Database for storing information
- Painting program for drawing pictures
- Calculator
- Calendar for your appointments
- Clock to place on the computer screen
- Games
- Notebook for jotting quick notes
- Screensaver to save CRT screen phosphors
- Virus protection (included in MS-DOS)

Windows can record and play back sound and video if you have the correct adapter cards installed. Because of its high graphics utilization, a Windows accelerator card is used to speed up screen displays and keep it from slowing down your system.

Windows 3.xx is a very powerful system of programs. It has been the standard for a number of years, but in 1995 Windows 95 was introduced. Windows 95 was more intuitive and easier to use. Windows 95 was upgraded to Windows 98, which is more Internet–Web page compatible. All versions of Windows use extended memory, so if you have used F5 or F8 to bypass CONFIG.SYS and/or AUTOEXEC.BAT, Windows will not run. The error message may just be an unreadable flash on the screen—after typing WIN at the DOS prompt, you see a flash and end up at the DOS prompt again.

To protect against viruses in Windows, DOS 6.22 has two programs in addition to its DOS antivirus programs, VSAFE and MSAV. The Windows versions are MWAV.EXE and MWAVTSR.EXE. MWAV can be run from Windows through File Manager, indicated by the little file cabinet icon. When run, it scans for viruses and cleans. MWAVTSR is a Terminate and Stay Resident program, so you do not exit it but minimize it and allow it to run constantly. You must run VSAFE from DOS before starting Windows, then run MWAVTSR through File Manager after starting Windows.

The program is called Windows because it allows you to keep a number of programs running at the same time, running each one in a window or box on the screen. You can change between the windows, control their size, move data between them, and do many other handy things. Windows has a graphical user interface (GUI), so it uses little pictures called icons to represent programs. You point at these icons using a mouse or trackball (pointing devices) to move a pointer (usually an arrow-shaped cursor), and you can click the mouse on a choice in a menu bar to drop down a menu.

### GETTING STARTED

#### Installing Windows 3.xx

Windows 3.xx comes on about six or more 1.44 MB disks. To install Windows, insert the disk labeled 1 into drive A:, then type A: to make it the default drive. The screen prompt should now show A>. Type the command SETUP and press the Enter key. Windows will lead you through a series of questions and almost automatically install itself on your hard disk. It changes your AUTOEXEC.BAT and CONFIG.SYS files to optimize them for running Windows.

## Mouse Functions

You interface with Windows through the mouse and keyboard, so you must understand some terms relating to the mouse.

| | |
|---|---|
| Click | You click on icons, menus, commands, and so on, by using the mouse to move the arrow cursor to the item and momentarily pressing a mouse button, usually the left. |
| Double-click | You double-click by clicking the left mouse button twice in quick succession. |
| Drag | You drag by using the mouse to move the arrow cursor to an item, then you press and hold the left mouse button while you move the mouse to move the arrow cursor to another location. There you release the mouse button you have been holding. The item moves with the cursor, giving the appearance of dragging. |

## Windows Tutorial

Windows is quite a complex system, but luckily the Windows program has its own tutorial included. The following steps will let you run the tutorial to become familiar with Windows.

1. Boot the system. At the DOS prompt type WIN. This will cause Windows to run.
2. Hold down Alt and press F for File and R for Run.
3. Type in the name of the file you want to run, in this case, WINTUTOR, and press Enter.

This should cause the Windows tutorial to run. Follow its instructions, and you should gain a lot of knowledge about Windows. You also can run the tutorial by typing WIN at the DOS prompt to run Windows, then clicking on HELP under the Program Manager menu bar, and then clicking on Windows Tutorial in the drop-down menu.

## Starting the Windows Program

Turn the computer on. You must be at a DOS prompt such as C:\. If you have a menu program, you must exit it to DOS. You start Windows from DOS by typing WIN and pressing the Enter key. A window should appear on the screen with a title bar stating Program Manager. If you always want Windows to run when you turn on your computer, add the two lines PATH=C:\WINDOWS;%PATH% and WIN to your AUTOEXEC.BAT file, and reboot.

## Exiting the Windows Program

Close all open application windows except Program Manager, then choose Close from the Program Manager menu.

## MANIPULATING A WINDOW

## Window Elements

Examine Figure 10.1 to learn to identify the elements of a window. In the top left corner, each window has a negative sign in a box called the **control menu box.** This is used to "control" the window, as we shall see later. In the top middle of each window is a **title bar** so you know what window you are in. In the right top corner of the window are little up and down arrows or triangles called **sizing buttons.** Windows **borders** on each side and top and bottom are used to change the size of the window. The inside area of a window is called the **work area. A vertical scroll bar** on the side of some windows allows you to scroll through

## FIGURE 10.1 Elements of a WIN 3.11 window

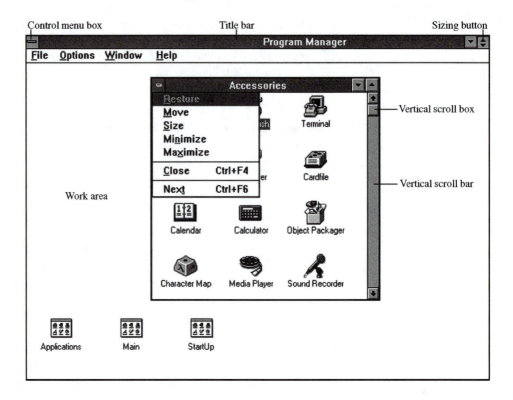

the contents that cannot be shown all at once in the window's current size. You can scroll by using the up and down arrows or by grabbing and dragging the **vertical scroll box.** Some windows have a **horizontal scroll bar** at the bottom.

## Opening and Closing a Window

A window is opened by double-clicking on the icon representing a program. You should see a rectangular window appear on the screen. You would now work in that window and, when done, close it. To close the window, click once on the control menu box. The menu to control the window appears in a rectangle, giving command choices of Restore, Move, Size, Minimize, Close, and Next or Switch To. If a command is in gray print instead of black, it is not allowed at the current time. You close the window by clicking once on the control menu box, and once on the Close command. You may also close the window by just double-clicking on the control menu box.

## Maximizing a Window

Open a window by double-clicking on an icon, such as the Clipboard Viewer. The window opens to a certain size, in a certain location on the screen. Click on the window control menu box to bring up the window control menu. Clicking on the Maximize option of this menu makes the window fill the entire display screen, so some work is easier to see and the window is less cluttered. You can also maximize the window using the sizing buttons at the top right corner of the window. The up arrow maximizes the window.

## Minimizing a Window

In an open window, click on the control menu box in the top left corner. From the menu select the Minimize command. The down arrow on the double arrow sizing button minimizes the window to a small icon, shrinking it temporarily out of the way and positioning it outside of the Program Manager. It is still open, but it is in the background, and you can switch to it quickly as needed. It will retain any data you were working on. The sizing buttons in the top right corner of the window can also minimize the window. The single down arrow minimizes the window also.

### Restoring a Window

If a program is already minimized, clicking the mouse once on the minimized icon will bring up the control menu. The Restore option of this control menu makes the window go back to the size it was before it was minimized. You can also double-click on the minimized icon to restore the window to its former size.

If a window is running maximized, the top right corner has a little double arrow, pointing up and down in the same square. The little down arrow is the Restore arrow. Clicking on it restores the window to the size it was before maximizing. You can also click once on the control menu box of a maximized window, then click on the Restore command.

### Moving a Window

The Move option lets you relocate a window by dragging it where you want it. When you release the mouse button, the window will stay at that position.

You may also move the window by just dragging on the title bar. Move the mouse, and you will see that you can move the whole window to any location you like. Let go of the button to leave the window where you want it.

### Sizing a Window

If you move the cursor to the bottom right corner and get the cursor arrow point exactly on the border corner, the cursor changes from an arrow to a two-headed arrow at a 45-degree angle. Now click and hold the left mouse button. This is like grabbing the corner of the window. Drag this corner and watch the size of the window change, being trimmed or expanded from the bottom right corner. The top and bottom and both side borders work exactly the same way to allow you to change the window to any size you like.

### Switching Windows

Switching between windows is done by clicking anywhere in the window you want to run. That window becomes active. You can also click on the control menu box and choose the Next or Switch To command. The Next option chooses the next group in the Program Manager. The Switch To option of the control menu gives you a choice of which active program to switch to. If a window gets hidden behind others, holding down the Alt key and pressing Tab repeatedly cycles through the active programs. When you release the Alt key, the program shown will be the active program.

## USING THE MENU BAR

You have already used the control menu box and commands. Each program also has a horizontal menu bar. Somewhere in the opening window of Windows, at least four icons should appear—Main, Accessories, Games, and Startup. These are group windows for holding groups of similar programs.

Double-click on the Main program icon. In the Main group window, double-click on the File Manager program (its icon is a two-drawer file cabinet). A window should appear with a menu bar under the title bar that displays File, Disk, Tree, View, Options, Window, and Help commands. To select a menu, hold down the Alt key and press the underlined letter you choose, or just click once on your choice. A drop-down menu of commands should appear beneath the subject. To leave that menu without doing anything, either select another, or press ESC. To select from the menu that appeared, click on your selection or press the underlined letter; no Alt key is needed this time.

To leave File Manager at this time, Press ESC until you are at the beginning File Manager window, and then press Alt-F for file, and X for exit, or click on File and then Exit. You can also double-click on the control menu box or click on the control menu box once, then click on Close. This should bring you back to the original screen you had when Windows started.

## RUNNING MULTIPLE PROGRAMS

For an example of opening multiple programs in different windows, open Main by double-clicking on its icon, then double-click on the File Manager's icon to open it. Size this window to about 2 × 3 inches by dragging the borders. Now click and hold on the title bar in the top row and drag the window to any location you like.

To open a second program at the same time, double-click on README.WRI. This is a small information file for you to read. For now we are just using it as another window. Grab the bottom right corner and make it smaller. Grab the title bar, and drag this window so that both it and File Manager are visible (they may overlap).

Now double-click on Control Panel. If it is hidden under other windows, move those windows aside. When it appears, make it smaller and move it where you want it. Now click anywhere in any of these windows, and watch the title bar become blue or dark. This means this window is active now. You can click in any window and work in it, then click in another and work in it.

## PROGRAM MANAGER

Now that you can get into and out of Windows, let's look at a few more things we can do. Windows opens in the Program Manager. Program Manager has at least four groups: Main, Accessories, Startup, and Games. The Program Manager menu bar shows choices of File, Options, Window, and Help. Figure 10.2 shows the Program Manager window.

Since Help may save you time and frustration, let's look at it first.

### Help Menu

Clicking on Help brings a drop-down menu with commands of Contents, Search for Help On, How to Use Help, Windows Tutorial, and About Program Manager. Click and read all of these commands so you will be familiar with them when you really need them. If you are new to Windows, the Tutorial is very useful. It includes a mouse tutorial.

### File Menu

The Program Manager File menu includes Open, Move, Copy, Delete, Properties, Run, and Exit.

    Open        The Open command runs a highlighted program. For example, Open the Games group by double-clicking on the Games icon. Now click once on the Minesweeper icon, then on the Program Manager File Open

**FIGURE 10.2    The Program Manager window**

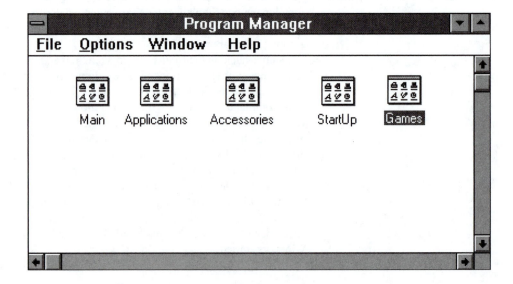

|            | command. This runs the game. Exit Minesweeper like all windows, by double-clicking on the games control menu box. You can open the game or any program more quickly by just double-clicking on its icon. |
|------------|---|
| Move       | The Program Manager File Move command moves a highlighted program icon from the group it is in to another group. You can also just click and drag the icon. |
| Copy       | The Copy command is similar to Move, but it leaves a copy in the original position. (Click and drag will not work to copy.) |
| Delete     | The Delete command removes a highlighted program from a group, but it will ask for verification first. |
| Properties | The Properties command shows information on the highlighted program. It allows you to pick or change an icon, and it includes many to choose from. From the Program Manager File menu, choose Properties. In the Properties window choose Change Icon, then Browse. A filename will be shown. Press the backspace key to delete this filename, type C:\WINDOWS\MORICONS.DLL, and press Enter. Icons should appear in the workspace. You can click on the arrow on the right of the horizontal scroll bar to scroll through the available icons. You may choose an icon not used for your program. |
| Run        | The Run command runs a highlighted program. |
| Exit       | The Exit Windows command closes things down properly and exits Windows. It does ask for verification first. |

## Options Menu

The Program Manager Options menu includes: Autoarrange, Minimize on Use, and Save Settings on Exit.

| Autoarrange          | This automatically arranges the icons in the active window in a tile pattern. |
|----------------------|---|
| Minimize on Use      | This makes Program Manager minimize to an icon at the bottom of the screen when you start another program. |
| Save Settings on Exit | This will remember what you've set up, and Windows will open with those settings the next time. |

## Window Menu

The Program Manager Window menu includes Cascade, Title, and Arrange Icons.

| Cascade       | This arranges the groups one behind the other with only the top left corner showing, like a hand of cards stacked behind one another. |
|---------------|---|
| Tile          | This arranges the groups side by side like the ceramic tiles on a kitchen floor. |
| Arrange Icons | This automatically neatens the arrangement of the icons in a group when you have moved one. |

## GROUPS

Windows groups programs together by function for example, all games in one group or all Financial programs in a Financial group. The Program Manager File menu allows you to add a program to a group and to make new groups. If you have a number of astronomy programs, you might add a group called Astronomy, then add those programs to that group. If you get a Windows game, you might add that program item to the existing Game group. Windows assigns programs to different groups. To run programs you must know how to work with groups.

## Opening Groups

To open a group you can click once on the group icon, then click on Restore, or you can double-click on the group icon.

### Running a Program in a Group

To run a program in a group, open the group, then double-click on the program.

### Exiting a Program in a Group

To exit a program, double-click on the program's menu control box.

### Adding a Program to a Group

To add a program to a group, first double-click on the File Manager icon in the Main group. Click on the drive the program is on. Click in the left workspace on the directory. In the right workspace, click and drag the filename wanted to the group you want it in. The file must be a Windows program that ends in an extension of .EXE, .COM, or .BAT.

### Deleting a Program from a Group

To delete a program from a group, click once on the program icon to highlight it, then click on the Program Manager File menu and choose the Delete command. It will ask for verification before deleting.

### Moving a Program from One Group to Another

To move a program to a different group, click once on the program icon to highlight it, then click on the Program Manager File menu and choose the Move command. You can also just drag the program icon to another group.

### Copying a Program from One Group to Another

To place a copy of a program in another group, click once on the program icon to highlight it, then click on the Program Manager File menu and choose the Copy command.

### Closing a Group

You close a group as you close any window, by clicking on the menu control box and then on Close or by just double-clicking on the menu control box.

## 10.10 WINDOWS MAIN GROUP

The Windows Main group has the programs more commonly used to set up and manipulate the system. Figure 10.3 shows the Main window.

### FILE MANAGER

The file cabinet icon is for File Manager, the program that can manipulate your files. The File Manager menu bar contains File, Disk, Tree, View, Options, Window, and Help. There are pictures of drives to allow you to select any drive that has the files you are interested in working with. A graphical tree, like a family tree, is shown. The arrows at the top and bottom and left and right of the scroll bars allow you to scroll through the files if there are too many to fit in the window.

### Selecting Files

File Manager can work on one or more files. You can select one file by clicking on it; it will be highlighted. To select two or more files in different positions (names not contiguous), click on one, move to another, hold down Ctrl, and click on that file. Move the cursor to another filename, hold Ctrl, and click on that file. Now click on Delete. You will be asked to confirm before it deletes anything.

FIGURE 10.3    The Main group window

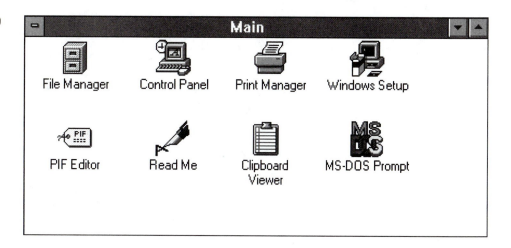

To select a contiguous group of files, click on the first, hold Shift, and click on the last of the group. All files between the two files will be selected. If you change your mind, you can deselect by clicking on a different file. To select all files in a list, page up to the top of the file list, hold Shift, then page down to the bottom of the list.

### File Manager Menu Bar

The File Manager File menu has commands to Open, Move, Copy, Delete, Rename, Run, Print, and Search files and to see their Properties. You may click on one filename, then on Delete to erase the selected file. The Disk menu allows you to Copy, Label (name the disk), Format, and Transfer the System to floppies. The Tree and View menus allow you to change how these filenames are graphically displayed. This program has many functions, and you need to spend time playing with it.

## CONTROL PANEL

The Control Panel has many options. They are all related to controlling how Windows interfaces with your computer system. You can control the colors the windows are seen in, add fonts for character display, set up your serial ports, and customize how your mouse and keyboard work. You can change your date and time and how they are displayed.

Desktop allows you to change the colors of everything you see in Windows, to select a wallpaper background design, and to select a screen saver to blank out the screen after a time you choose. Printers lets you install printer drivers and set which printer is the default printer. 386 Enhanced allows you to change the settings for enhanced mode running, including setting the size of your virtual memory (using the hard drive to hold information when there is too much for your RAM). Drivers allows you to set up drivers for your audio and video. Sound allows you to set what sounds will be heard for certain events, such as errors and starting and exiting Windows.

## PRINT MANAGER

Print Manager allows you to see what printer is set as the default and what files are in the print queue.

## CLIPBOARD VIEWER

Clipboard Viewer allows you to see what has been cut or copied from the Edit menu of a window to the clipboard. Pressing Alt+PrintScrn will copy the current active window into the clipboard.

## MS-DOS PROMPT

MS-DOS Prompt allows you to run a DOS, non-Windows program from Windows. Holding down Alt and pressing Tab repeatedly cycles you through the open windows. It stops at the window you are in when you release the Alt key. Alt+Enter creates a partial window for the DOS program, making it easy to move between windows.

## WINDOWS SETUP

In Windows Setup you can change the settings for your display, keyboard, mouse, and network. This is used when you install Windows or new devices.

## PIF EDITOR

The PIF (Program Information File) Editor allows you to examine these files and change the graphic mode, memory allocated (conventional and extended), and include any parameters the program needs.

## READ ME

This file has extra information about Windows that is not in the manual.

## 10.11 WINDOWS ACCESSORIES GROUP

One group in Program Manager that comes with Windows and has many helpful programs is Accessories. Figure 10.4 shows the Accessories window. Some of the more interesting or useful programs are briefly explained below.

## WRITE

Double-clicking on the fountain pen icon starts a word processor. It is far too complex to explain well here, so this is just an introduction. Figure 10.5 shows the Write window.

You can just start typing in the words you want in your document. As you approach the end of a line, just keep typing. The program will automatically jump the cursor to the next line at the right time. This is called word wrap. Menu bar topics are File, Edit, Find, Character, Paragraph, Document, and Help. Just type something and experiment with the

**FIGURE 10.4    The Accessories group window**

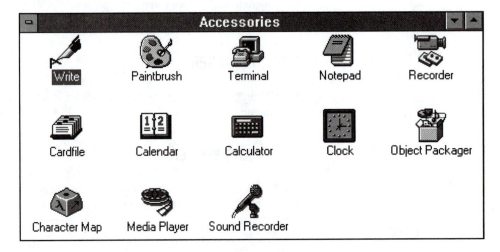

**FIGURE 10.5** The Write word processor window

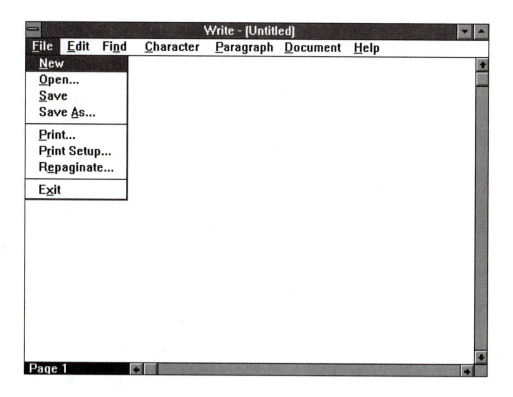

choices. When you're finished typing, click on File, then on Save As and type a name. This is only needed the first time you save the file. After that, choose File, Save.

To modify or add to an existing file, use File, Open, and choose the filename. To start a new document, choose File, New, but be sure you save the old file first.

To print to a printer, choose Print Setup from the File Menu and then Print from the File menu.

Under the Character menu, the Font choice allows you to choose letter style (font) and size. You can set top, bottom, and side margins using the Document menu Page Layout command.

You may choose to number pages under the Document menu Footer command. Select Insert Page#. While this window is open, select the Paragraph menu, then left, centered, or right for your page number position. Finally, select Return to Document.

Under the Document menu Ruler option, you can display icons for tabs, single or double spacing, justifying, or centering. Experiment with the other topics and choices.

## Inserting a Drawing in a Write Document

In Write, position the cursor where you want the picture. Pictures take whole lines, no matter how small they are. In the menu bar, click on Edit, Insert Object, Paintbrush Picture, and OK. The Paint program will automatically start.

Draw a small picture as explained in the Paint program description. (*Note:* Even a small picture takes many kilobytes of file space.) Now, use the square scissor tool to drag a square around your drawing, selecting it. From the Paint menu bar, select Edit, Copy, then File, Exit. Say No to the update request. Paint will close, and you will be back in Write. From Write's menu bar, select Edit, Paste. The picture should appear.

If you use the arrow keys to move up and down your document's lines, note that the picture darkens when selected and that it takes up the whole line. To later modify the picture, double-click on it. Paint opens with the drawing loaded. Modify the picture, then click on File, Update, File Exit, and Return. You will be back in Write, with your modified drawing.

If you wanted to, you could save your drawing as a separate drawing while in Paint by selecting File, Save As, and naming it, before exiting and returning to Write. In Write, you can move or change the size (and proportions) of your drawing by clicking once on it to select it, then on Edit and Move or Size Picture.

### Inserting a Sound in a Write Document

If your computer system has a sound card and speakers (and a microphone), you can place a sound in your file. It appears as an icon in the document, but when you double-click on the icon, the sound is heard.

Start by positioning the cursor where you want the sound icon to appear. From the Edit menu, select Insert Object. In the Insert Object window, arrow down on the vertical scroll bar to Sound. Click on Sound and OK. This automatically opens the Sound Recorder, which is just like a little audiocassette recorder. Click on the microphone icon to record, and speak into the real microphone. Click on the black square to the left of the mic icon to Stop the recorder. By clicking on Rewind, and Play, you can listen and repeat to get it right. With the Edit and Effects menus, you can increase or decrease volume or speed and add "echo."

When your sound is right, select File, Update, and File, Exit. The Sound Recorder window should close, and you are back in Write with a mic icon in the document. Double-click on the icon to hear your sound.

### Inserting a Video in a Write Document

A video file with the extension .AVI can be inserted just like a sound file, except it uses the media player. Test your system with the video file and the media player before trying to insert it. In Write, position the cursor where you want the video to appear. Choose Edit, Insert Object, then click on Media Clip and on OK. The Media Player opens. Choose an .AVI file, Edit, Update, Edit, Exit, and Return. The video's first frame will be seen in the document. Double-clicking on the picture will run the video, playing any sounds recorded with it.

### Inserting a Write Document in a Write Document

In Write, position the cursor where you want the Write document to appear. Click on Edit, Insert Object, Package, OK. This brings up the Object Packager. Click on Contents, File, Import, and type the name of the file you want to insert. It should end in .WRI.

Now click on File, Update, File, Exit. The pen icon for Write should appear in the document. Double-clicking on the icon opens a second Write program with the old document presented. To leave the second Write program, File, Exit. This returns you to the first document.

If you draw your own icon, you could use this feature to have your own help button (icon) in your document.

### Creating Your Own Icon for an Object Package in a Write Document

You can create your own icon for an object package. Alt+Tab to the Program Manager and open Sound Recorder. Record a phrase, such as "Have a nice day." Save it under a filename with the .WAV extension, such as HAVEA.WAV. Exit the Sound Recorder.

Open Paint. Draw a small picture, perhaps a smiley face. Use the square scissor tool to select it, and Edit, Copy it so it is in the clipboard. File, Exit Paint.

Now open the Object Packager. Click on Content, File, Import, and type the sound file, FILENAME.WAV. This should appear to the right in words. Now click on Appearance, Edit, Paste. The drawing you made for an icon should now be on the left, and the words for the sound file on the right. Now Edit, Copy Package, File, Exit.

Finally, open Write, position the cursor where you want the object with your custom-made icon, then click on Edit, Paste. Double-clicking on your custom icon (smiley face) should make it say "Have a nice day."

### PAINTBRUSH

Clicking on the artist's palette runs the Paintbrush program. This creates bitmapped image files with the extension .BMP. Figure 10.6 shows the Paintbrush window.

**FIGURE 10.6  The Paintbrush window**

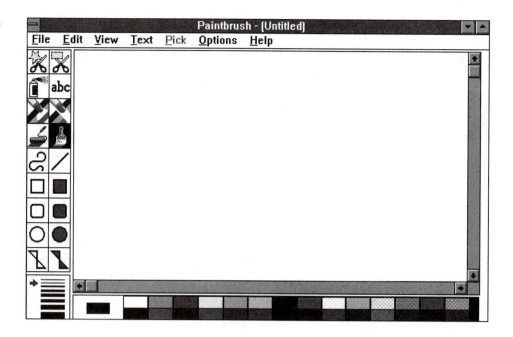

## Choosing Colors

Click the left mouse button on a color in the color bar at the bottom of the screen. The inner square to the right turns that color as the primary or foreground color. Click on another color with the right mouse button, and the outer square turns that color as the background color.

## Choosing Line Width

Click on the fat line at the bottom of the column of different width lines to choose the cursor or brush width. The line width selections in the bottom left affect curves, lines, erasers, the perimeters of the shapes, the brush, and the airbrush.

## Drawing Figures

Click on the Filled Square icon, in the second column, sixth row. Now go onto the painting area and drag. When you release the button, you should have a square filled with the foreground color and framed by the background color. The Unfilled Square draws a frame only and in the foreground color, opposite of the Filled Square. The Round-Cornered Square and the Circle work just the same.

To use the Polygon at the bottom, click and drag the first line, then either click and drag or just move and click the other sides. You may have as many sides as you like, and it will finish automatically when you finally click on the first point again, creating a closed polygon.

Above the Filled Square icon is the Line icon. To draw a line, just click and drag. To the left of the Line icon is the Curve icon. Click and drag, as you did for the line, but then go somewhere on that line and click and drag one or two more times. The line curves toward the cursor. It stops after two clicks or when you click on the end of the line.

## Painting

Above the Curve is the Paint Roller. Placing the bottom tip of it in a closed figure and clicking totally paints the inside of the figure the foreground color.

To the right of the Paint Roller is the Brush. Just drag it around and it leaves a trail of paint. Above the Color Eraser is the Airbrush, or spray paint can. As with the Brush, you just drag it around and it sprays. Staying in one area longer or more often will make it darker.

### Erasing

Above the Paint Roller is the Color Eraser, and to the right of that is the Eraser. They both leave behind the background color, but the Eraser erases all colors, while the Color Eraser only erases the foreground color selected.

### Writing Text

The ABC icon to the right of the Airbrush is Text. After clicking on the ABC icon, click on Text from the menu bar. This gives you a number of striking effects. Under Fonts you can choose the font, style, and size, creating very large or small print.

### Moving, Copying, Cutting, and Pasting

The Scissors icon with the star allows you to cut a jagged piece of the painting. Just make sure you cross the start point of the cut to get a closed figure. Then click on Edit in the menu bar, and you can Cut, Copy, or Paste that cut piece.

Pasting always puts the cut piece in the top left corner, but you can move it by dragging. Clicking outside the area finishes the process. You can paste as often as you want; the figure is stored in the clipboard.

To the right of the scissors is the Pick icon, another pair of scissors with a square. It is used in the same way. The area you picked can be manipulated by choosing Pick from the menu bar. It allows you to flip horizontally or vertically, inverse the colors, shrink or grow, and clear the background.

### The Menu Bar

The **File** menu gives the same choices as other Windows programs: a New file, Open an old file, Save the file already named, Save As a new filename, Print the file, Setup the printer, or Exit the program.

The **Edit** menu allows you to Undo what you just did, Cut out an area selected with the Pick or Scissors icon, Copy that area without erasing it, Paste what was cut or copied from the clipboard, and Copy To or Paste From files.

The **View** menu allows you to Zoom In to the pixel level and change each dot of the picture. Zoom Out is the opposite of this, getting you back to normal. Picture displays the picture on the full screen. A mouse click brings you back to normal. The last three commands are to show or not show the Tools and Linesize, color Palette, and cursor position.

The **Text** menu allows you to choose Regular, Bold, Italic, Underlined, Outlined, or Shadowed characters. You may choose from a dozen or more Fonts, and select the point size of the characters.

The **Pick** command allows you to Flip the cutout made with the Pick or Scissors icon horizontally or vertically, inverse the colors, shrink or grow, and clear the background.

The **Options** command allows you to change attributes such as picture size or whether to use black-and-white or color, to change the brush shapes, and even to change the colors in the color bar.

Commands on the **Help** menu are Contents, Search, How to Search, and About. These are the same as with other programs.

### TERMINAL

The Terminal program in the Accessories group works with your modem, allowing you to connect over the phone lines with other computers. The menu bar includes File, Edit, and Help, which are the same as in all Windows programs. Settings, Phone, and Transfers are unique.

Settings lets you set your modem up to communicate properly with the one on the other end. Baud speed is set here, as well as what kind of terminal your computer will act like. The Phone menu has commands to Dial and to Hang Up. The Transfers menu allows you to send and receive text files or binary (program) files.

## NOTEPAD

The Notepad program is a tiny word editor. It is very limited in function. The menu bar includes File, Edit, Search, and Help with the standard commands. One command to be aware of is Word Wrap, under the Edit menu. Without this on, each line you type keeps getting longer until you press Enter. With Word Wrap on, the lines automatically break on the right and start new lines.

## RECORDER

The Recorder is a Macro recorder. This means it memorizes what you do in Windows. From the menu bar titles of File, Macro, Options, and Help, choose Macro. Now choose the command record. Type in a name. The program minimizes, and the icon blinks to show you it is recording. Now run the Calculator, and exit, and perhaps the game Solitaire. Now exit and click on the blinking Recorder icon. Click on OK to Save the macro. Now click on Macro, Run, and watch a replay of what you did. The macro is saved in a file with an extension of .REC.

## CARDFILE

The Cardfile program under Accessories creates a database that can be displayed as a 3 × 5 card file. It can be used to store recipes, class members, or anything.

The menu bar choices in all Windows programs are quite similar. File lets you Open, Save, and Print files, and Exit the program. Edit allows Cut and Paste and more. View gives a List, or a cardfile look. Search lets you find a card or data record quickly by searching for a word, name, ZIP code, phone number, or anything. Help is the same as always.

## CALENDAR

The Calendar program allows you to keep track of appointments, birthdays, and so on. It even includes a reminder alarm. You can set the calendar under the View menu to show the hours of one day only or the days of the month. The Alarm command lets you Set the alarm, or Control the alarm to beep or not and to tell it how long of a warning to give you.

## CALCULATOR

Double-clicking on the calculator icon runs the Calculator program. Menu topics are Edit, View, and Help. Edit allows you to copy the number in the calculator display to the clipboard. This is a temporary holding place. You can switch windows, for example to the Write word processor, then Paste from the clipboard to where the cursor is located. This allows you to move the numbers you calculate from the calculator to another application. You also can copy a number from another application, and paste it in the calculator's display.

The View menu allows Scientific and Standard views. If you are unfamiliar with scientific calculators, use the Standard view at first. Click on numbers and operation signs (+, =, -, *, /) just as you would on a normal calculator. For example, 35 + 27 = will give 62 in the display.

The C button is for clear all. The CE is for clearing only the last number typed; the previous numbers remain. The Back button allows you to back up the display if you enter a wrong digit. MS is memory save, and it is not affected by Clear. When you save a number to memory, an M appears in the empty box by Back to show you the memory has a value in it. MR is memory recall, and it does not affect memory. M+ means to add the display value to the number in memory. MC clears the memory and removes the M from the box.

Sqrt gives the square root of a number, 1/x gives the inverse or reciprocal of a number, and % uses the number as a percent. For example, 4.00 * 75% gives 3.00, showing 75% of $4.00 is $3.00.

The Help menu has Contents, Search, How to Use Help, and About Calculator Choices. Contents gives some general topics, Search allows you to type the subject of your question and searches for it through a list of topics, which you can also scan. How to Use Help tells just that, and About gives a little information on the program.

## CLOCK

When you double-click on the Clock icon, a window with a clock appears. There is only one menu, Settings. Clicking on this will drop down a menu including Analog, Digital, Set Font, No Title, Seconds, Date, and About Clock.

Analog chooses a clock with hour and minute hands. The date is displayed in the title bar. Set Font is not active in this mode. Digital chooses a numeric display, including date. Set Font is active and can be chosen to display different letter styles.

No Title removes the title block to give a cleaner, less busy look to the clock display. A double-click in the window brings back the title bar. Seconds works on either display mode and adds or removes seconds for a cleaner display. Date adds or removes the date for a cleaner display also. In Analog mode, the date is in the title bar.

About Clock tells information about the program. As in all windows, the sizing buttons (arrows) are at the top right corner of the window. If minimized, click on the icon, and there is an Always On Top choice that keeps the time always visible. Click and drag by the title bar to an unused portion of the screen. This is a good choice to add to your Startup Group, so you always have the time and date on screen.

## OBJECT PACKAGER

The Object Packager program allows you to take sounds, pictures, or documents and associate them with icons. Then you can just double-click on the icon in your document, and the object appears. For instance, in Write, you can embed a sound object, another document object, and a paint object. When you then double-click on the icon in your Write document, the sound sounds, the document appears, and the drawing appears.

First identify or create the sound file (.WAV), another short Write document, or a Paintbrush picture file (such as PICTURE.BMP). Then double-click on the Object Packager icon. Click on File, Import, then type in the filename of the file you created. Now click on Edit, Copy Package. The object is now in the clipboard.

Start Write, Open the document to receive the object, place the cursor where you want the object to be, and Edit, Paste. Don't forget to File, Save. The icon from Object Packager should appear. Double-click on it, and Paintbrush starts and the picture appears. You must Exit Paintbrush to end the object presentation.

## CHARACTER MAP

Character Map allows you to see special characters that are available and to put them in a document. For example, start Write. Now start Character Map. Scan through fonts, and see the characters available, such as fractions and Greek letters. Click on a character, then Select, and Copy. This gets it in the clipboard. Now click on Write, and Edit, Paste. If the font in Write is different from the one you chose in Character Map, select the character, and change the font under Character, Font. The character will then appear as it did in the Character Map.

## MEDIA PLAYER

Media Player is like an audiotape player. You must have a sound card for audio and a MIDI device for the MIDI files. Media Player plays sound files (.WAV), video files (.AVI), and MIDI files (.MID). The Play, Pause, and Stop buttons work like those on a normal tape recorder.

## SOUND RECORDER

Sound Recorder is also like a tape recorder. Click on the microphone icon, and talk into the microphone connected to your sound card. The menu bar has File, Edit, Effects, and Help menus. Effects lets you Echo or Reverse the sound.

## 10.12 WINDOWS STARTUP, APPLICATIONS, AND GAMES GROUPS

### STARTUP GROUP

Placing a program in this group makes that program come up when Windows is started. Clock may be a good one to always have running and visible.

### APPLICATIONS GROUP

When you install new Windows programs, place them here or create new groups.

### GAMES GROUP

Windows comes with two games already installed in the Games group: Solitaire and Minesweeper. You open them as you do all Windows, by double-clicking on the icon. From there on, you can figure these out. They of course have the standard Windows Help menu.

### TROUBLESHOOTING WINDOWS 3.X

The preceding pages were just an introduction to using Windows. For those who choose to become familiar with its inner workings and learn to troubleshoot Windows problems, CyberMedia has a product called First Aid for Windows Users. It detects and corrects many software problems that occur when using different programs in Windows.

## 10.13 WINDOWS 95

Installing: To install Windows 95 on a system you have just built, your system must have a CD-ROM drive, and you must install the drivers to run the drive to access the Windows CD.

### TROUBLESHOOTING WITH WINDOWS 95

If the system will load Windows, clicking Start, Help, Index, and Troubleshooting will give about twenty help topics on troubleshooting various components of your system.

Microsoft Diagnostics (MSD) is missing, but Start, Settings, Control Panel, System (see below) give all the same information and more, although not as directly and succinctly as MSD did.

Press F8 when "Starting Windows 95," and choose selection 4 to go step by step through the CONFIG.SYS and AUTOEXE.BAT files to see problems.

Press F8 and choose selection 2, "logged," to boot and write each event to a log named BOOTLOG.TXT in the root directory C:\. Then Notepad can be used to read the file, and Notepad, Search, Find "failed" to find any event that failed.

Start, Run, Sysedit gives access to the CONFIG.SYS AUTOEXEC.BAT, CONFIG.SYS, WIN.INI, SYSTEM.INI, and PROTOCOL.INI files. Start, Run, Sysedit gives access to the registries.

To find and download drivers for various video cards, CD drives, and so on, go to www.microsoft.com for the Windows driver library. Private vendors also have drivers on their Web sites for downloading their own products. The handy DOS HELP command is missing in Win95. Familiar and handy files under the Windows directory include:

```
DEFRAG   EXE      241,600 07-11-95  9:50a DEFRAG.EXE
DRVSPACE EXE      336,736 07-11-95  9:50a DRVSPACE.EXE
EMM386   EXE      125,495 07-11-95  9:50a EMM386.EXE
SCANDSKW EXE        4,608 07-11-95  9:50a SCANDSKW.EXE
SETDEBUG EXE       36,352 01-31-97  8:24a SETDEBUG.EXE
SETVER   EXE       18,939 10-23-97  6:51a SETVER.EXE
SMARTDRV EXE       45,145 10-23-97  6:33a SMARTDRV.EXE
WINVER   EXE        3,632 07-11-95  9:50a WINVER.EXE
```

Files under Windows\command directory include:

```
ANSI     SYS        9,719 07-11-95  9:50a ANSI.SYS
ATTRIB   EXE       15,252 07-11-95  9:50a ATTRIB.EXE
CHKDSK   EXE       27,248 07-11-95  9:50a CHKDSK.EXE
CHKLIST  MS          999 11-22-97  2:28p CHKLIST.MS
CHOICE   COM        5,175 07-11-95  9:50a CHOICE.COM
COUNTRY  SYS       27,094 07-11-95  9:50a COUNTRY.SYS
DBLSPACE SYS       15,831 07-11-95  9:50a DBLSPACE.SYS
DEBUG    EXE       20,522 07-11-95  9:50a DEBUG.EXE
DELTREE  EXE       19,019 07-11-95  9:50a DELTREE.EXE
DISKCOPY COM       21,959 07-11-95  9:50a DISKCOPY.COM
DISPLAY  SYS       17,175 07-11-95  9:50a DISPLAY.SYS
DOSKEY   COM       15,431 07-11-95  9:50a DOSKEY.COM
DRVSPACE BIN       71,287 07-11-95  9:50a DRVSPACE.BIN
DRVSPACE SYS       15,831 07-11-95  9:50a DRVSPACE.SYS
EDIT     COM       69,886 07-11-95  9:50a EDIT.COM
EDIT     HLP       10,790 07-11-95  9:50a EDIT.HLP
EGA      CPI       58,870 07-11-95  9:50a EGA.CPI
EXTRACT  EXE       46,656 10-23-97  6:33a EXTRACT.EXE
FC       EXE       20,494 07-11-95  9:50a FC.EXE
FDISK    EXE       59,128 07-11-95  9:50a FDISK.EXE
FIND     EXE        6,658 07-11-95  9:50a FIND.EXE
FORMAT   COM       40,135 07-11-95  9:50a FORMAT.COM
KEYB     COM       19,927 07-11-95  9:50a KEYB.COM
KEYBOARD SYS       34,566 07-11-95  9:50a KEYBOARD.SYS
KEYBRD2  SYS       31,942 07-11-95  9:50a KEYBRD2.SYS
LABEL    EXE        9,260 07-11-95  9:50a LABEL.EXE
MEM      EXE       32,082 07-11-95  9:50a MEM.EXE
MODE     COM       29,191 07-11-95  9:50a MODE.COM
MORE     COM       10,471 07-11-95  9:50a MORE.COM
MOVE     EXE       27,235 07-11-95  9:50a MOVE.EXE
MSCDEX   EXE       25,473 07-11-95  9:50a MSCDEX.EXE
NLSFUNC  EXE        6,940 07-11-95  9:50a NLSFUNC.EXE
SCANDISK EXE      134,738 10-23-97  6:33a SCANDISK.EXE
SCANDISK INI        7,270 07-11-95  9:50a SCANDISK.INI
SHARE    EXE       10,304 07-11-95  9:50a SHARE.EXE
SORT     EXE       25,802 07-11-95  9:50a SORT.EXE
START    EXE        9,216 07-11-95  9:50a START.EXE
SUBST    EXE       17,904 07-11-95  9:50a SUBST.EXE
SYS      COM       13,239 07-11-95  9:50a SYS.COM
XCOPY    EXE        3,878 07-11-95  9:50a XCOPY.EXE
XCOPY32  EXE       40,960 07-11-95  9:50a XCOPY32.EXE
```

Luckily commandname /? still works for help, such as "DIR /?" to see help for the directory command.

The Windows directory has a number of text files (*.TXT) that contain valuable information. For example, the DISPLAY.TXT file has information on finding and installing drivers.

## GENERAL WINDOWS 95 INTRODUCTION

Windows 95 gave Windows a new look and feel. There is a task bar across the bottom of the screen, with the time at the right. In pull-down menus, the command is sometimes followed on the right by keyboard shortcut commands. For example, Alt-F4 is Exit, Ctrl-x is Cut, Ctrl-c is Copy, Ctrl-v is Paste, and Ctrl-z is Undo. Ctrl-Esc pops up the Start Menu from the taskbar, as shown in Figure 10.7.

Let's glance at each selection on the Start Menu.

## START—SHUT DOWN

Click on Start, then Shut Down to get three choices: Shut down the computer, Restart the computer, Restart the computer in DOS mode. When you are ready to turn off your computer, always use the Shut down the computer choice, and wait for Windows to say your system can be turned off safely.

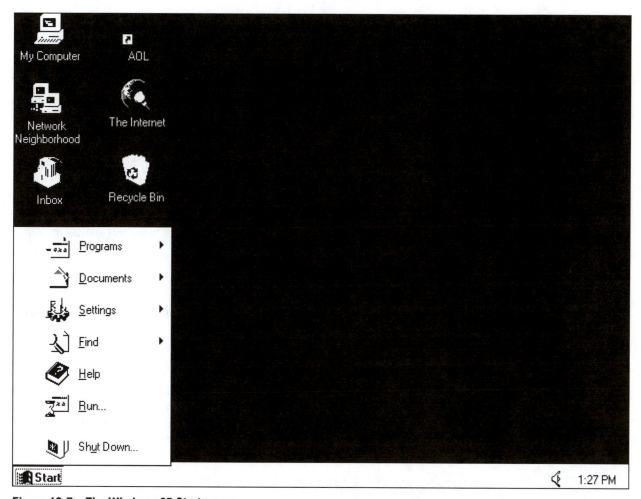

**Figure 10.7   The Windows 95 Start menu**

## START—RUN

Click on Start, then Run, and fill in a filename or browse a hard drive or the CD drive to find a program you want to run. Usually, this will be when installing a program from the CD player or floppy drives.

## START—HELP

Clicking on Start, then Help gives you a help screen with choices for Contents, Index, Find. Contents gives general topics. Index lets you type and finds letter by letter the topic you want, if it is in the index. Find goes a little deeper; it does a word search for any word you type.

## START—FIND

Click on Start, then Find to enter a search program that allows you to find out if a certain file or folder (directory) exists, and where it is on your system.

## START—SETTINGS

Click on Start, then Settings to access the Control Panel, Printers, and Taskbar.

## SETTINGS—CONTROL PANEL

Similar to previous versions, Control Panel gives you some control and insight into your hardware and software drivers. Selections include: 32bit ODBC, Add New Hardware, Add/Remove Programs, Date/Time, Display, Fonts, Internet, Joystick, Keyboard, Modems, Mouse, Multimedia, Network, Passwords, Printers, Regional Settings, Sounds, and System (Figure 10.8).

Control Panel includes many functions useful in setting up, maintaining, and repairing your system. A brief glance at each follows.

## CONTROL PANEL—32BIT ODBC

This is a data base connection interface with information on how to connect to data providers.

## CONTROL PANEL—ADD NEW HARDWARE

This program allows you to install drivers for hardware manually, or it will attempt to do it automatically.

## CONTROL PANEL—ADD/REMOVE PROGRAMS

Installing and especially uninstalling programs is much easier with this program. It keeps track of those little control files and lines added to many files, and it removes them when you uninstall a program.

## CONTROL PANEL—DATE/TIME

As you can guess, this lets you change the date and time. It also allows you to select a time zone.

**Figure 10.8 Windows 95 Control Panel**

### CONTROL PANEL—DISPLAY

Under Control Panel, a double-click on Display allows you to see and adjust the background Pattern and Wallpaper. The screensaver tab allows you to set your screensaver. The Appearance tab allows changing the colors of your windows. The Monitor refresh allows you to change refresh rates, and the Settings tab allows changing number of colors and resolution. Of course that is limited by your hardware.

### CONTROL PANEL—FONTS

Shows installed fonts and allows you to install new fonts.

### CONTROL PANEL—INTERNET

This gives six tabs for setting parameters: General, Connection, Navigation, Programs, Security, and Advanced.

### CONTROL PANEL—JOYSTICK

A double-click on Joystick will allow you to test or calibrate your joystick.

### CONTROL PANEL—KEYBOARD

This window allows you to control the key repeat delay, the repeat rate, and the cursor blink rate.

## CONTROL PANEL—MODEMS

Here you can see type and brand of modem, the COM port it is attached to, adjust your modem's speaker volume (usually only used to hear the call connect), set the maximum baud speed, set data and stop bits, set parity, see the port settings, and choose flow control such as X on – X off.

## CONTROL PANEL—MOUSE

Use this window to choose whether you are using the mouse with your right or left hand, set the double-click speed, choose the pointer type and speed, and choose whether to have the pointer leave a trail behind it. This trail sometimes makes it easier to find, especially on laptop screens.

## CONTROL PANEL—MULTIMEDIA

Multimedia Properties in Control Panel gives tabs for the Audio, Video, MIDI, CD Music, and Advanced Controls. Under Audio, you can set the playback and record volumes. If you check the box for showing the volume control on the taskbar, a Speaker icon appears at the lower right side of the screen, by the time. One click on this gives overall volume control. Double-clicking gives a full Master Control window. Recording quality can also be set here.

Under Video, you may choose if video clips are shown full screen or in a window and at what size. As you can guess, enlarging a video reduces its quality.

The MIDI tab allows you to set up your system for a MIDI device. The CD tab allows you to set the earphone volume from your CD player. The Advanced tab shows your multimedia drivers in the same format as the Device Manager.

## CONTROL PANEL—NETWORK

Shows network components installed.

## CONTROL PANEL—PASSWORDS

Allows different users of the system to login under their own preferences and desktop settings by setting a password.

## CONTROL PANEL—PRINTERS

Use this to see what printers are installed and what jobs they are working on. Under File, you may Pause or Purge (delete) a print job. You select one printer as the default printer and can then see Properties of that printer. This is the same program accessed by clicking on Start, Settings, Printer. Single-clicking on one of the available printers to select it, you may click on File, Properties to gain full control of the printer settings. Under File, Properties on the General tab you see the type of printer and can print a test page. Under Details you see the port (usually LPT1:) and the installed driver. Under copies you see the number of copies and the orientation of the paper. Under Size and Source you choose the size of paper and whether it is manually or automatically fed. Under Graphics and Color you can choose automatic color, manual color, or gray scale for black-and-white printing. Under Quality you can choose the resolution of your printing. Of course all of the above settings depend on the type of printer you have installed. See more below under Settings—Printers.

## CONTROL PANEL—REGIONAL SETTINGS

In this window you may customize how your time and date will be printed on the screen.

## CONTROL PANEL—SOUNDS

This allows you to choose sounds to accompany system actions.

## CONTROL PANEL—SYSTEM

System Properties includes General, Device Manager, Hardware Profiles, and Performance.

## SYSTEM PROPERTIES—GENERAL

The General tab shows you the operating system, the registration information, the type of computer, and the RAM installed.

## SYSTEM PROPERTIES—DEVICE MANAGER—SEEING YOUR HARDWARE DEVICES

Click on Start, Settings, Control Panel, System, Device Manager to see your hardware devices. Click on the little plus sign (+) by any one to see a tree of the drivers or devices. Click on the minus sign (–) to collapse the tree. Click on the first Computer Device and properties to see the IRQs, I/O Addresses, DMAs, and memory usage for your system. Click on the hard disk controller plus sign to expand the tree. Click on Primary IDE controller and click on properties. Click on Resources to see the I/O Address or addresses, and the IRQ. Your CD-ROM drive will have a Settings tab to click on. Click on your sound card to see tabs for the Driver and Resources. This time you can see I/O addresses, IRQ, and DMA. By clicking off the Use Automatic Settings check mark, you can force the setting if you are having a problem. The Ports Device will expand to serial communications and parallel printer devices. In addition to the Driver and Resources tabs, this will also have a Port Settings tab. This is where you can set the baud speed, and data and stop bits. The Modem Device properties will show the Driver and Resources tabs, plus a Connection and Modem tab. You may adjust the modem's speaker volume by clicking on this tab and see the serial port the modem is using. The mouse device, oddly enough, doesn't show the port being used by a serial mouse. See Figure 10.9 for System Properties.

## SYSTEM PROPERTIES—HARDWARE PROFILES

Hardware profiles allows you to define different profiles: for instance, if an old DOS program needs a different driver to run the sound card, or if you do not want to load the device drivers for a network you are not going to use this time.

## SYSTEM PROPERTIES—PERFORMANCE

Performance gives the amount of RAM, system resources in use, disk compression if used, whether a PCMCIA card is in use, and advanced settings. Under advanced settings, there are buttons for File System, Graphics, and Virtual Memory.

Performance—File system lets you choose Hard Disk, CD-ROM, or Troubleshooting tabs. Under Hard Disk, the hard drive will read ahead to speed up hard drive access. This is

**Figure 10.9   Windows 95
System Properties Window**

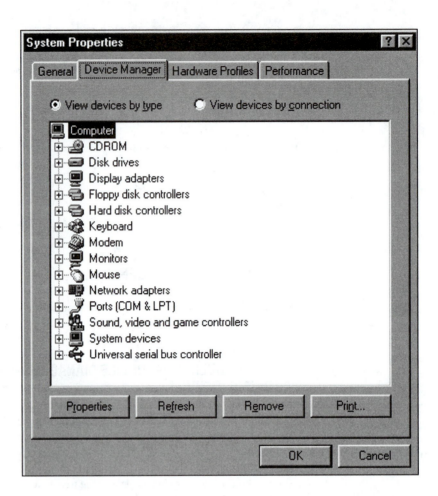

normally set to full. Under CD-ROM, you set the cache size and select the drive speed. Under Troubleshooting, you can disable settings to test your system and see which setting was causing problems.

Performance—Graphics, you may choose your graphics acceleration. This is also normally set to full.

Performance—Virtual Memory, you may set this, or let Windows control the size automatically.

## SETTINGS—PRINTERS

This takes you to the same program as Settings—Control Panel—Printers. It gives you quick access to controlling and monitoring you printers. Clicking on the printer icon once, then File, Properties, allows setting colors, print quality, orientation, and so on, and choosing one printer as the default printer. You may also click on Print Test Page to do just that. See Figure 10.10 for the Graphics and Color tab window of the Printer program.

## SETTINGS—TASKBAR

The Taskbar Properties has two tabs: Taskbar Options and Start Menu Programs as shown in Figure 10.11. Taskbar Options allows you to keep the taskbar on top of the desktop and to have the time always displayed. The Start Menu Programs tab allows you to add or remove programs from the Start menu. When you click on Start, then Programs, a list of these programs appears, giving easy access to them. You can still run other programs not on this list by using the Start, Run selections, or finding them with Windows Explorer and double-clicking on them. You may also clear the Documents list described below by clicking on that button.

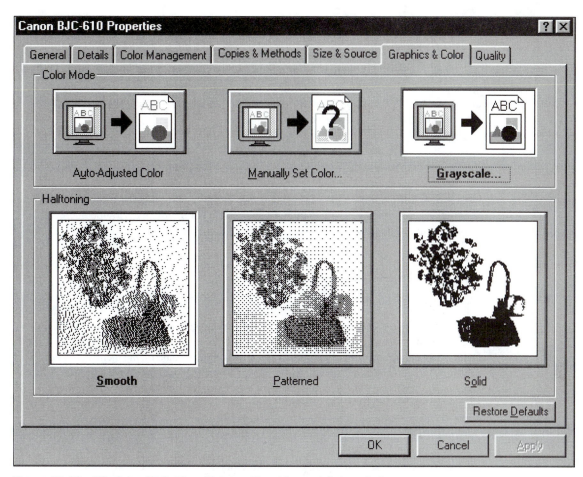

Figure 10.10 Windows 95 Settings/Printers/Graphics and Color window

Figure 10.11 Setting up the
Windows 95 Taskbar and Start
menu

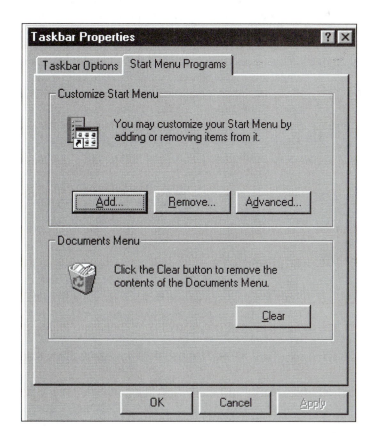

## START—DOCUMENTS

This is another easy way to start a program you have recently used, to continue on the same job. Click on Start, Documents, then the document desired. Windows automatically runs the application program for the document or data file you want to work with. This is a very quick way to return to a job worked on recently. This list can be cleared under the Taskbar, Start Menu Programs, as described above.

## START—PROGRAMS

Clicking on this displays a list of programs you have created with the Taskbar, Start Menu Programs. Most programs you install on your system automatically are added here. It is the standard way to run most of you programs.

## TO FIND FILES—WINDOWS EXPLORER

Click on Start, Programs, Windows Explorer. This is like File Manager in Windows 3.1. The command on the menu bar is File, and using this you can create a new directory by clicking on New, then Folder. A folder is a directory. Under the File drop-down menu, you can also Create, Delete, or see the Properties of a folder or file. The next command on the menu bar is Edit. The drop down menu includes the Undo Delete, Cut, Copy, Paste, and Select All commands. These function as they did in older versions, except Undo Delete, which does just what it says. This is because of the Recycle Bin, which holds deleted items until it is emptied by double-clicking on it, then on File, Empty Recycle Bin. Undo Delete just takes files out of the Recycle Bin.

The View command drop-down menu includes Tool Bar, Status Bar, Large and Small Icons, List, Details, and Options.

Toolbar gives you icons for Up One Directory Level (like CD.. in DOS), Cut, Copy, Paste, Undo Delete, and Properties. These are icons for most of the choices in the File, Edit, and View menus.

Status Bar displays a horizontal bar at the bottom of the window with the number of objects (files and folders), size of the currently selected object, and free disk space.

Large and Small Icons allows you to choose just that.

List gives nothing but names, to cram more filenames into the window. Details gives name, size, type, and date modified. Clicking at the top of one of these columns sorts by that parameter. This is handy and is the same as the "DIR /O:P" DOS command, where P is either D for date or S for size.

Options gives a card file with two tabs. The View tab allows you to show or hide files. The File Type tab shows file details, such as which application program this data file was created with.

The Tools drop-down menu gives Find and Goto commands. The Find command runs a search for folder or filename. The Goto command does just that.

Files may be moved from one folder to another by just clicking on it in the right window and dragging it to a folder in the left window.

## ACCESSORIES

Start, Programs, Accessories, wil give you most of the same programs as Windows 3.1, except for the calendar, and many new features. Depending on the features you added, Accessories may include:

Folders (subdirectories with a number of programs in them)
Fax              Compose new fax
                 Cover page editor
                 Request a fax

| | |
|---|---|
| Games | Free cell |
| | Hearts |
| | Minesweeper |
| | Solitaire |
| Internet Tools | Your browser |
| Multimedia | CD player |
| | Media player for movies |
| | Sound recorder |
| | Volume controls |
| System Tools | Backup |
| | Compression utilities |
| | Defragmentation |
| | Drivespace compression |
| | Resource bar meter to show system usage |
| | ScanDisk FAT repair |
| | System monitor graph of system usage |
| Programs | Calculator |
| | Character map |
| | Notepad ASCII text word processor (less than 64K) |
| | Hyperterminal |
| | Paint |
| | Wordpad word processor Microsoft Word DOC file output |

## 10.14   WINDOWS 98

Windows 98 is an upgraded or improved version of Windows 95. The Start button is still at the bottom left of the screen and is a good place to "Start." There is a "tour" included, which may be started by clicking on Start, Run, and typing in "tour98". It may request that you insert your CD in the drive. The tour includes Computer Essentials, Windows 98 Overview, What's New, and More Windows 98 Resources. Also you may click on Start, Help, Contents, and Getting Started Book Online Version to learn more about Windows 98. For more technical detail, clicking on Start, Programs, Accessories, System Tools, System Information gives a window filled with lots of data on your hardware, components, and software. This more or less takes the place of MSD in DOS.

The layout of the desktop may be made to look "Web-like," and the whole operating system is designed to work in close integration with the Internet. Microsoft FrontPage Express is included in Windows 98 and allows you to create HTML (Hyper Text Markup Language) documents for display on the Web. Microsoft has its own network (MSN), and if you choose it as an Internet Service Provider (ISP), Windows sets your system up on the Internet almost automatically.

Windows 98 supports Digital Video Disk (DVD) technology and smoother multimedia display for "live" video. The "troubleshooters" are greatly improved, and other aids are added for hardware problem solving. Plug-and-Play seems to have many of the bugs worked out, and many installations are truly automatic. Universal Serial Bus (USB) is supported allowing peripherals to be "hot swappable." They may be plugged in while the system is running and the system will notice and start responding to them.

In fact, installing Windows 98 in a new system with a CD-ROM drive is often as easy as installing the Windows 98 CD in the drive, inserting the Windows 98 boot-up disk in the floppy, and turning on the system. I have been amazed at the results. Following are only a few of the features, noting mostly the new features or features that will help the technician.

### START—RUN

This choice is the same as in Windows 95. It allows you to run a program from the floppy or CD drive, or browse those drives to find a program to run.

## START—HELP

For Windows 98, help has been improved considerably from Windows 95. Figure 10.12 shows the main help window. The main feature I like is the addition of "troubleshooters." This helps the technician diagnose system problems. To use the troubleshooters, click on Start, Help, Contents, Troubleshooting, Windows 98 Troubleshooters, and any of the topics that pop up. Troubleshooters lead you through a series of questions to isolate the probable cause of your trouble. Regretfully, they don't cover all topics, but you can see the topics covered in Figure 10.13.

Help is also available at http://support.microsoft.com/support/. This is a really useful site. Also http://www.microsoft.com/windows98 has useful information.

## START—SETTINGS—CONTROL PANEL

The Windows 98 Control Panel shown in Figure 10.14 is similar to Windows 95, but notice the "Web page" feel. The System icon can be accessed for system properties to aid in checking hardware problems.

## SYSTEM INFORMATION

Click on Start, Programs, Accessories, System Tools, and System Information for a very handy source of hardware, component, and software information. It's all here: DMAs, IRQs, I/O addresses, memory usage, which IRQs are being shared or conflicting, everything. Under

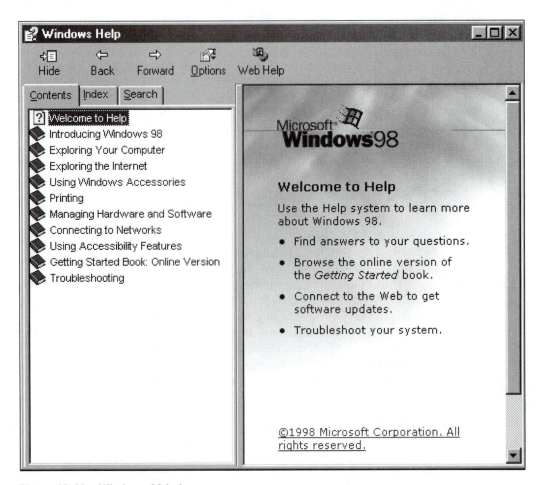

**Figure 10.12   Windows 98 help screen**

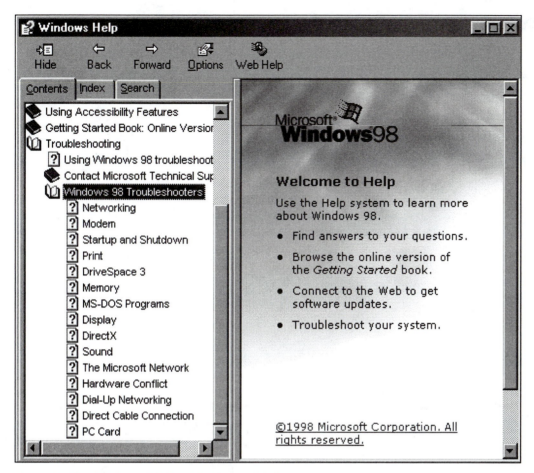

Figure 10.13    Windows 98 "Troubleshooters" help screen

Components you may select Multimedia, Display, Infrared, Input, Modem, Network, Ports, Storage (drives), Printing, Problem Devices (if you are lucky, this says None!), USB, History, and System. This is more information than you can believe; this screen is a real winner. You really should spend some time exploring all the information available with this tool.

## MY COMPUTER

Even though similar to Windows 95, clicking on My Computer on the desktop gives some quick information. A single-click on any drive gives the type, capacity used, and free of the drive. Then a click on File, Properties gives a pie chart showing usage, as shown in Figure 10.15, and a tab for the tools Scandisk, Backup, and Defrag. This will also show if your hard drive is FAT32 or FAT16. Windows 98 has a FAT32 converter program under Accessories, System Tools. A double-click on a drive icon will give the directory of that drive. For the printer icon, this gives another path to access properties and print a test page.

My Computer is also a fast way to get to System Properties. Right click on My Computer and then Properties. The System Properties window comes up. You may click on Device Manager to check for hardware conflicts. A yellow exclamation point shows a problem.

## TROUBLESHOOTING

Many of the problems that caused General Protection Faults in Windows 95 seem to have been resolved in Windows 98. With enhanced Plug-and-Play, hardware conflicts are less frequent with newer hardware. Sometimes getting a newer modem or other device with improved Plug-and-Play takes care of the problem much more easily than fighting with

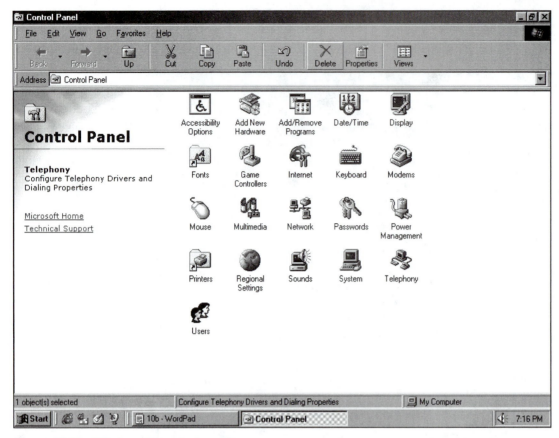

**Figure 10.14    Windows 98 control panel**

**Figure 10.15    Windows 98 Desktop/C:/Files/Properties**

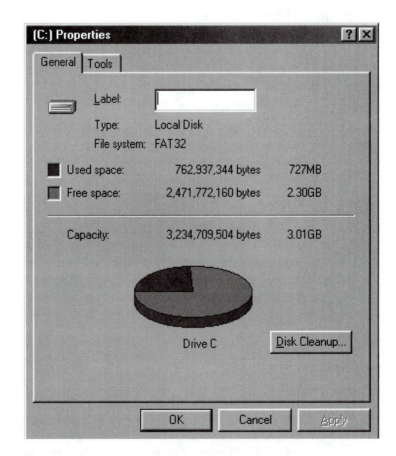

settings and drivers. Many of the INI files of Windows 3.1 have been eliminated, and in Windows 95 and 98 the information is stored in a Registry. If you go to the DOS prompt and type "DIR*.DAT" you see a few, but not all the Data files. Typing "ATTRIB*.DAT" shows the USER.DAT, SYSTEM.DAT, and HWINFO.DAT as hidden, read-only files. Windows backs these up each time it successfully boots. First it checks them with a program called Registry Checker. This can be run by clicking on Start, Programs, Accessories, System Tools, System Information, Tools, Registry Checker. If you want to make your own copies of these, you must first restart in DOS mode, for the SYSTEM.DAT file will cause a sharing violation if you try to use it while running Windows. Once you have re-booted, you may use the DOS ATTRIB command to remove hidden and read-only qualities, and then copy the files. Use ATTRIB to reset the hidden and read-only settings for these files, then type WIN to start Windows. You may edit the Registry directly yourself by clicking Start, Run, and typing REGEDIT. ScanDisk is available under System Tools to fix bad FAT tables. As before Windows 98, if the software seems totally flaky, running a virus scan may find and clean a virus and fix the software problem. If worse comes to worse, you may re-FDISK and reformat the hard drive, and reinstall Windows.

If the system refuses to boot, troubleshooting is the same as before Windows 98. If the system is working well enough to run Windows, the Windows 98 troubleshooters and the system information window are a great help in solving system problems. Clicking on Start, Settings, Control Panel, System, Device Manager may show a yellow exclamation point where hardware conflicts. Uninstall the conflicting hardware and reinstall, letting Windows search for it. If Windows does not see it, you may try to install it manually. Remember, it is a good idea to keep a Windows 98 startup disk. You create this by clicking Start, Settings, Control Panel, Add/Remove Programs, Startup tab, Create Disk. This disk may be used to start the system from the floppy drive similarly to the old DOS boot disk.

## 10.15 CHAPTER REVIEW/QUIZ QUESTIONS

1. DIR
   a. Changes the direction of the drive.
   b. Does interlaced reads.
   c. Shows the direction of the path.
   d. Gives a listing of files.

2. CHKDSK
   a. Allows you to format any drive.
   b. Allows you to partition the hard drive only.
   c. Examines the File Allocation Table for errors.
   d. Changes disk letters.

3. The DOS RECOVER program
   a. Recovers information lost when the power fails.
   b. Recovers information in "unreadable" files.
   c. Recovers a file that you've written over.
   d. Recovers what you keyed before you hit Q.

4. The DOS TYPE command
   a. Shows the type of hard drives in the system.
   b. Shows the type of memory in the system.
   c. Shows the type of video controller and display.
   d. Prints a file on the display.

5. The DOS BACKUP command
   a. Backs up to the last DOS command given.
   b. Backs up a program to the beginning.
   c. Saves hard drive data to floppies.
   d. Saves floppy drive data to hard drive.

6. The DOS RESTORE command
   a. Finds data lost when the power fails.
   b. Repairs a damaged program back to normal.
   c. Transfers information from floppies to the hard drive.
   d. Sets system memory back to normal.

7. The DOS HIMEM command in the CONFIG.SYS file
   a. Writes the higher memory of the system.
   b. Erases all higher memory for future use.
   c. Manages the system's extended memory.
   d. Places DOS in higher memory.

8. The line DOS=HIGH in the CONFIG.SYS file
   a. Allows DOS to check the high memory.
   b. Loads DOS in the high-memory area.
   c. Allows other programs to use the high-memory area.
   d. Uses a higher level of DOS.

9. The line DEVICE=SMARTDRV.SYS in the CONFIG.SYS file
   a. Allows the drive to spin faster.
   b. Allows the CPU to calculate faster.
   c. Creates a cache in memory to speed hard drive access.
   d. Writes data on the hard drive in compressed form.

10. The line DEVICE=RAMDRIVE.SYS in the CONFIG.SYS file
    a. Makes the floppy drive run like a RAMjet.
    b. Makes the floppy run faster.
    c. Creates a cache in memory for the floppy.
    d. Creates a RAM or electronic disk drive.

11. The line INSTALL=C:\DOS\FASTOPEN.EXE C:=40 in CONFIG.SYS
    a. Allows the CPU to operate at forty times the clock speed, speeding system performance.
    b. Creates a cache to hold forty file names and locations, speeding system performance.
    c. Allows the video controllers to print forty characters to the screen at a time, speeding system performance.
    d. Allows the CPU to process forty programs at the same time, speeding system performance.

12. A batch file
    a. Is a data file.
    b. Is a binary file.
    c. Is a program.
    d. Is not shown with the DIR command.

13. The DOS SORT command
    a. Alphabetically sorts a file line by line.
    b. Alphabetizes the files on the disk.
    c. Prints the sort of hard drive you have on the screen.
    d. Prints the sort of monitor you have on the screen.

14. The PAUSE batch command
    a. Prints "Strike a key when ready" and stops the batch file.
    b. Pauses the program for ten seconds each time it's used.
    c. Pauses the program until a warm boot.
    d. Is now antiquated.

15. The > symbol indicates
    a. Piping information from one program through another.
    b. Redirecting output of a program to a file or device.
    c. Redirecting input of a program from a file or device.
    d. Selection of a higher level of DOS.

16. The >> symbol
    a. Writes to a file, destroying original data.
    b. Writes to a file, appending to original data.
    c. Points to the next command in a batch file.
    d. Precedes a remark in a batch file.

17. The ¦ symbol
    a. Erases all files.
    b. Stands for programs only.
    c. Pipes information from one program through another.
    d. Redirects input or output.

18. The ECHO OFF batch command
    a. Stops all audio echoes.
    b. Stops the commands from being shown on the screen.
    c. Stops the CPU from repeating commands.
    d. Copies the command to a file automatically.

19. The PROMPT command
    a. Can change the visual prompt.
    b. Can change the color of the screen.
    c. Can change the position of the cursor.
    d. All of the above, and more.

20. The AUTOEXEC.BAT file
    a. Runs DOS automatically.
    b. Is run by DOS automatically when DOS is booted.
    c. Automatically executes DOS and/or Windows.
    d. Is not very useful.

21. You load Windows into a computer by
    a. Inserting Windows disk #1 in drive A:, typing A:, then typing SETUP.
    b. Inserting Windows disk #1 in drive A:, then typing INSTALL.
    c. Inserting any Windows disk in any drive, then typing SETUP.
    d. Typing WINDOWS in any version of DOS.

22. Windows has at least three groups, the most commom of which are:
    a. Programs, Files, and Data.
    b. Spreadsheets, Databases, and Word Processors.
    c. Applications, Accessories, and Main.
    d. Windows, Icons, and Pointers.

23. Under the Accessories group Windows has which of the following?
    a. A word processor, a calendar, a calculator, and an ASCII character map
    b. A drawing program, a database (cardfile), a clock, and a modem terminal program
    c. A macro recorder, a sound recorder, a media browser for video and sound, and a notepad
    d. Amazingly, all of the above

24. Under the Main group Windows has which of the following?
    a. A file manager, a print manager, and Windows setup
    b. An MS-DOS prompt, a clipboard, and a control panel
    c. A PIF editor and Read Me file with more Windows help
    d. Amazingly, all of the above

25. Under the Applications group
    a. You install third-party programs for your work.
    b. Windows comes with many applications preinstalled.
    c. You can store your résumé.
    d. All of the above.

# 11

# Symptoms and Repairs

**OBJECTIVES**

After completing this chapter, you should be able to

- Differentiate between software and hardware problems
- Isolate a hardware problem to a specific subassembly
- Correct or replace a failing subassembly

To assist the novice in developing a troubleshooting strategy for a failed system, this chapter presents some common problems and possible causes. It is not meant to be exhaustive, but to give some general guidance. General rules are

1. Be positive your DOS and test disks are good.
2. Confirm the drive parameters.
3. Document/archive every technical document, jumper setting, and software setup you see.

The following hints are in order of severity of problem, in the general order you would need to follow to ensure a properly operating system, from major to minor. This chapter deals with generalities. For deeper details see each chapter's Basic Trouble-shooting section.

## 11.1  COMPUTER DEAD

Figure 11.1 shows a flowchart for diagnosing a dead computer. Use it and the following symptoms and troubleshooting strategies to solve the problem.

### NO LIGHTS, NO SOUNDS, NO FAN—NOTHING

Confirm that the 110 V AC from wall line is good. This may be done by observing the pilot light of the display. No pilot means no 110 or that the display low-voltage power supply is out. Having a display pilot confirms 110 and displays LVPS. No fan indicates no +12 V (sometimes no –12 V), which probably means the whole power supply is shut down. Check power supply for all correct voltages per Chapter 2.

**FIGURE 11.1 Diagnosing a dead computer**

*Note:* Numbers correspond to sections in text.

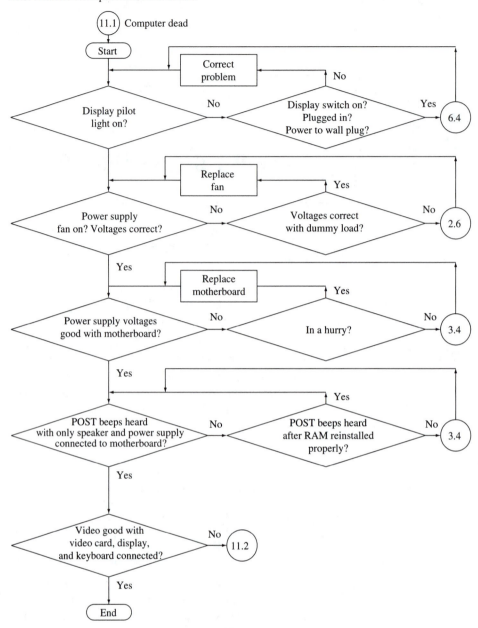

## FAN WORKS, NO LIGHTS, NO BOOT

Proper power to the motherboard, but no video on the display and no booting of any drive may indicate a motherboard problem. Remove all adapter cards, front panel connectors, keyboard, and so on. Front panel connectors installed improperly can put system into constant reset or can short and shut down power supply. Leave only the speaker and power supply attached. If you remove a memory card, check memory switches on motherboard for new setting. Improperly installed RAM can make a good motherboard appear totally dead. Check the clock to the CPU, the reset signal to the CPU, and for activity on the data and address bus. The RAM, ROM, CPU, and keyboard ROM are usually socketed and easily replaceable. Check the motherboard per Chapter 3.

One beep on power-up means OK. Many beeps give an error report, but rightfully so on a stripped system. This means the power supply, CPU, main RAM, ROM, and most of the rest of the motherboard are working. You can plug in the keyboard next, always turning the system off before adding anything. Press Alt+Ctrl+Delete for a warm boot. If the system resets and sounds the beeps after each reboot, the motherboard is 90 percent cleared of major problems.

**FIGURE 11.2  Diagnosing
video display problems**

*Note:* Numbers correspond to sections in text.

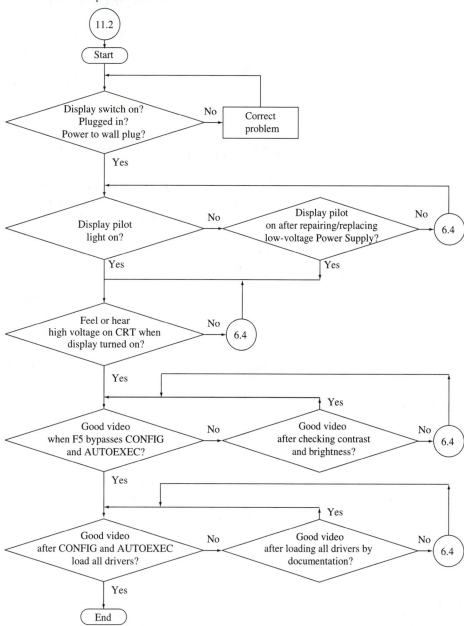

## 11.2  COMPUTER OK, NO VIDEO

Figure 11.2 shows a flowchart for diagnosing problems that result in no video display. Use it and the following symptoms and troubleshooting strategies to solve the problem.

### NO VIDEO, NO HIGH VOLTAGE, DISPLAY PILOT LIGHT OFF

Good high voltage will hold a sheet of paper to the screen with static electricity. A 4 × 4 sheet is sufficient, but any size will prove the voltage is high enough for visible video. On most monitors or displays, the pilot light is powered from the display's low-voltage power supply. Check 110 V AC from wall to display, and check display fuse if accessible. If there is still no pilot, the display low-voltage power supply is either bad or loaded down by a failure in the rest of the display. Check silk screen on LVPS board for output voltages and measure. These are usually switching supplies and may require a dummy load to operate. Check this per Chapter 6.

## NO VIDEO, NO HIGH VOLTAGE, DISPLAY PILOT LIGHT ON

While the pilot shows that some of the low-voltage supply is operating, check all display power supply voltages on input and output connectors. If the low-voltage supply is good, track down the horizontal signal from the oscillator to the flyback. The horizontal output transistor and the flyback transformer are two high failure points. See Chapter 6.

## NO VIDEO, HIGH VOLTAGE OK, PILOT ON

With a pilot and high voltage on, the display has a 90 percent chance of operating. Check brightness and contrast controls for center setting. Suspicion is thrown on the video adapter card or software. Confirm output signals from the adapter card, pins 1, 2, 3, 13, and 14; with an oscilloscope. If you have these signals, the display is definitely bad.

A quick switch of the display with a known working display of the same kind will confirm whether the display or the adapter card is failing. If the new display works, attempt to repair the old one per Chapter 6. If the adapter card does not drive the new monitor, change the card. Adapter cards can sometimes be repaired by reseating or replacing the adapter RAM.

An incorrect video driver program can blank a screen. Press F5 when DOS starts loading to bypass the CONFIG.SYS and AUTOEXEC.BAT files that may be loading the incorrect driver. See Chapter 6.

## BAD, WEIRD, OR FLAKY VIDEO

Note that improperly installed software drivers can totally blank out a screen. On booting the system, press F5 to bypass CONFIG.SYS and AUTOEXEC.BAT and come up in raw DOS, with no drivers loaded. The display should default to standard operation. This will eliminate the possibility of software problems. Once standard DOS text video is established, install the drivers that came with the video card, and retest.

## 11.3 COMPUTER OK, VIDEO OK, KEYBOARD ERROR

Figure 11.3 shows a flowchart for diagnosing problems that cause keyboard errors. Use it and the following symptoms and troubleshooting strategies to solve the problem.

### KEYBOARD DEAD

Check cable from keyboard to computer. Check signals at motherboard keyboard connector. If good, switch with known good keyboard. If bad, check motherboard keyboard connector. May need resoldering. If signals there are bad, repair or replace motherboard.

### A FEW KEYS DEAD OR INTERMITTENT

If dead keys are widely separated, pop the plastic cap from each and blow out each one with compressed air or vacuum out. If dead keys are in same general area or in a pattern, remove back and do a continuity test on copper foil traces. See Chapter 5.

**FIGURE 11.3  Diagnosing keyboard errors**

*Note:* Numbers correspond to sections in text.

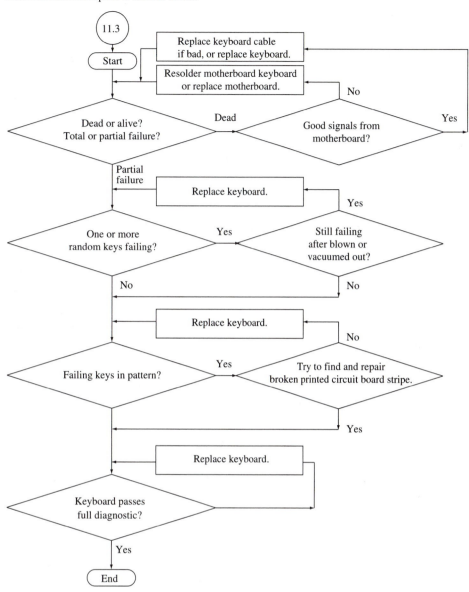

## 11.4  COMPUTER OK, VIDEO OK, KEYBOARD OK, MEMORY TEST FAILS

Figure 11.4 shows a flowchart for diagnosing problems that cause memory test failure. Use it and the following symptoms and troubleshooting strategies to solve the problem.

### FAILS MEMORY TEST CONSISTENTLY

Consider yourself lucky if the POST memory test catches a bad RAM. Reseat or replace the RAM per Chapter 4.

### INTERMITTENT MEMORY FAILURE

Often, the most difficult step of troubleshooting intermittent memory failure is duplicating the problem to observe symptoms. The second difficulty is positively eliminating software or TSR program collisions that appear memory related.

FIGURE 11.4    Diagnosing
memory test failure

*Note:* Numbers correspond to sections in text.

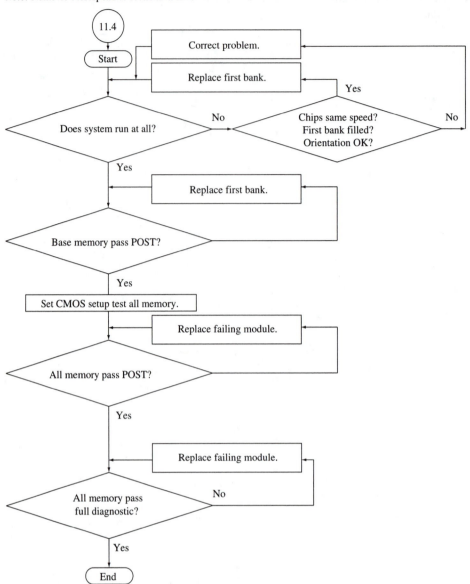

Boot with F5 or F8 for no AUTOEXEC.BAT or CONFIG.SYS, run memory check program only. If it is a stubborn, irritating problem, just replace the RAM for spite and thoroughly test system for an extended time. If the problem remains with minimal DOS and new RAM, a motherboard stripe (actual copper foil trace) may be broken. It is probably most cost efficient to just replace the motherboard. See Chapter 4.

## 11.5   COMPUTER, VIDEO, KEYBOARD, MEMORY OK, FLOPPY DRIVE BAD

Figure 11.5 shows a flowchart for diagnosing a bad floppy drive. "No ROM basic, system halted" means the system could not find a boot device. This means that neither the hard drive nor the floppy drive is working. Most systems now have one drive controller adapter card to control both the floppy and hard drives. You will need a bootable DOS disk for the A: drive to test it. Many a technician has wasted time on a drive, to later find out his trusted DOS boot disk was a bad disk. Always check your boot disk first to be absolutely positive it is a good bootable copy. Bootable copies of DOS are created with the FORMAT A:/S command to transfer the system (assuming drive A:) from the hard drive to the floppy.

**FIGURE 11.5 Diagnosing a bad floppy drive**

*Note:* Numbers correspond to sections in text.

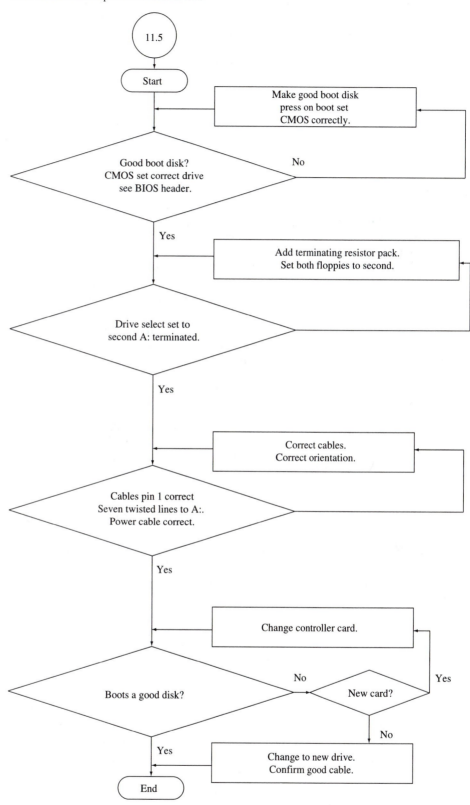

Check power to drives, control cables to all drives with pin 1 in correct position, drive select settings, terminating resistor, if applicable, and switches or jumpers on the controller card. Female dual headers are often misplaced, missing an entire row of pins. Examine them closely.

Check the CMOS setup to verify floppy types (see Chapter 3). If the CMOS setup is incorrect, weird symptoms can occur. For example, if a 1.44-MB drive is set up as a 1.2-MB drive, you may be able to read a directory, but you get the error message "Sector Not Found" when trying to load a program. The DRIVPARM command may be useful in the CONFIG.SYS if problems persist.

Just because the floppy/hard drive combination controller adapter card boots the hard drive does not mean the floppy portion of the card is good. A quick change of the floppy drive with a known good drive of the proper capacity will settle suspicion regarding the card or cables. New cables or card should confirm which unit had failed. Verify the CMOS setup and the drive jumper settings. If you change settings, document the current settings and configuration before changing. See Chapter 7.

## 11.6 COMPUTER, VIDEO, KEYBOARD, MEMORY OK, HARD DRIVE BAD

Watch for drive or controller LED to come on. If controller LED does not come on, replace controller card. If drive light LED does not come on, check cabling (and drive jumper settings, if customer has been inside case).

Can you boot with floppy with the same DOS version and access hard drive with a DIR C: command? If so, attempt to boot hard drive. If either "No System" or "Track Zero Bad" is displayed, you may need to low-level format the hard drive. If the error message says "No System Found," boot from a DOS floppy, enter the SYS command to transfer back to the hard drive, and attempt to boot from it again. If no error message and no boot, use FDISK to check (not change yet) partitions. Boot from a system floppy and run FDISK. From the menu, choose to look at the partition, and make sure it is active. If it is not active, it won't boot. Set it to active and try to boot again.

If the problem is getting two hard drives to work together, check factory jumper settings for master and slave settings. Even though two IDE drives should work together as master and slave, it does not always work if they are different brands. Try switching the master and slave, changing the settings, and trying again. Two of the same brand usually do work together. See Chapter 8.

## 11.7 COMPUTER MAKES ERRORS INTERMITTENTLY

### RANDOM ERRORS, TOTALLY UNRELATED

Protect power line from transient spiking with filters. Carefully observe long-term environment for heavy equipment switching on and off, extreme temperature fluctuations, line-voltage variations from time-dependent usage (such as the whole building turning on air conditioners at 2:00 o'clock each hot day).

Check AUTOEXEC.BAT and CONFIG.SYS for TSR programs that may cause conflicts. Pressing F5 or F8 during the loading of DOS will bypass or single-step through the CONFIG.SYS and AUOTEXEC.BAT files, allowing you to either skip both entirely or to pick and choose what will load.

### RANDOM ERROR WHEN RUNNING SAME SOFTWARE

Check for memory usage and other programs running at the same time. Memory conflict can cause failure in different routines of program. Check AUTOEXEC.BAT and CONFIG.SYS for TSR programs that may conflict. Try new copy of program, or copy all files to a floppy and run on different machine.

## 11.8 EVERYTHING WAS OK UNTIL NEW DEVICE WAS ADDED

Peripheral devices usually need an IRQ, DMA, and I/O address space of their own. If another device already is using the same one, conflicts occur. You must solve the conflict or forget the device. The following steps should help.

1. Remove all cards except video, hard drive, and floppy controllers. This eliminates all nonessential cards and allows you to customize installation of all cards.
2. Do MSD or diagnostic program, and write down or print out current settings of these three controllers.
3. Arrange remaining peripheral cards to be installed in order of worst to best. The worst has no documentation and no adjustable settings. The best has much documentation and many adjustable settings.
4. Now install the worst card first, hardware and software drivers, and run a diagnostic program to see the IRA, DMA, and I/O address settings. Add these to the list you are making.
5. Continue this, moving from worst through best card. Test all peripherals every time. If you cannot find a free IRQ, you can attempt to share one with a device that will not be used at the same time as the new device. At the end, test all devices thoroughly.

# APPENDIX A
# POST Error Codes

## POST AUDIO ERROR CODES

These beeps from the system speaker are given by many BIOS during POST (Power-On Self-Test).

| Audible Beeps from Speaker | Possible Error |
|---|---|
| 1 short beep, system works | All OK |
| No beep, dead unit | Power supply |
| Continuous beep | Power supply |
| Repeating short beep | Power supply |
| 1 short, continuous | First bank of RAM |
| 1 long, 1 short | System motherboard |
| 1 long, 2 short | Display |
| 1 long, 3 short | Display |
| 1 short, 1 long, bad video | Display |
| 1 short, bad video | Display |
| 1 short, video good, no boot | Disk drive |

## POST VIDEO ERROR CODES

If the system is working enough to give a legible display, these numeric error indicators are shown at the top left of the video display if the POST finds an error. If the video is out, these are unseen. But even this short list is sometimes helpful. Much more extensive lists can be obtained from the motherboard manufacturer.

| Error Code | General Failure Block |
|---|---|
| 1xx | System motherboard |
| 100 | Option configuration error |
| 101 | System interrupt circuitry |
| 102 | System timer circuitry |
| 103 | System timer interrupt |
| 104 | System protected mode |
| 105 | System 8042 command |

| Error Code | General Failure Block |
| --- | --- |
| 106 | System converting logic test |
| 107 | System NMI test |
| 108 | System timer bus test |
| 109 | DMA test error |
| 121 | Unexpected hardware interrupt |
| 131 | Cassette port error |
| 161 | System option error, battery? |
| 162 | System option error, run SETUP |
| 163 | Time, date not set, run SETUP |
| 164 | Memory size error, run SETUP |
| 2xx | Memory |
| 201 | Memory test failure |
| 202 | Memory address error |
| 203 | Memory address error |
| 3xx | Keyboard |
| 301 | Keyboard not responding |
| 302 | User error or keyboard locked |
| 303 | Keyboard or system error |
| 304 | Keyboard error, CMOS config error |
| 4xx | Display (Mono) |
| 401 | Mono memory, horizontal synchronization, or video |
| 408 | User indicated attribute error |
| 416 | User indicated character set error |
| 424 | User indicated 80x25 mode error |
| 432 | Mono adapter parallel port error |
| 5xx | Display (Color) |
| 501 | Color memory, horizontal synchronization, or video |
| 508 | User indicated attribute error |
| 516 | User indicated character set error |
| 524 | User indicated 80x25 mode error |
| 532 | User indicated 40x25 mode error |
| 540 | User indicated 320x200 mode error |
| 548 | User indicated 640x200 mode error |
| 556 | Light pen test |
| 564 | User indicated screen paging test |
| 6xx | Disk drives |
| 601 | Disk POST error, interface |
| 602 | Disk test error |
| 603 | Disk size error |
| 606 | Disk verify error |
| 607 | Write protected |
| 608 | Disk status returned |
| 610 | Disk initialization error |
| 611 | Timeout |
| 612 | Cable or adapter error |
| 613 | Bad DMA |
| 614 | DMA boundary error |
| 621 | Bad seek |
| 622 | Bad CRC |
| 623 | Record not found |
| 624 | Bad address mark |
| 625 | Bad controller seek |
| 626 | Data compare error |
| 627 | Disk change line error |
| 628 | Disk removed |

| Error Code | General Failure Block |
| --- | --- |
| 630 | A: index stuck high |
| 631 | A: index stuck low |
| 632 | A: track 0 stuck off |
| 633 | A: track 0 stuck on |
| 640 | B: index stuck high |
| 641 | B: index stuck low |
| 642 | B: track 0 stuck off |
| 643 | B: track 0 stuck on |
| 650 | Drive speed error |
| 651 | Format failure |
| 652 | Verify failure |
| 653 | Read failure |
| 654 | Write failure |
| 655 | Controller |
| 656 | Drive |
| 657 | Write protect stuck protected |
| 658 | Change line stuck changed |
| 659 | Write protect stuck unprotected |
| 660 | Change line stuck unchanged |
| 7xx | Math coprocessor |
| 702 | Exception errors test |
| 703 | Rounding test |
| 704 | Math test 1 |
| 705 | Math test 2 |
| 706 | Math test 3 |
| 707 | Combination test |
| 708 | Integer store test |
| 709 | Equivalent expressions |
| 710 | Exceptions |
| 711 | Save state |
| 712 | Protected mode test |
| 713 | Sensitivity test |
| 9xx | Parallel printer card |
| 901 | Data register latch |
| 902 | Control register latch |
| 903 | Register address decode |
| 904 | Address decode |
| 910 | Status line wrap connector |
| 911 | Status line bit 8 wrap |
| 912 | Status line bit 7 wrap |
| 913 | Status line bit 6 wrap |
| 914 | Status line bit 5 wrap |
| 915 | Status line bit 4 wrap |
| 916 | Interrupt wrap |
| 917 | Unexpected interrupt |
| 11xx | COM serial card |
| 1101 | Adapter test failure |
| 1102 | Card selected feedback |
| 1103 | Port 102 register test |
| 1106 | Serial option |
| 1107 | Com cable |
| 1108 | IRQ3 |
| 1109 | IRQ4 |
| 1110 | Modem status register not clear |
| 1111 | Ring indicate |

| Error Code | General Failure Block |
|---|---|
| 1112 | Trailing edge ring indicate |
| 1113 | Delta receive line signal detect |
| 1114 | Receive line signal detect |
| 1115 | Delta receive signal detect |
| 1116 | Line control register, bits not set |
| 1117 | Line control register, bits can't be reset |
| 1118 | Transmit holding/shift register stuck |
| 1119 | Data ready stuck on |
| 1120 | Interrupt enable register, bits not set |
| 1121 | Interrupt enable register, bits can't be reset |
| 1122 | Interrupt pending stuck on |
| 1123 | Interrupt ID register stuck on |
| 1124 | Modem control register bits not set |
| 1125 | Modem control register bits can't be reset |
| 1126 | Modem status register, bits not set |
| 1127 | Modem status register, bits can't be reset |
| 1128 | Interrupt ID |
| 1129 | Can't force overrun error |
| 1130 | No modem status interrupt |
| 1131 | Invalid interrupt pending |
| 1132 | No data ready |
| 1133 | No data available interrupt |
| 1134 | No transmit holding interrupt |
| 1135 | No interrupts |
| 1136 | No received line status interrupt |
| 1137 | No receive data available |
| 1138 | Transmit holding register not empty |
| 1139 | No modem status interrupt |
| 1140 | Transmit holding register not empty |
| 1141 | No interrupts |
| 1142 | No IRQ4 interrupt |
| 1143 | No IRQ3 interrupt |
| 1144 | No data transferred |
| 1145 | Maximum baud rate |
| 1146 | Minimum baud rate |
| 1148 | Timeout error |
| 1149 | Invalid data returned |
| 1150 | Modem status register error |
| 1151 | No DSR and delta DSR |
| 1152 | No DSR |
| 1153 | No delta DSR |
| 1154 | Modem status register not clear |
| 1155 | No CTS and delta CTS |
| 1156 | No CTS |
| 1157 | No delta CTS |
| 13xx | Game control card |
| 1301 | Adapter failure |
| 1302 | Joystick test |
| 17xx | Hard drive card |
| 1701 | Fixed disk or adapter test |
| 1702 | Time out |
| 1703 | Drive |
| 1704 | Controller or drive error |
| 1705 | No record found |
| 1706 | Write fault |

| Error Code | General Failure Block |
|---|---|
| 1707 | Track 0 error |
| 1708 | Head select error |
| 1709 | Bad ECC |
| 1710 | Read buffer overrun |
| 1711 | Bad address mark |
| 1712 | Undetermined |
| 1713 | Data compare error |
| 1714 | Drive not ready |
| 1730 | Adapter |
| 1731 | Adapter |
| 1733 | Adapter |
| 1750 | Drive verify |
| 1751 | Drive read |
| 1752 | Drive write |
| 1753 | Random read test |
| 1754 | Drive seek test |
| 1755 | Controller |
| 1756 | Controller ECC test |
| 1757 | Controller head select |
| 1780 | Drive C: fatal time out |
| 1781 | Drive D: fatal time out |
| 1782 | Controller no IPL from hardfile |
| 1790 | Drive C: nonfatal |
| 1791 | Drive D: nonfatal |

# APPENDIX B
# Helpful Products and Aids

A number of companies offer diagnostic disks to aid in troubleshooting. Of course, the system must be running well enough to boot a floppy and show the results on the display. However, some of the most perplexing problems are those small glitches that hinder an operating system's usefulness. The very best aid I have seen is Howard W. Sams's *Computerfacts,* which has extensive specific component level repair information on numerous systems.

*Note:* Some suppliers' primary means of contact is via modem—modem numbers are labeled as such. A few of these suppliers are accessible via the Internet, and their online addresses are listed here.

## DIAGNOSTIC AIDS

### Software

Ami Diags, a diagnostic program available from Software Addons, (800) 822-8088 (modem).

Check-It Pro Wincheckit, and Check-it Set-Up Advisor. Diagnostic programs available from many sources. The producer is TouchStone Software, P.O. Box 1311, Huntington Beach, CA 92647, (800) 932-5566.

Conner Low-Level Formatter. Blocks out bad sectors. From TCE Company, 63 Douglas Avenue, Elgin, IL 60120, (800) 383-8001.

CyberMedia, First Aid for Windows Users. For Windows problems. Also, for boot-up or configuration problems, PC 911. Available from Micro Warehouse, 1720 Oak Street, P.O. Box 3014, Lakewood NJ 08701-9823, (800) 367-7080.

Disk Technician Gold. A disk alignment package available from Software Addons, (800) 822-8088 (modem).

FLOPPY DRIVE ALIGNMENT DISK and POST code master card. Micro Systems Development, 4100 Moorpark Ave., Ste. #104, San Jose, CA 95117, (408) 296-4000.

Hard Drive Diagnostic Programs. Paul Mace, Inc., 123 N. First St., Ashland, OR 97520, (503) 488-0224. Also accessible via modem at (503) 482-7435.

Hard Drive Reference Manual on 2000 drives, 350 controllers. From TCE Company, 63 Douglas Avenue, Elgin, IL 60120, (800) 383-8001.

Micro House Technical Library. Configurations, settings, diagrams of 1500 motherboards, available from JDR Microdevices, 1850 South 10th St., San Jose, CA 95112-4108, (800) 538-5000.

Optune. A hard disk optimizer, available from Software Addons, (800) 822-8088 (modem).

PC Doctor. Diagnoses, configures, and gives information about your system. Available from JDR Microdevices, 1850 South 10th St., San Jose, CA 95112-4108, (800) 538-5000.

Professional QAPlus. 5.0 Diagnostics QuapPlus Win-Win from Diagsoft, Inc., 5615 Scotts Valley Drive, Suite 140, Scotts Valley, CA 95066, (408) 438-8247.

Service Diagnostics. Landmark Research International Corporation, 5770 Roosevelt Blvd., #400, Clearwater, FL 34620, (813) 443-6603. They have many related software tools for technicians.

Spinrite. A nondestructive hard disk low-level reformatter that also optimizes interleave. Gibson Research, 22991 LaCadena, Laguna Hills, CA 92653, (800) 736-0637.

The Troubleshooter from Forefront is the best diagnostic I have seen. (800) 475-5831, www.ffg.com

Win Check-It. A Windows diagnostic tool to set up, clean up, or repair Windows files. Available from JDR Microdevices, 1850 South 10th St., San Jose, CA 95112-4108.

Xidex floppy alignment disks. Xidex/Dysan, 1244 Ramwood Ave., Sunnyvale, CA 94089, (408) 988-3472 (modem).

### Hardware

POST Code Display Card. Plugs into an expansion slot and gives the POST code when the system is incapable of doing so. JDR Microdevices, 2233 Samaritan Drive, San Jose, CA 95124.

SIMM tester for 30- and 72-pin SIMMS. Made by Chroma. Available from JDR Microdevices, 1850 South 10th St., San Jose, CA 95112–4108.

## MAGAZINES, TRADE, AND VENDOR

Circuit Specialists, Inc.
  P.O. Box 3047
  Scottsdale, AZ 85271-3047
  (800) 528-1417
  (602) 464-2485

Computer Hotline
  15400 Knoll Trail
  Dallas, TX 75248
  (800) 999-5131

Computer Shopper
  P.O. Box 52568
  Boulder, CO 80322-2568
  (800) 274-6384
  http://www.ziff.com

Digi Key
P.O. Box 677
Thief River Falls, MN 56701-0677
(800) 344-4539

JDR Microdevices
2233 Samaritan Drive
San Jose, CA 95124
(800) 538-5000

Mouser Electronics
2401 Hwy. 287 N.
Mansfield, TX 76063
(800) 992-9943

Nuts & Volts
430 Princeland Court
Corona, CA 91719
(800) 783-4624

## PARTS

## Batteries

Battery-Biz
31352 Via Colinas, Unit 104
Westlake Village, CA 91362
(800) 848-6782

## Card Replacements, Video, Drive Controllers, Motherboards

Altex Electronics (Also power supplies, keyboards, cases, drives)
11342 IH-35 North
San Antonio, TX 78233
(800) 531-5369

Computer Parts & Pieces
301 W. Abram St.
Arlington, TX 76010
(800) 235-0096

JDR Microdevices (Also rails, bezels, frames, hard-to-find supplies)
2233 Samaritan Dr.
San Jose, CA 95124
(800) 538-5000

Shecom Computers (Also drives, keyboards, cases)
22755 Savi Ranch Parkway, Unit G
Yorba Linda, CA 92687
(800) 366-4433

USA Electronics, Inc. (Also cases, power supplies, drives)
9090 N. Stemmons Freeway
Dallas, TX 75247
(800) 332-8434

## Cases and Power Supplies

Jinco Computer
5122 Walnut Grove Ave.
San Gabriel, CA 91776
(800) 253-2531

## Flybacks

International Components Marketing (ICM)
    11953 Kiowa
    Losa, CA
    (800) 748-6232

MCM Electronics
    858 E. Congress Park Dr.
    Centerville, OH 45459
    (800) 543-4330

## General Parts

Jameco Electronics
    1355 Shoreway Rd.
    Belmont, CA 94002
    (415) 592-8097
    (800) 831-4242

## Hard Drive Replacements

Insight (Also memory)
    1912 W. Fourth St.
    Tempe, AZ 85281
    (800) 736-3475
    http://www.insight.com

MegaHaus (Also controllers)
    1110 Nasa Rd. 1
    Suite 306
    Houston, TX 77058
    (800) 426-0560

Quantum Corp. (The hardcard, a slot hard drive)
    500 McCarthy Blvd.
    Milpitas, CA 95035
    (800) 624-5545

## Keyboards

PC Discount Center (Also cases, power supplies, motherboards, video cards)
    4341 Dipaolo Center
    Glenview, IL 60025
    (800) 245-7453

## Memory, RAM

MicroTech (Also processors)
    7304 15th Ave. N.E.
    Seattle, WA 98115
    (800) 521-9035

Worldwide Technologies (Also motherboards, video cards)
    21 S. Fifth St.
    Suite 900
    Philadelphia, PA 19106
    (800) 457-6937

## Monitors

Ralin Wholesalers, Inc. (Also cases, motherboards, video cards)
    P.O. Box 450
    Orchard Park, NY 14127
    (800) 752-9512

## ROM Replacements

Phoenix Technologies, Ltd. (Replacement ROMs)
    846 University Ave.
    Norwood, MA 02062
    (617) 551-4000

## Schematics and Documentation

Chips and Technologies, Inc. (Pinouts of VLSI SMT chipsets)
    2950 Zanker Rd.
    San Jose, CA 95117
    (800) 782-8148

Computer Component Source (Schematics)
    135 Eileen Way
    Syosset, NY 11791
    (800) 356-1227

Conner
    1650 Sun Flower Ave.
    Costa Mesa, CA 92626
    (800) 626-6637

C.R.C. Components, Inc. (Schematics for monitors, terminals, system boards)
    186 University Parkway
    Pomona, CA 91768
    (714) 468-9711

Eagan Technical Services
    1408 Northland Dr., #304
    Mendota Heights, MN 55120
    (612) 688-0098

Fujitsu
    2904 Orchard Parkway
    San Jose, CA 95134-2009
    (800) 626-4686

Howard W. Sams and Co., Inc. (Sams *Computerfacts* schematics
    and troubleshooting guides)
    4300 West 62nd St.
    P.O. Box 7092
    Indianapolis, IN 46206
    (800) 428-7267

Landmark Research International Corp. (System Speed Test; Alignit for floppies;
    Kickstart, a diagnostic card; and Service Diagnostics, PC Probe, a diagnostic program;
    and more)
    703 Grand Central St.
    Clearwater, FL 34616
    (800) 683-6696

Maxtor
2190 Miller Dr.
Longmont, CO 80501
(303) 651-6000 voice
(303) 678-2222 modem
(800) 262-9867
http://www.maxtor.com

Micropolis
21211 Nordhoff
Chatsworth, CA 91311
(800) 395-3748
http://www.micropolis.com

Samsung
105 Challenger Rd.
Richfield Park, NJ 07660
(800) 229-2239
http://www.samsung.com

SeaGate Technologies
920 Disk Dr.
Scotts Valley, CA 95066-4544
(800) 468-3472
http://www.seagate.com

Teac Corp. of America
7733 Telegraph Rd.
Montebello, CA 90640
(213) 726-0303

VLSI Engineering
1240 McKay Dr.
San Jose, CA 95131
(408) 434-7641

Western Digital Corp.
2445 McCabe Way
Irvine, CA 92714
(714) 932-5000
(800) 832-4778

# APPENDIX C
# Repair Depots

Many of the following repair all kinds of equipment, but each is listed under its specialty.

## Floppy Drives

Computer Repair Connection, Ltd.
    CRC Systems
    6900 W. 117th Ave.
    Broomfield, CO 80020
    (800) 231-0743

MicroNational, Inc. (repair, exchange)
    (800) 722-9917

Uptime Computer Support Services
    23633 Via Primero
    Valencia, CA 91355
    (805) 254-3384
    http://www.scv.net.com\~uptime

## Hard Drives

AA Computech, Inc.
    2810 Crocker Ave., #306
    Valencia, CA 91355
    (805) 257-6801

Advanced Technologies and Services, Inc.
    12567 Crenshaw Blvd.
    Hawthorne, CA 90250
    (310) 676-0487

Compusol, Inc.
    17344 Eastman St.
    Irvine, CA 92714
    (714) 253-9533

Data Solutions
    22699 Old Canal Rd.
    Yorba Linda, CA 92687
    (714) 637-5060

Independent Technology Service
    9146 Jordan Ave.
    Chatsworth, CA 91311
    (800) 874-6898

Maxdata Technologies
    3001 Redhill Ave.
    Building 3, Suite 108
    Costa Mesa, CA 92626
    (800) 354-2952

Winchester Service
    6485 Almar St.
    Simi Valley, CA 93063
    (805) 584-6409

## Keyboards

Technical Equipment Services
    256 Great Rd.
    Littleton, MA 01460
    (508) 486-0600

## Monitors

*Note:* Monitor flat rates do *not* include CRTs or flybacks.

Compusol, Inc.
    17344 Eastman St.
    Irvine, CA 92714
    (714) 253-9533

Computers, Service, & Support
    1206 Northeast Fourth Ave.
    Fort Lauderdale, FL 33304
    (800) 669-2311

Kennsco Engineering
    600 Ansin Blvd.
    Hallandale, FL 33009
    (800) 966-9279

Midwest Monitor Service
    2709 South K St.
    Belleville, KS 66935
    (913) 527-2602

3E Corp.
    165 Front Street
    Chicopee, MA 01013
    (800) 682-5175

## Motherboards (System Boards)

Computer Best Service
    472 11th St.
    Palisade Park, NJ 07650
    (201) 461-0479

Micro Repair Center
    1139 Stone Gate Rd.
    Shrub Oak, NY 10588
    (800) 829-6671

## Power Supplies

Compusol, Inc.
    17344 Eastman St.
    Irvine, CA 92714
    (714) 253-9533

## Printer Boards

Microtrace, Inc.
    8802 Basil St., Suite F
    Indianapolis, IN 46256-1288
    (800) 666-5611

## Printer Heads

Computer Terminal Repair
    8971 Quiet Canyon Rd.
    Placerville, CA 95667
    (916) 621-2888

# APPENDIX D
# The ASCII Code

| ASCII Value | Control Character | Key | ASCII Value | Key | ASCII Value | Key | ASCII Value | Key |
|---|---|---|---|---|---|---|---|---|
| 00 | NUL | Ctrl+@ | 32 | SPC | 64 | @ | 96 | ` |
| 01 | SOH | Ctrl+A | 33 | ! | 65 | A | 97 | a |
| 02 | STX | Ctrl+B | 34 | " | 66 | B | 98 | b |
| 03 | ETX | Ctrl+C | 35 | # | 67 | C | 99 | c |
| 04 | EOT | Ctrl+D | 36 | $ | 68 | D | 100 | d |
| 05 | ENQ | Ctrl+E | 37 | % | 69 | E | 101 | e |
| 06 | ACK | Ctrl+F | 38 | & | 70 | F | 102 | f |
| 07 | BEL | Ctrl+G | 39 | ' | 71 | G | 103 | g |
| 08 | BS | Ctrl+H | 40 | ( | 72 | H | 104 | h |
| 09 | HT | Ctrl+I | 41 | ) | 73 | I | 105 | i |
| 10 | LF | Ctrl+J | 42 | * | 74 | J | 106 | j |
| 11 | VT | Ctrl+K | 43 | + | 75 | K | 107 | k |
| 12 | FF | Ctrl+L | 44 | , | 76 | L | 108 | l |
| 13 | CR | Ctrl+M | 45 | - | 77 | M | 109 | m |
| 14 | SO | Ctrl+N | 46 | . | 78 | N | 110 | n |
| 15 | SI | Ctrl+O | 47 | / | 79 | O | 111 | o |
| 16 | DLE | Ctrl+P | 48 | 0 | 80 | P | 112 | p |
| 17 | DC1 | Ctrl+Q | 49 | 1 | 81 | Q | 113 | q |
| 18 | DC2 | Ctrl+R | 50 | 2 | 82 | R | 114 | r |
| 19 | DC3 | Ctrl+S | 51 | 3 | 83 | S | 115 | s |
| 20 | CD4 | Ctrl+T | 52 | 4 | 84 | T | 116 | t |
| 21 | NAK | Ctrl+U | 53 | 5 | 85 | U | 117 | u |
| 22 | SYN | Ctrl+V | 54 | 6 | 86 | V | 118 | v |
| 23 | ETB | Ctrl+W | 55 | 7 | 87 | W | 119 | w |
| 24 | CAN | Ctrl+X | 56 | 8 | 88 | X | 120 | x |
| 25 | EM | Ctrl+Y | 57 | 9 | 89 | Y | 121 | y |
| 26 | SUB | Ctrl+Z | 58 | : | 90 | Z | 122 | z |
| 27 | ESC | Ctrl+[ | 59 | ; | 91 | [ | 123 | { |
| 28 | FS | Ctrl+\ | 60 | < | 92 | \ | 124 | ¦ |
| 29 | GS | Ctrl+] | 61 | = | 93 | ] | 125 | } |
| 30 | RS | Ctrl+^ | 62 | > | 94 | ^ | 126 | ~ |
| 31 | US | Ctrl+_ | 63 | ? | 95 | _ | 127 | Del |

# THE EXTENDED ASCII CODE

Codes 1–26 are Ctrl+A through Crtl+Z. Graphic codes 127–255 can be entered by holding down the Alt key and pressing the three-digit code on the keypad, then releasing the Alt key.

| | | | | | | | | | | | | | |
|---|---|---|---|---|---|---|---|---|---|---|---|---|---|
| 1 | ☺ | 2 | ☻ | 3 | ♥ | 4 | ♦ | 5 | ♣ | 6 | ♠ | 7 | • |
| 8 | ◘ | 9 | ○ | 10 | ◙ | 11 | ♂ | 12 | ♀ | 13 | ♪ | 14 | ♫ |
| 15 | ☼ | 16 | ► | 17 | ◄ | 18 | ↕ | 19 | ‼ | 20 | ¶ | 21 | § |
| 22 | ▬ | 23 | ↨ | 24 | ↑ | 25 | ↓ | 26 | → | 27 | ← | 28 | ∟ |
| 29 | ↔ | 30 | ▲ | 31 | ▼ | 32 | | 33 | ! | 34 | " | 35 | # |
| 36 | $ | 37 | % | 38 | & | 39 | ' | 40 | ( | 41 | ) | 42 | * |
| 43 | + | 44 | , | 45 | - | 46 | . | 47 | / | 48 | 0 | 49 | 1 |
| 50 | 2 | 51 | 3 | 52 | 4 | 53 | 5 | 54 | 6 | 55 | 7 | 56 | 8 |
| 57 | 9 | 58 | : | 59 | ; | 60 | < | 61 | = | 62 | > | 63 | ? |
| 64 | @ | 65 | A | 66 | B | 67 | C | 68 | D | 69 | E | 70 | F |
| 71 | G | 72 | H | 73 | I | 74 | J | 75 | K | 76 | L | 77 | M |
| 78 | N | 79 | O | 80 | P | 81 | Q | 82 | R | 83 | S | 84 | T |
| 85 | U | 86 | V | 87 | W | 88 | X | 89 | Y | 90 | Z | 91 | [ |
| 92 | \ | 93 | ] | 94 | ^ | 95 | _ | 96 | ` | 97 | a | 98 | b |
| 99 | c | 100 | d | 101 | e | 102 | f | 103 | g | 104 | h | 105 | i |
| 106 | j | 107 | k | 108 | l | 109 | m | 110 | n | 111 | o | 112 | p |
| 113 | q | 114 | r | 115 | s | 116 | t | 117 | u | 118 | v | 119 | w |
| 120 | x | 121 | y | 122 | z | 123 | { | 124 | ¦ | 125 | } | 126 | ~ |
| 127 | | 128 | Ç | 129 | ü | 130 | é | 131 | â | 132 | ä | 133 | à |
| 134 | å | 135 | ç | 136 | ê | 137 | ë | 138 | é | 139 | ï | 140 | î |
| 141 | ì | 142 | Ä | 143 | Å | 144 | É | 145 | æ | 146 | Æ | 147 | ô |
| 148 | ö | 149 | ò | 150 | û | 151 | ù | 152 | ij | 153 | Ö | 154 | Ü |
| 155 | ¢ | 156 | £ | 157 | ¥ | 158 | ₧ | 159 | ƒ | 160 | á | 161 | í |
| 162 | ó | 163 | ú | 164 | ñ | 165 | Ñ | 166 | ª | 167 | º | 168 | ¿ |
| 169 | ⌐ | 170 | ¬ | 171 | ½ | 172 | ¼ | 173 | ¡ | 174 | << | 175 | >> |
| 176 | ░ | 177 | ▒ | 178 | ▓ | 179 | │ | 180 | ┤ | 181 | ╡ | 182 | ╢ |
| 183 | ╖ | 184 | ╕ | 185 | ╣ | 186 | ║ | 187 | ╗ | 188 | ╝ | 189 | ╜ |
| 190 | ╛ | 191 | ┐ | 192 | └ | 193 | ┴ | 194 | ┬ | 195 | ├ | 196 | ─ |
| 197 | ┼ | 198 | ╞ | 199 | ╟ | 200 | ╚ | 201 | ╔ | 202 | ╩ | 203 | ╦ |
| 204 | ╠ | 205 | ═ | 206 | ╬ | 207 | ╧ | 208 | ╨ | 209 | ╤ | 210 | ╥ |
| 211 | ╙ | 212 | ╘ | 213 | ╒ | 214 | ╓ | 215 | ╫ | 216 | ╪ | 217 | ┘ |
| 218 | ┌ | 219 | █ | 220 | ▄ | 221 | ▌ | 222 | ▐ | 223 | ▀ | 224 | α |
| 225 | β | 226 | Γ | 227 | π | 228 | Σ | 229 | σ | 230 | µ | 231 | τ |
| 232 | φ | 233 | θ | 234 | Ω | 235 | δ | 236 | ∞ | 237 | ø | 238 | ε |
| 239 | ∩ | 240 | ≡ | 241 | ± | 242 | ≥ | 243 | ≤ | 244 | ⌠ | 245 | ⌡ |
| 246 | ÷ | 247 | ≈ | 248 | ° | 249 | • | 250 | · | 251 | √ | 252 | ⁿ |
| 253 | ² | 254 | ■ | 255 | | | | | | | | | |

# APPENDIX E
# Hard Drive Parameters

| Model | Interface | Form Factor | Capacity | Cylinders | Heads | Read/Write Cylinders | Write/Precomp Cylinders | Encoding Cylinders | Sectors | Other |
|---|---|---|---|---|---|---|---|---|---|---|
| **Alps Electric** | | | | | | | | | | |
| DRND10A | ST506 | 3.5" | 10 | 615 | 2 | 616 | 616 | MFM | 17 | |
| DRND20A | ST506 | 3.5" | 2 | 615 | 4 | 616 | 616 | MFM | 17 | |
| RPO20A | ST506 | 3.5" | 20 | 615 | 2 | 616 | 616 | RLL | 26 | |
| DRPO20D | ST506 | 3.5" | 20 | 615 | 2 | 616 | 616 | RLL | 26 | |
| **Ampex** | | | | | | | | | | |
| PYXIS7 | ST506 | 5.25" FH | 5 | 320 | 2 | 132 | 132 | MFM | 17 | |
| PYXIS13 | ST506 | 5.25" FH | 10 | 320 | 4 | 132 | 132 | MFM | 17 | |
| PYXIS20 | ST506 | 5.25" FH | 15 | 320 | 6 | 132 | 132 | MFM | 17 | |
| PYXIS27 | ST506 | 5.25" FH | 20 | 320 | 8 | 132 | 132 | MFM | 17 | |
| **Atasi** | | | | | | | | | | |
| 3020 | ST506 | | 17 | 645 | 3 | 320 | 320 | MFM | 17 | |
| 3033 | ST506 | | 28 | 645 | 5 | 320 | 320 | MFM | 17 | |
| 3046 | ST506 | | 39 | 645 | 7 | 320 | 320 | MFM | 17 | |
| 3051 | ST506 | | 43 | 704 | 7 | 359 | 350 | MFM | 17 | |
| 3053 | ST506 | | 44 | 733 | 7 | 350 | | MFM | 17 | |
| 3075 | ST506 | | 67 | 1024 | 8 | 1025 | 1025 | MFM | 17 | |
| 3085 | ST506 | | 67 | 1024 | 8 | 512 | | MFM | 17 | |
| **BASF** | | | | | | | | | | |
| 6185 | ST506 | | 23 | 440 | 6 | 220 | 220 | MFM | 17 | |
| 6186 | ST506 | | 15 | 440 | 4 | 220 | 220 | MFM | 17 | |
| 6187 | ST506 | | 8 | 440 | 2 | 220 | 220 | MFM | 17 | |

| Model | Interface | Form Factor | Capacity | Cylinders | Heads | Read/Write Cylinders | Write/Precomp Cylinders | Encoding Cylinders | Sectors | Other |
|---|---|---|---|---|---|---|---|---|---|---|
| **Bull** | | | | | | | | | | |
| D530 | ST506 | | 25 | 987 | 3 | 988 | 988 | MFM | 17 | |
| D550 | ST506 | | 43 | 987 | 5 | 988 | 988 | MFM | 17 | |
| D570 | ST506 | | 59 | 987 | 7 | 988 | 988 | MFM | 17 | |
| **C. Itoh** | | | | | | | | | | |
| YD3530 | ST506 | | 32 | 731 | 5 | 732 | 732 | MFM | 17 | |
| YD3540 | ST506 | 5.25" HH | 45 | 731 | 7 | 732 | 732 | MFM | 17 | |
| YD3042 | SCSI | 5.25" FH | 43 | 788 | 4 | 789 | 789 | RLL | | |
| YD3082 | SCSI | 5.25" FH | 87 | 788 | 8 | 789 | 789 | RLL | | |
| **Cardiff** | | | | | | | | | | |
| F3053 | ST506 | 3.5" | 44 | 1024 | 5 | | | MFM | 17 | |
| F3080 | ESDI/SCSI | 3.5" | 68 | 1024 | 5 | | | R/N | 26 | |
| F3127 | ESDI/SCSI | 3.5" | 109 | 1024 | 5 | | | R/N | 35 | |
| **CDC** | | | | | | | | | | |
| 94244219 | AT IDE | 5.25" HH | 19 | 1747 | 4 | 1748 | 1748 | RLL | | |
| 94244383 | AT IDE | 5.25" HH | 383 | 1747 | 7 | 1748 | 1748 | RLL | | |
| 94354126 | AT IDE | | 111 | 1072 | 7 | 1073 | 1073 | | 29 | |
| 94354160 | AT IDE | | 143 | 1072 | 9 | 1073 | 1073 | | 29 | |
| 94354200 | AT IDE | | 177 | 1072 | 9 | 1073 | 1073 | | 36 | |
| 94354230 | AT IDE | | 311 | 1272 | 9 | 1273 | 1273 | | 36 | |
| BJ7D5A 77731614 | AT IDE | 5.25" FH | 23 | 67 | 4 | 671 | 128 | MFM | 17 | |
| BJ7D5A 77731608 | AT IDE | 5.25" FH | 29 | 670 | 5 | 671 | 128 | MFM | 17 | |
| BJ7D5A 77731613 | AT IDE | | | 733 | 5 | 734 | 128 | | 17 | |
| 9415521 WREN1 | AT IDE | 5.25" FH | 21 | 697 | 3 | 698 | 128 | MFM | 17 | |
| 9415525 | AT IDE | | 24 | 697 | 4 | 698 | 128 | MFM | 17 | |
| 9415528 | AT IDE | | 24 | 697 | 4 | 698 | 128 | MFM | 17 | |
| 9415536 WREN1 | AT IDE | 5.25" FH | 36 | 697 | 5 | 698 | 128 | MFM | 17 | |
| 9415538 | AT IDE | | 31 | 733 | 5 | 734 | 128 | MFM | 17 | |
| 9415548 WREN1 | AT IDE | 5.25" FH | 36 | 697 | 5 | 698 | 128 | MFM | 17 | |
| 9415538 | AT IDE | | 31 | 733 | 5 | 734 | 128 | MFM | 17 | |
| 9415548 WREN2 | AT IDE | 5.25" FH | 40 | 925 | 5 | 926 | 128 | MFM | 17 | |
| 9429551 WREN2 | AT IDE | 5.25" FH | 43 | 989 | 5 | 990 | 128 | MFM | 17 | |
| 9415557 WREN2 | AT IDE | 5.25" FH | 48 | 925 | 6 | 926 | 128 | MFM | 17 | |
| 9415567 WREN2 | ST506 | 5.25" FH | 56 | 925 | 7 | 926 | 128 | MFM | 17 | |
| 9415577 WREN2 | ST506 | | 64 | 925 | 8 | 926 | 128 | MFM | 17 | |
| 9415585 WREN2 | ST506 | 5.25" FH | 71 | 1024 | 8 | 1025 | 128 | MFM | 17 | |
| 9415586 WREN2 | ST506 | 5.25" FH | 72 | 925 | 9 | 926 | 128 | MFM | 17 | |
| 9420551 | ST506 | 5.25" HH | 43 | 989 | 5 | 990 | 128 | MFM | 17 | |
| 9433555 | ST506 | 3.5" | 46 | | 5 | | | MFM | 17 | |
| 94335100 | ST506 | 3.5" | 83 | | 9 | | | MFM | 17 | |
| 9435555 SWIFT2 | ST506 | 3.5" | 46 | | 5 | | | MFM | 17 | |
| 94355100 SWIFT | ST506 | 3.5" | 88 | 1072 | 9 | 1073 | 300 | MFM | 17 | |

| Model | Interface | Form Factor | Capacity | Cylinders | Heads | Read/Write Cylinders | Write/Precomp Cylinders | Encoding Cylinders | Sectors | Other |
|---|---|---|---|---|---|---|---|---|---|---|
| **CDC (*continued*)** | | | | | | | | | | |
| 94155135 WREN2 | ST506 | 5.25" HH | 115 | 960 | 9 | 961 | 128 | RLL | 26 | |
| 9420577 WREN2 | ST506 | 5.25" HH | 63 | 989 | 5 | 990 | 128 | RLL | 26 | |
| 94335150 | ST506 | 3.5" | 128 | | 9 | | | RLL | 26 | |
| 94355150 | ST506 | 3.5" | 133 | 1072 | 9 | 1073 | 300 | RLL | 28 | |
| 9415648 WREN2 | ESDI | | 40 | 925 | 5 | 926 | 128 | N | | |
| 9415667 WREN2 | ESDI | | 56 | 925 | 7 | 926 | 128 | N | | |
| 9415686 WREN2 | ESDI | | 72 | 925 | 9 | 926 | 128 | N | | |
| 9416610 WREN3 | ESDI | 5.25" FH | 86 | 969 | 5 | 97 | 128 | N | | |
| 94166141 WREN3 | ESDI | 5.25" FH | 121 | 969 | 7 | 970 | 128 | N | | |
| 94166182 WREN3 | ESDI | 5.25" FH | 155 | 969 | 9 | 970 | 128 | N | | |
| 94181702 WRENV | ESDI | 5.25" FH | 702 | 1549 | 15 | 1550 | 128 | Z | | |
| 94186265 WRENV | ESDI | 5.25" FH | 265 | 1412 | 9 | 1413 | 128 | | | |
| 94186324 WRENV | ESDI | 5.25" FH | 324 | 1412 | 11 | 1413 | 128 | | | |
| 94186383 WRENV | ESDI | 5.25" FH | 383 | 1412 | 13 | 1413 | 128 | R/N | | |
| 94186383H WRENV | ESDI | 5.25" FH | 383 | 1224 | 15 | 1225 | 128 | R/N | | |
| 94196766 WRENVI | ESDI | 5.25" FH | 766 | 1632 | 15 | | | RLL | | |
| 94216106 WREN3 | ESDI | 5.25" FH | 91 | 1024 | 5 | 1025 | 128 | R/N | | |
| 94246182 WRENVI | ESDI | 5.25" HH | 182 | 1453 | 4 | 1454 | 128 | RLL | | |
| 94246383 WRENVI | ESDI | 5.25" HH | 383 | 1747 | 7 | 1748 | 128 | RLL | | |
| 94356111 SWIFT | ESDI | 3.5" | 98 | 1072 | 5 | 1073 | 1073 | R/N | 34/36 | |
| 94356155 SWIFT | ESDI | 3.5" | 138 | 1072 | 7 | 1073 | 1073 | R/N | 34/36 | |
| 94356200 SWIFT | ESDI | 3.5" | 177 | 1072 | 9 | 1073 | 1073 | R/N | 34/36 | |
| WRENIII | ESDI | 5.25" HH | 106 | 969 | 5 | 970 | 123 | R/N | | |
| 916186 WREN3 | SCSI | 5.25" FH | 86 | 969 | | 970 | 970 | | | |
| 94161121 WREN3 | SCSI | 5.25" FH | 121 | 969 | | 970 | 970 | | | |
| 94171300 WREN4 | SCSI | 5.25" FH | 300 | 1365 | 9 | 1366 | 1366 | RLL | | |
| 94171344 WRENV | SCSI | 5.25" FH | 344 | 1549 | 9 | 1550 | 1550 | Z | | |
| 94171350 WRENIV | SCSI | 5.25" FH | 350 | 1412 | 9 | 1550 | 1550 | Z | | |
| 94171376 WRENIV | SCSI | 5.25" FH | 376 | 1549 | 9 | 1550 | 1550 | Z | | |
| 94181574 WRENV | SCSI | 5.25" FH | 574 | 1549 | 15 | 1550 | 1550 | Z | | |
| 94181702 WRENV | SCSI | 5.25" FH | 702 | 1549 | 15 | 1550 | 128 | Z | | |
| 94186383 WRENV | SCSI | 5.25" FH | 383 | 1412 | 13 | 1413 | 128 | R/N | | |
| 94186383H WRENV | SCSI | 5.25" FH | 383 | 1224 | 15 | 1225 | 128 | R/N | | |
| 94186442 WRENV | SCSI | 5.25" FH | 442 | 1412 | 15 | 1413 | 128 | R/N | | |
| 94191766 WRENVI | SCSI | 5.25" FH | 766 | 1632 | 15 | | | RLL | | |
| 9421191 WREN3 | SCSI | 5.25" FH | 91 | 969 | | 970 | 970 | | | |
| 94211106 WREN3 | SCSI | 5.25" HH | | 1024 | 5 | 1025 | 1025 | RLL | | |
| 94221190 WRENV | SCSI | 5.25" HH | 190 | 1547 | 5 | 1548 | 1548 | RLL | | |
| 94211209 WRENV | SCSI | 5.25" HH | | 1547 | 5 | 1548 | 1548 | R/Z | | |
| 94351128 SWIFT | SCSI | 3.5" | 111 | 1068 | 7 | 1069 | 1069 | | 36 | |
| 94351134 SWIFT | SCSI | 3.5" | 134 | | 7 | | | | | |
| 94351172 SWIFT | SCSI | 3.5" | 172 | | 9 | | | | | |
| 94351160 SWIFT | SCSI | 3.5" | 142 | 1068 | 9 | 1069 | 1069 | | 36 | |
| 94351200 SWIFT | SCSI | 3.5" | 177 | 1068 | 9 | 1069 | 1069 | | 36 | |
| 94351200S SWIFT | SCSI | 3.5" | 177 | 1068 | 9 | 1069 | 1069 | | 36 | |

| Model | Interface | Form Factor | Capacity | Cylinders | Heads | Read/Write Cylinders | Write/Precomp Cylinders | Encoding Cylinders | Sectors | Other |
|---|---|---|---|---|---|---|---|---|---|---|
| **CDC (*continued*)** | | | | | | | | | | |
| 94351230S SWIFT | SCSI | 3.5" | 211 | 1272 | 9 | 1273 | 1273 | | 36 | |
| 9720368 SABRE | SCSI | 8.0" FH | 316 | 1217 | 10 | 1218 | 1218 | R/N | | |
| 9720500 SABRE | SCSI | 8.0" FH | 427 | 1217 | 10 | 1218 | 1218 | R/N | | |
| 9720736 SABRE | SCSI | 8.0" FH | 637 | 1635 | 15 | 1636 | 1636 | R/N | | |
| 9720850 SABRE | SCSI | 8.0" FH | 727 | 1381 | 15 | 1382 | 1382 | R/N | | |
| 97201230 SABRE | SCSI | 8.0" FH | 1056 | 1635 | 15 | 1636 | 1636 | R/N | | |
| WREN III | SCSI | 5.25" HH | 106 | 969 | 5 | 970 | 970 | R/N | | |
| **Century Data** | | | | | | | | | | |
| CAST10203E | ST506 | 5.25" FH | 55 | | 3 | 1051 | 1051 | R/N | 35 | |
| CAST10304 | ST506 | 5.25" FH | 75 | 1050 | 4 | 1051 | 1051 | R/N | 35 | |
| CAST10305 | ST506 | 5.25" FH | 94 | 1050 | 5 | 1051 | 1051 | R/N | 35 | |
| CAST14404 | ST506 | 5.25" HH | 114 | 1590 | 4 | 1591 | 1591 | R/N | 35 | |
| CAST14405 | ST506 | 5.25" HH | 140 | 1590 | 5 | 1591 | 1591 | R/N | 35 | |
| CAST14406 | ST506 | 5.25" HH | 170 | 1590 | 6 | 1591 | 1591 | R/N | 35 | |
| CAST24509 | ST506 | 5.25" FH | 258 | 1599 | 9 | 1600 | 1600 | R/N | 35 | |
| CAST24611 | ST506 | 5.25" FH | 315 | 1599 | 11 | 1600 | 1600 | R/N | 35 | |
| CAST24713 | ST506 | 5.25" FH | 372 | 1599 | 13 | 1600 | 1600 | R/N | 35 | |
| CAST10203S | SCSI | 5.25" FH | 55 | 1050 | 3 | 1051 | 1051 | | 35 | |
| CAST10304S | SCSI | 5.25" FH | 75 | 1050 | 4 | 1051 | 1051 | RLL | 35 | |
| CAST10205S | SCSI | 5.25" FH | 94 | 1050 | 5 | 1051 | 1051 | RLL | 35 | |
| CAST14404S | SCSI | 5.25" HH | 114 | 1590 | 4 | 1591 | 1591 | RLL | 35 | |
| CAST14405S | SCSI | 5.25" HH | 140 | 1590 | 5 | 1591 | 1591 | RLL | 35 | |
| CAST14406S | SCSI | 5.25" HH | 170 | 1590 | 6 | 1591 | 1591 | RLL | 35 | |
| CAST24509S | SCSI | 5.25" FH | 258 | 1599 | 9 | 1600 | 1600 | RLL | 35 | |
| CAST24611S | SCSI | 5.25" FH | 315 | 1599 | 11 | 1600 | 1600 | RLL | 35 | |
| CAST24713S | SCSI | 5.25" FH | 372 | 1599 | 13 | 1600 | 1600 | RLL | 35 | |
| **CMI** | | | | | | | | | | |
| 3426 | ST506 | | 20 | 615 | 4 | 616 | 256 | MFM | 17 | |
| 5206 | ST506 | | 5 | 306 | 2 | 307 | 256 | MFM | 17 | |
| 5205 | ST506 | | 4 | 256 | 2 | 128 | 128 | MFM | 17 | |
| 5410 | ST506 | | 8 | 256 | 4 | 128 | 128 | MFM | 17 | |
| 5412 | ST506 | | 10 | 306 | 4 | 307 | 128 | MFM | 17 | |
| 5616 | ST506 | | 13 | 256 | 6 | 257 | 257 | MFM | 17 | |
| 5619 | ST506 | | 15 | 306 | 6 | 307 | 128 | MFM | 17 | |
| 6213 | ST506 | | 11 | 640 | 2 | 641 | 256 | MFM | 17 | |
| 6426 | ST506 | | 21 | 640 | 4 | 641 | 256 | MFM | 17 | |
| 6640 | ST506 | | 33 | 615 | 6 | 616 | 256 | MFM | 17 | |
| 7660 | ST506 | | 50 | 960 | 6 | 961 | 450 | MFM | 17 | |
| 7880 | ST506 | | 67 | 960 | 8 | 961 | 450 | MFM | 17 | |

| Model | Interface | Form Factor | Capacity | Cylinders | Heads | Read/Write Cylinders | Write/Precomp Cylinders | Encoding Cylinders | Sectors | Other |
|---|---|---|---|---|---|---|---|---|---|---|
| **COGITO** | | | | | | | | | | |
| CG906 | ST506 | | 5 | 306 | 2 | 128 | 128 | MFM | | |
| CG912 | ST506 | | 11 | 306 | 4 | 128 | 128 | MFM | | |
| PT912 | ST506 | | 11 | 612 | 2 | 307 | 307 | MFM | | |
| PT925 | ST506 | | 21 | 612 | 4 | 307 | 307 | MFM | | |
| **Conner Peripherals** | | | | | | | | | | |
| CP342 | AT IDE | 3.5" | 40 | 805 | 4 | | | RLL | | |
| CP344 | AT IDE | 3.5" | 42 | 805 | 4 | | | RLL | 26/27 | Voice |
| CP3022 | AT IDE | 3.5" | 21 | 636 | 2 | | | RLL | | |
| CP3024 | AT IDE | 3.5" | 21 | 636 | 2 | | | RLL | 33/34 | Voice |
| CP3044 | AT IDE | 3.5" | 42 | 1047 | 2 | | | RLL | 41 | Voice |
| CP3102 | AT IDE | 3.5" | 104 | 776 | 8 | | | RLL | | |
| CP3104 | AT IDE | 3.5" | 104 | 776 | 8 | | | RLL | 33/34 | Voice |
| CP2304 | AT IDE | 3.5" | 209 | 1348 | 8 | | | RLL | 39 | Voice |
| CP340 | SCSI | 3.5" | 42 | 788 | 4 | | | RLL | 26/27 | Voice |
| CP3020 | SCSI | 3.5" | 21 | 636 | 2 | | | RLL | 33/34 | Voice |
| CP3040 | SCSI | 3.5" | 42 | 1047 | 2 | | | RLL | 41 | Voice |
| CP3100 | SCSI | 3.5" | 104 | 776 | 8 | | | RLL | 33/34 | Voice |
| CP3200 | SCSI | 3.5" | 209 | 1348 | 8 | | | RLL | 39 | Voice |
| **Core International** | | | | | | | | | | |
| AT 32 | ST506 | 5.25" HH | 31 | 733 | 5 | | | MFM | 17 | |
| AT 30 | ST506 | 5.25" FH | 31 | 733 | 5 | | | MFM | 17 | |
| AT 40 | ST506 | 5.25" FH | 40 | 924 | 5 | | | MFM | 17 | |
| AT 63 | ST506 | 5.25" FH | 42 | 988 | 5 | | | MFM | 17 | |
| AT 72 | ST506 | 5.25" FH | 72 | 924 | 9 | | | MFM | 17 | |
| OPTIMA 30 | ST506 | 5.25" HH | 31 | 733 | 5 | | | MFM | 17 | |
| OPTIMA 40 | ST506 | 5.25" HH | 41 | 963 | 5 | | | MFM | 17 | |
| OPTIMA 70 | ST506 | 5.25" FH | 71 | 918 | 9 | | | MFM | 17 | |
| RLLAT 32 | ST506 | 5.25" HH | 48 | 733 | 5 | | | RLL | 26 | |
| AT 30 | ST506 | 5.25" FH | 48 | 733 | 5 | | | RLL | 26 | |
| AT 40 | ST506 | 5.25" FH | 61 | 924 | 5 | | | RLL | 26 | |
| AT 63 | ST506 | 5.25" FH | 65 | 988 | 5 | | | RLL | 26 | |
| AT 72 | ST506 | 5.25" FH | 107 | 924 | 9 | | | RLL | 26 | |
| OPTIMA 30 | ST506 | 5.25" HH | 48 | 733 | 5 | | | RLL | 26 | |
| OPTIMA 40 | ST506 | 5.25" HH | 64 | 963 | 5 | | | RLL | 26 | |
| OPTIMA 70 | ST506 | 5.25" FH | 109 | 918 | 9 | | | RLL | 26 | |
| HC 40 | ESDI | 5.25" FH | 40 | 564 | 4 | | | R/N | 35 | |
| HC 90 | ESDI | 5.25" HH | 91 | 969 | 5 | | | R/N | 35 | |
| HC 150 | ESDI | 5.25" FH | 156 | 969 | 9 | | | R/N | 35 | |
| HC 260 | ESDI | 5.25" FH | 260 | 1212 | 12 | | | R/N | 35 | |
| HC 310 | ESDI | 5.25" FH | 311 | 1582 | 12 | | | R/N | 35 | |

| Model | Interface | Form Factor | Capacity | Cylinders | Heads | Read/Write Cylinders | Write/Precomp Cylinders | Encoding Cylinders | Sectors | Other |
|-------|-----------|-------------|----------|-----------|-------|---------------------|-------------------------|--------------------|---------|-------|
| **Disctron** | | | | | | | | | | |
| D503 | ST506 | | 3 | 153 | 2 | | | MFM | 17 | |
| D504 | ST506 | | 4 | 215 | 2 | | | MFM | 17 | |
| D506 | ST506 | | 5 | 153 | 4 | | | MFM | 17 | |
| D507 | ST506 | | 5 | 306 | 2 | 128 | 128 | MFM | 17 | |
| D509 | ST506 | | 8 | 215 | 4 | 128 | 128 | MFM | 17 | |
| D512 | ST506 | | 11 | 153 | 8 | | | MFM | 17 | |
| D513 | ST506 | | 11 | 215 | 6 | 128 | 128 | MFM | 17 | |
| D514 | ST506 | | 11 | 306 | 4 | 128 | 128 | MFM | 17 | |
| D518 | ST506 | | 15 | 215 | 8 | 128 | 128 | MFM | 17 | |
| D519 | ST506 | | 16 | 306 | 6 | 128 | 128 | MFM | 17 | |
| D526 | ST506 | | 21 | 306 | 8 | 128 | 128 | MFM | 17 | |
| **DMA** | | | | | | | | | | |
| 306 | ST506 | | 11 | 612 | 2 | 612 | 400 | MFM | 17 | Removable media |
| **Elcoh** | | | | | | | | | | |
| DISCACHE10 | ST506 | | 10 | 320 | 4 | 321 | 321 | MFM | 17 | |
| DISCACHE20 | ST506 | | 20 | 320 | 8 | 321 | 321 | MFM | 17 | |
| **Fuji** | | | | | | | | | | |
| FK30526 | ST506 | 3.5" | 21 | 615 | 4 | | 616 | MFM | 17 | |
| FK30539 | ST506 | 3.5" | 32 | 615 | 6 | | 616 | MFM | 17 | |
| FK30926 | ST506 | 3.5" | 21 | 615 | 4 | | 616 | MFM | 17 | |
| FK301 | ST506 | | 10 | 306 | 4 | 307 | 128 | MFM | 17 | |
| FK30213 | ST506 | | 10 | 612 | 2 | 613 | 307 | MFM | 17 | |
| FK30226 | ST506 | | 21 | 612 | 4 | 613 | 307 | MFM | 17 | |
| FK30239 | ST506 | | 32 | 612 | 6 | 613 | 307 | MFM | 17 | |
| FK30352 | ST506 | 3.5" | 40 | 615 | 8 | | 616 | MFM | 17 | |
| FK30539R | ST506 | 3.5" | 32 | 615 | 4 | | 616 | RLL | 26 | |
| FK30558R | ST506 | 3.5" | 49 | 615 | 6 | | 616 | RLL | 26 | |
| FK30939R | ST506 | | 32 | 615 | 4 | | 616 | RLL | 26 | |
| FK308S58R | SCSI | 3.5" | 45 | 615 | 6 | | | RLL | | |
| FK308S39R | SCSI | 3.5" | 32 | 615 | 4 | | 616 | RLL | | |
| FK309S50R | SCSI | 3.5" | 41 | 615 | 4 | | | RLL | | |
| **Fujitsu** | | | | | | | | | | |
| M2611T | AT IDE | 3.5" | 45 | 1334 | 2 | | | RLL | | |
| M2612T | AT IDE | 3.5" | 90 | 1334 | 4 | | | RLL | | |
| M2613T | AT IDE | 3.5" | 135 | 1334 | 6 | | | RLL | | |
| M2614T | AT IDE | 3.5" | 180 | 1334 | 8 | | | RLL | | |
| 2230 AS | ST506 | | 5 | 320 | 2 | 321 | 321 | MFM | 17 | |
| 2233 AS | ST506 | | 10 | 320 | 4 | 321 | 321 | MFM | 17 | |
| 2234 AS | ST506 | | 15 | 320 | 6 | 321 | 321 | MFM | 17 | |

| Model | Interface | Form Factor | Capacity | Cylinders | Heads | Read/Write Cylinders | Write/Precomp Cylinders | Encoding Cylinders | Sectors | Other |
|---|---|---|---|---|---|---|---|---|---|---|
| **Fujitsu (continued)** | | | | | | | | | | |
| 2235 AS | ST506 | | 20 | 320 | 8 | 321 | 321 | MFM | 17 | |
| 2241 AS | ST506 | | 26 | 754 | 4 | 755 | 755 | MFM | 17 | |
| M2226D2 | ST506 | 3.5" | 30 | 615 | 6 | | | MFM | 17 | |
| MSSS7D2 | ST506 | 3.5" | 42 | 615 | 8 | | | MFM | 17 | |
| M2242AS2 | ST506 | 5.25" FH | 43 | 754 | 7 | | | MFM | 17 | |
| MSS43AS2 | ST506 | 5.25" FH | 67 | 754 | 11 | | | MFM | 17 | |
| M2243T | ST506 | 5.25" HH | 68 | 1186 | 7 | | | MFM | 17 | |
| M2225DR | ST506 | 3.5" | 32 | 615 | 4 | | | RLL | 26 | |
| M2226DR | ST506 | 3.5" | 49 | 615 | 6 | | | RLL | 26 | |
| M2227DR | ST506 | 3.5" | 65 | 615 | 8 | | | RLL | 26 | |
| M2243R | ST506 | 5.25" HH | 110 | 1186 | 7 | | | RLL | 26 | |
| 2244E | ESDI | 5.25" FH | 73 | 823 | 5 | | | R/N | 35 | |
| 2245E | ESDI | 5.25" FH | 120 | 823 | 7 | | | R/N | 35 | |
| M2246E | ESDI | 5.25" FH | 172 | 823 | 10 | | | R/N | 35 | |
| M2249 | ESDI | 5.25" FH | 389 | 1243 | 15 | | | RLL | | |
| M2263 | ESDI | 5.25" FH | 778 | 1658 | 15 | | | RLL | | |
| 2244SA | SCSI | | 73 | 823 | 5 | | | RLL | 35 | 16K |
| M2245SA | SCSI | | 120 | 823 | 7 | | | RLL | 35 | 16K |
| M2246SA | SCSI | 5.25" FH | 171 | 823 | 10 | | | RLL | 35 | 16K |
| M2249 | SCSI | 5.25" FH | 389 | 1243 | 15 | | | RLL | | |
| M2263 | SCSI | 5.25" FH | 778 | 1658 | 15 | | | RLL | | |
| M2344KS | SCSI | | 690 | 624 | 27 | | | RLL | | |
| M2611S | SCSI | 3.5" | 45 | 1334 | 2 | | | RLL | | |
| M2612S | SCSI | 3.5" | 90 | 1334 | 4 | | | RLL | | |
| M2613S | SCSI | 3.5" | 136 | 1334 | 6 | | | RLL | | |
| M2614S | SCSI | 3.5" | 182 | 1334 | 8 | | | RLL | | |
| **Hitachi** | | | | | | | | | | |
| DK3011 | ST506 | 3.5" | 10 | 306 | 4 | | | MFM | 17 | |
| DK3012 | ST506 | 3.5" | 15 | 306 | 6 | | | MFM | 17 | |
| DK5113 | ST506 | 5.25" FH | 28 | 699 | 5 | | | MFM | 17 | |
| DK5115 | ST506 | 5.25" FH | 40 | 699 | 7 | | | MFM | 17 | |
| DK5118 | ST506 | 5.25" FH | 67 | 823 | 10 | | | MFM | 17 | |
| DK5215 | ST506 | 5.25" HH | 51 | 823 | 6 | | | MFM | 17 | |
| DK5128 | ESDI | 5.25" FH | 67 | 823 | 5 | | | R/N | | |
| DK51212 | ESDI | 5.25" FH | 94 | 823 | 7 | | | R/N | | |
| DK51217 | ESDI | 5.25" FH | 134 | 823 | 10 | | | R/N | | |
| DK51438 | ESDI | 5.25" FH | 330 | 903 | 14 | | | R/N | 51 | |
| DK52210 | ESDI | 5.25" HH | 103 | 823 | 6 | | | R/N | 36 | |
| DK512C8 | SCSI | 5.25" FH | 67 | 823 | 5 | | | RLL | | |
| DK512C12 | SCSI | 5.25" FH | 94 | 823 | 7 | | | RLL | | |
| DK512C17 | SCSI | 5.25" FH | 134 | 819 | 10 | | | RLL | 35 | |
| DK522C10 | SCSI | 5.25" HH | 88 | 819 | 6 | | | RLL | 35 | |

| Model | Interface | Form Factor | Capacity | Cylinders | Heads | Read/Write Cylinders | Write/Precomp Cylinders | Encoding Cylinders | Sectors | Other |
|---|---|---|---|---|---|---|---|---|---|---|
| **IMI** | | | | | | | | | | |
| 5006 | ST506 | | 5 | 306 | 2 | 307 | 214 | MFM | 17 | |
| 5012 | ST506 | | 10 | 306 | 4 | 307 | 214 | MFM | 17 | |
| 5018 | ST506 | | 15 | 306 | 6 | 307 | 214 | MFM | 17 | |
| **JCT** | | | | | | | | | | |
| 100 | ST506 | 5.25" HH | 5 | | | | | MFM | 17 | |
| 105 | ST506 | 5.25" HH | 7 | | | | | MFM | 17 | |
| 110 | ST506 | 5.25" HH | 14 | | | | | MFM | 17 | |
| 120 | ST506 | 5.25" HH | 20 | | | | | MFM | 17 | |
| **Kalok Corporation** | | | | | | | | | | |
| KL343 Octagon 40 | AT Bus | | 4 | 644 | 4 | 645 | | RLL | 30 | |
| KL320 Octagon 20 | ST506 | 3.5" | 20 | 615 | 4 | 616 | 300 | MFM | 17 | |
| KL330 Octagon 30 | ST506 | | 32 | 616 | 4 | 617 | 617 | RLL | 26 | |
| KL332 Octagon 30 | PS/2 | | 40 | 615 | 4 | | | RLL | 30 | |
| KL341 Octagon 40 | SCSI | | 40 | 644 | 4 | | | RLL | 30 | |
| **Kyocera** | | | | | | | | | | |
| KC20A/KC20B | ST506 | 3.5" | 20 | 616 | 4 | | | MFM | 17 | |
| KC30A/KC30B | ST506 | 3.5" | 30 | 615 | 4 | | | RLL | 26 | |
| **La Pine** | | | | | | | | | | |
| 3522 | ST506 | | 10 | 306 | 4 | 307 | | MFM | 17 | |
| LT 10 | ST506 | | 10 | 615 | 2 | 616 | | MFM | 17 | |
| LT 20 | ST506 | | 20 | 615 | 4 | 616 | | MFM | 17 | |
| LT 200 | ST506 | | 20 | 614 | 4 | 615 | | MFM | 17 | |
| LT 2000 | ST506 | | 20 | 614 | 4 | 615 | | MFM | 17 | |
| LT 300 | ST506 | | 32 | 614 | 4 | 615 | | RLL | 26 | |
| **Maxtor** | | | | | | | | | | |
| XT1065 | ST506 | 5.25" FH | 56 | 918 | 7 | 919 | 919 | MFM | 17 | |
| XT1085 | ST506 | 5.25" FH | 71 | 1024 | 8 | 1025 | 1025 | MFM | 17 | |
| XT1105 | ST506 | 5.25" FH | 87 | 918 | 11 | 919 | 919 | MFM | 17 | |
| XT1140 | ST506 | 5.25" FH | 119 | 918 | 15 | 919 | 919 | MFM | 17 | |
| XT2085 | ST506 | 5.25" FH | 74 | 1224 | 7 | 1225 | 1225 | MFM | 17 | |
| XT2140 | ST506 | 5.25" FH | 117 | 1224 | 11 | 1225 | 1225 | MFM | 17 | |
| XT2190 | ST506 | 5.25" FH | 159 | 1224 | 15 | 1225 | 1225 | MFM | 17 | |
| XT1120R | ST506 | 5.25" FH | 104 | 1024 | 8 | | | RLL | 25 | |
| XT1240R | ST506 | 5.25" FH | 196 | 1024 | 15 | | | RLL | 25 | |
| XT4170E | ESDI | 5.25" FH | 157 | 1224 | 7 | | | R/N | 35/36 | |
| XT4175 | ESDI | 5.25" FH | 150 | 1224 | 7 | | | R/N | 35 | |
| ST4380E | ESDI | 5.25" FH | 338 | 1224 | 15 | | | R/N | 35/36 | |
| XT8380E | ESDI | 5.25" FH | 360 | 1632 | 8 | | | R/N | 54 | |
| XT8760E | ESDI | 5.25" FH | 676 | 1632 | 15 | | | R/N | 54 | |

| Model | Interface | Form Factor | Capacity | Cylinders | Heads | Read/Write Cylinders | Write/Precomp Cylinders | Encoding Cylinders | Sectors | Other |
|---|---|---|---|---|---|---|---|---|---|---|
| **Maxtor (*continued*)** | | | | | | | | | | |
| XT3170 | SCSI | 5.25" FH | 146 | 1224 | 9 | | | RLL | 48 | |
| XT3280 | SCSI | 5.25" FH | | 1224 | 15 | | | RLL | | |
| XT3380 | SCSI | 5.25" FH | | | 15 | | | RLL | | |
| XT4170S | SCSI | 5.25" FH | 157 | 1224 | 7 | | | RLL | 36 | |
| XT4280S | SCSI | 5.25" FH | 338 | 1224 | 11 | | | RLL | 36 | |
| XT4380S | SCSI | 5.25" FH | 337 | 1224 | 15 | | | RLL | 36 | |
| XT8380S | SCSI | 5.25" FH | 360 | 1632 | 8 | | | RLL | 54 | |
| XT8760S | SCSI | 5.25" FH | 676 | 1632 | 15 | | | RLL | 54 | |
| LXT200 | SCSI | 3.5" | | 201 | 7 | | | | | |
| LXT100 | SCSI | 3.5" | | 96 | 8 | | | | | |
| RXT800S | SCSI | 5.25" FH | 786 | | 2 | | | | | |
| TAHITI | SCSI | 5.25" FH | 650 | | 2 | | | | | |
| **Memorex** | | | | | | | | | | |
| 321 | ST506 | | 5 | 320 | 2 | 321 | 128 | MFM | 17 | |
| 322 | ST506 | | 10 | 320 | 4 | 321 | 128 | MFM | 17 | |
| 323 | ST506 | | 15 | 320 | 6 | 321 | 128 | MFM | 17 | |
| 324 | ST506 | | 20 | 320 | 6 | 321 | 128 | MFM | 17 | |
| 450 | ST506 | | 10 | 612 | 2 | 321 | 350 | MFM | 17 | |
| 512 | ST506 | | 25 | 961 | 3 | 321 | 480 | MFM | 17 | |
| 513 | ST506 | | 41 | 961 | 5 | 321 | 480 | MFM | 17 | |
| 514 | ST506 | | 58 | 961 | 7 | 321 | 480 | MFM | 17 | |
| **Micropolis** | | | | | | | | | | |
| 17435 | AT IDE | 3.5" | 112 | 1140 | 5 | 1141 | 1141 | | 28–48 | |
| 17446 | AT IDE | 3.5" | 135 | 1140 | 6 | 1141 | 1141 | | 28–48 | |
| 17447 | AT IDE | 3.5" | 157 | 1140 | 7 | 1141 | 1141 | | 28–48 | |
| 17458 | AT IDE | 3.5" | 180 | 1140 | 8 | 1141 | 1141 | | 28–48 | |
| 17459 | AT IDE | 3.5" | 202 | 1140 | 9 | 1141 | 1141 | | 28–48 | |
| 1302 | AT IDE | 5.25" FH | 21 | 830 | 3 | 831 | 831 | MFM | 17 | |
| 1303 | AT IDE | 5.25" FH | 36 | 830 | 5 | 831 | 831 | MFM | 17 | |
| 1304 | AT IDE | 5.25" FH | 43 | 830 | 6 | 831 | 831 | MFM | 17 | |
| 1323 | AT IDE | 5.25" FH | 35 | 1024 | 4 | 1025 | 1025 | MFM | 17 | |
| 1323A | AT IDE | 5.25" FH | 44 | 1024 | 5 | 1025 | 1025 | MFM | 17 | |
| 1324 | AT IDE | 5.25" FH | 53 | 1024 | 6 | 1025 | 1025 | MFM | 17 | |
| 1324A | AT IDE | 5.25" FH | 62 | 1024 | 7 | 1025 | 1025 | MFM | 17 | |
| 1325 | AT IDE | 5.25" FH | 71 | 1024 | 8 | 1025 | 1025 | MFM | 17 | |
| 1333A | AT IDE | 5.25" FH | 44 | 1024 | 5 | 1025 | 1025 | MFM | 17 | |
| 1334 | AT IDE | 5.25" FH | 53 | 1024 | 6 | 1025 | 1025 | MFM | 17 | |
| 1335 | AT IDE | 5.25" FH | 71 | 1024 | 8 | 1025 | 1025 | MFM | 17 | |
| 1352 | ESDI | | | 1024 | 2 | 1025 | 1025 | R/N | 36 | |
| 1352A | ESDI | | 41 | 1024 | 3 | 1025 | 1025 | R/N | 36 | |
| 1353 | ESDI | 5.25" FH | 75 | 1024 | 4 | 1025 | 1025 | R/N | 36 | (HS) |
| 1353A | ESDI | 5.25" FH | 94 | 1024 | 5 | 1025 | 1025 | R/N | 36 | (HS) |

## Micropolis (*continued*)

| Model | Interface | Form Factor | Capacity | Cylinders | Heads | Read/Write Cylinders | Write/Precomp Cylinders | Encoding Cylinders | Sectors | Other |
|---|---|---|---|---|---|---|---|---|---|---|
| 1354 | ESDI | 5.25" FH | 113 | 1024 | 6 | 1025 | 1025 | R/N | 36 | (HS) |
| 1354A | ESDI | 5.25" FH | 132 | 1024 | 7 | 1025 | 1025 | R/N | 36 | (HS) |
| 1355 | ESDI | 5.25" FH | 150 | 1024 | 8 | 1025 | 1025 | R/N | 36 | (HS) |
| 151610S | ESDI | 5.25" FH | 678 | 1840 | 10 | 1841 | 1841 | R/N | 72 | |
| 151713 | ESDI | 5.25" FH | 922 | 1925 | 13 | 1926 | 1926 | R/N | 72 | |
| 151814 | ESDI | 5.25" FH | 993 | 1925 | 14 | 1926 | 1926 | R/N | 72 | |
| 155611 | ESDI | 5.25" FH | 1064 | 1925 | 15 | 1926 | 1926 | R/N | 72 | |
| 155712 | ESDI | 5.25" FH | 248 | 1224 | 11 | 1225 | 1225 | R/N | 36 | |
| 155713 | ESDI | 5.25" FH | 270 | 1224 | 12 | 1225 | 1225 | R/N | 36 | |
| 155714 | ESDI | 5.25" FH | 293 | 1224 | 13 | 1225 | 1225 | R/N | 36 | |
| 155715 | ESDI | 5.25" FH | 315 | 1224 | 14 | 1225 | 1225 | R/N | 36 | |
| 156611 | ESDI | 5.25" FH | 338 | 1224 | 15 | 1225 | 1225 | R/N | 36 | |
| 156712 | ESDI | 5.25" FH | 496 | 1632 | 11 | 1633 | 1633 | R/N | 54 | |
| 156713 | ESDI | 5.25" FH | 541 | 1632 | 12 | 1633 | 1633 | R/N | 54 | |
| 156814 | ESDI | 5.25" FH | 586 | 1632 | 13 | 1633 | 1633 | R/N | 54 | |
| 156815 | ESDI | 5.25" FH | 631 | 1632 | 14 | 1633 | 1633 | R/N | 54 | |
| 16534 | ESDI | 5.25" HH | 676 | 1632 | 15 | 1633 | 1633 | R/N | 54 | |
| 16535 | ESDI | 5.25" HH | 92 | 1249 | 4 | 1250 | 1250 | R/N | 36 | |
| 16546 | ESDI | 5.25" HH | 115 | 1249 | 5 | 1250 | 1250 | R/N | 36 | |
| 16547 | ESDI | 5.25" FH | 138 | 1249 | 6 | 1250 | 1250 | R/N | 36 | |
| 16634 | ESDI | 5.25" HH | 161 | 1249 | 7 | 1250 | 1250 | R/N | 36 | |
| 16635 | ESDI | 5.25" HH | 197 | 1780 | 4 | 1781 | 1781 | R/N | | |
| 16646 | ESDI | 5.25" HH | 246 | 1780 | 5 | 1781 | 1781 | R/N | | |
| 16647 | ESDI | 5.25" HH | 295 | 1780 | 6 | 1781 | 1781 | R/N | | |
| | ESDI | 5.25" HH | 345 | 1780 | 7 | 1781 | 1781 | R/N | | |
| 1373 | SCSI | 5.25" FH | 72 | 1016 | 4 | 1017 | 1017 | 1.25 | 3616KBCCS | |
| 1373A | SCSI | 5.25" FH | 91 | 1016 | 5 | 1017 | 1017 | 1.25 | 3616KBCCS | |
| 1374 | SCSI | 5.25" FH | 109 | 1016 | 6 | 1017 | 1017 | 1.25 | 3616KBCCS | |
| 1374A | SCSI | 5.25" FH | 127 | 1016 | 7 | 1017 | 1017 | 1.25 | 3616KBCCS | |
| 1375 | SCSI | 5.25" FH | 145 | 1016 | 8 | 1017 | 1017 | 1.25 | 3616KBCCS | |
| 157611 | SCSI | 5.25" FH | 243 | 1220 | 11 | 1221 | 1221 | 1.6 | 3616KBCCS | |
| 157712 | SCSI | 5.25" FH | 266 | 1220 | 12 | 1221 | 1221 | 1.6 | 3616KBCCS | |
| 157713 | SCSI | 5.25" FH | 287 | 1220 | 13 | 1221 | 1221 | 1.6 | 3616KBCCS | |
| 157814 | SCSI | 5.25" FH | 310 | 1220 | 14 | 1221 | 1221 | 1.6 | 3616KBCCS | |
| 157815 | SCSI | 5.25" FH | 332 | 1220 | 15 | 1221 | 1221 | 1.6 | 3616KBCCS | |
| 158611 | SCSI | 5.25" FH | 490 | 1628 | 11 | 1629 | 1629 | RLL | 54 | |
| 158712 | SCSI | 5.25" FH | 535 | 1628 | 12 | 1629 | 1629 | RLL | 54 | |
| 158713 | SCSI | 5.25" FH | 579 | 1628 | 13 | 1629 | 1629 | RLL | 54 | |
| 158814 | SCSI | 5.25" FH | 624 | 1628 | 14 | 1629 | 1629 | RLL | 54 | |
| 158815 | SCSI | 5.25" FH | 668 | 1628 | 15 | 1629 | 1629 | RLL | 54 | |
| 159610S | SCSI | 5.25" FH | 668 | 1834 | 10 | 1835 | 1835 | | 72 | |
| 159713 | SCSI | 5.25" FH | 909 | 1919 | 13 | 1920 | 1920 | | 72 | |
| 159814 | SCSI | 5.25" FH | 979 | 1919 | 14 | 1920 | 1920 | | 72 | |
| 159815 | SCSI | 5.25" FH | 1049 | 1919 | 15 | 1920 | 1920 | | 72 | |

| Model | Interface | Form Factor | Capacity | Cylinders | Heads | Read/Write Cylinders | Write/Precomp Cylinders | Encoding Cylinders | Sectors | Other |
|---|---|---|---|---|---|---|---|---|---|---|
| **Micropolis (*continued*)** | | | | | | | | | | |
| 16734 | SCSI | 5.25" HH | 90 | 1249 | 4 | 1250 | 1250 | RLL | | |
| 16735 | SCSI | 5.25" HH | 112 | 1249 | 5 | 1250 | 1250 | RLL | | |
| 16746 | SCSI | 5.25" HH | 135 | 1249 | 6 | 1250 | 1250 | RLL | | |
| 16747 | SCSI | 5.25" HH | 158 | 1249 | 7 | 1250 | 1250 | RLL | | |
| 16834 | SCSI | 5.25" HH | 193 | 1776 | 4 | 1777 | 1777 | | 54 | |
| 16835 | SCSI | 5.25" HH | 242 | 1776 | 5 | 1777 | 1777 | | 54 | |
| 16846 | SCSI | 5.25" HH | 291 | 1776 | 6 | 1777 | 1777 | | 54 | |
| 16847 | SCSI | 5.25" HH | 340 | 1776 | 7 | 1777 | 1777 | | 54 | |
| 17735 | SCSI | 3.5" | 112 | 1140 | 5 | 1441 | 1441 | | 28–48 | |
| 17746 | SCSI | 3.5" | 135 | 1140 | 6 | 1141 | 1141 | | 28–48 | |
| 17747 | SCSI | 3.5" | 157 | 1140 | 7 | 1141 | 1141 | | 28–48 | |
| 17758 | SCSI | 3.5" | 180 | 1140 | 8 | 1141 | 1141 | | 28–48 | |
| 17759 | SCSI | | 202 | 1140 | 9 | 1141 | 1141 | | 28–48 | |
| **Microscience** | | | | | | | | | | |
| 7040 | AT IDE | 3.5" | 47 | 855 | 3 | | | RLL | 36 | |
| 7100 | AT IDE | 3.5" | 110 | 855 | 7 | | | RLL | 36 | |
| HH312 | ST506 | | 10 | 306 | 4 | 307 | 307 | MFM | 17 | |
| HH315 | ST506 | | 10 | 306 | 4 | 307 | 307 | MFM | 17 | |
| HH325 | ST506 | | 21 | 612 | 4 | 613 | 613 | MFM | 17 | |
| HH612 | ST506 | | 11 | 306 | 4 | 307 | 307 | MFM | 17 | |
| HH625 | ST506 | | 21 | 612 | 4 | 613 | 613 | MFM | 17 | |
| HH712 | ST506 | 5.25" HH | 10 | 612 | 2 | 613 | 613 | MFM | 17 | |
| HH725 | ST506 | 5.25" HH | 21 | 612 | 4 | 613 | 613 | MFM | 17 | |
| HH825 | ST506 | 5.25" HH | 21 | 615 | 4 | 616 | 616 | MFM | 17 | |
| HH1050 | ST506 | 5.25" HH | 44 | 1024 | 5 | 1025 | 1025 | MFM | 17 | |
| HH1075 | ST506 | 5.25" HH | 65 | 1024 | 7 | 1025 | 1025 | MFM | 17 | |
| HH1090 | ST506 | 5.25" HH | 80 | 1314 | 7 | 1315 | 1315 | MFM | 17 | |
| 4050 | ST506 | 3.5" | 44 | 1024 | 5 | 1025 | 1025 | MFM | 17 | |
| HH330 | ST506 | | 32 | 612 | 4 | 613 | 613 | RLL | 26 | |
| HH738 | ST506 | 5.25" HH | 32 | 612 | 4 | 613 | 613 | RLL | 26 | |
| HH830 | ST506 | 5.25" HH | 38 | 615 | 4 | 616 | 616 | RLL | 26 | |
| HH1060 | ST506 | 5.25" HH | 66 | 1024 | 5 | 1025 | 1025 | RLL | 26 | |
| HH1095 | ST506 | 5.25" HH | 95 | 1024 | 7 | 1025 | 1025 | RLL | 26 | |
| HH1120 | ST506 | 5.25" HH | 122 | 1314 | 7 | 1315 | 1315 | RLL | 26 | |
| HH2120 | ESDI | 5.25" HH | 121 | 1024 | 7 | | | R/N | 33 | |
| 5100 | ESDI | 3.5" | 110 | 855 | 7 | | | R/N | 36 | |
| HH3120 | SCSI | 5.25" HH | 122 | 1314 | 7 | | | RLL | 26 | |
| 6100 | SCSI | 3.5" | 110 | 855 | 7 | | | RLL | 36 | |
| **Miniscribe** | | | | | | | | | | |
| 8225AT | AT IDE | 3.5" | 21 | 745 | 2 | | | | 28 | |
| 8051AT | AT IDE | 3.5" | 42 | 745 | 4 | | | | | |
| 8450AT | AT IDE | 3.5" | 42 | 745 | 4 | | | | 28 | |

## Miniscribe (*continued*)

| Model | Interface | Form Factor | Capacity | Cylinders | Heads | Read/Write Cylinders | Write/Precomp Cylinders | Encoding Cylinders | Sectors | Other |
|---|---|---|---|---|---|---|---|---|---|---|
| 8225XT | XT | 3.5" | 21 | 805 | 2 | | | | 26 | |
| 8425XT | XT | 3.5" | 21 | 615 | 4 | | | | | |
| 8438XT | XT | 3.5" | 31 | 615 | 4 | | | | | |
| 8450XT | XT | 3.5" | 42 | 805 | 4 | | | | 26 | |
| 1006 | ST506 | | 5 | 306 | 2 | 153 | 128 | MFM | 17 | |
| 1012 | ST506 | | 10 | 306 | 4 | 153 | 128 | MFM | 17 | |
| 2006 | ST506 | | 5 | 306 | 2 | 307 | 128 | MFM | 17 | |
| 2012 | ST506 | | 11 | 306 | 4 | 307 | 128 | MFM | 17 | |
| 3053 | ST506 | 5.25" HH | 44 | 1024 | 5 | 1025 | 512 | MFM | 17 | |
| 3085 | ST506 | 5.25" HH | 71 | 1170 | 7 | 1171 | 512 | MFM | 17 | |
| 3212 | ST506 | 5.25" HH | 11 | 612 | 2 | 613 | 128 | MFM | 17 | |
| 3212 PLUS | AT | 5.25" HH | 11 | 612 | 2 | 613 | 128 | MFM | 17 | |
| 3412 | AT | | 11 | 306 | 4 | 307 | 128 | MFM | 17 | |
| 3425 | AT | 5.25" HH | 21 | 612 | 4 | 613 | 128 | MFM | 17 | |
| 3425 PLUS | AT | 5.25" HH | 21 | 612 | 4 | 613 | 128 | MFM | 17 | |
| 3650 | AT | 5.25" HH | 42 | 809 | 6 | 810 | 128 | MFM | 17 | |
| 3650F | AT | 5.25" HH | 42 | 809 | 6 | 810 | 128 | MFM | 17 | |
| 4010 | AT | | 8 | 480 | 2 | 481 | 128 | MFM | 17 | |
| 4020 | AT | | 17 | 480 | 4 | 481 | 128 | MFM | 17 | |
| 5330 | AT | | 25 | 480 | 6 | 481 | 128 | MFM | 17 | |
| 5338 | AT | | 32 | 612 | 6 | 613 | 128 | MFM | 17 | |
| 5440 | AT | | 38 | 612 | 8 | 613 | 128 | MFM | 17 | |
| 5451 | AT | | 43 | 612 | 8 | 613 | 128 | MFM | 17 | |
| 6032 | AT | 5.25" FH | 26 | 1024 | 3 | 1025 | 512 | MFM | 17 | |
| 6053 | AT | 5.25" FH | 44 | 1024 | 5 | 1025 | 512 | MFM | 17 | |
| 6074 | AT | 5.25" FH | 62 | 1024 | 7 | 1025 | 512 | MFM | 17 | |
| 6085 | AT | 5.25" FH | 71 | 1024 | 8 | 1025 | 512 | MFM | 17 | |
| 6212 | AT | | 10 | 612 | 2 | 613 | 128 | MFM | 17 | |
| 7426 | AT | | 21 | 612 | 4 | 613 | 128 | MFM | 17 | |
| 8212 | AT | | 11 | 612 | 2 | 613 | 128 | MFM | 17 | |
| 8412 | AT | | 10 | 306 | 4 | 307 | 128 | MFM | 17 | |
| 8425 | AT | 3.5" | 21 | 615 | 4 | 616 | 128 | MFM | 17 | |
| 8425F | AT | 3.5" | 21 | 615 | 4 | 616 | 128 | MFM | 17 | |
| 3438 | ST506 | 5.25" HH | 32 | 612 | 4 | 613 | 128 | RLL | 26 | |
| 3438 PLUS | ST506 | 5.25" HH | 32 | 612 | 4 | 613 | 128 | RLL | 26 | |
| 3675 | ST506 | 5.25" HH | 63 | 809 | 6 | 810 | 128 | RLL | 26 | |
| 6079 | ST506 | 5.25" FH | 68 | 1024 | 5 | 1025 | 512 | RLL | 26 | |
| 6128 | ST506 | 5.25" FH | 110 | 1024 | 8 | 1025 | 512 | RLL | 26 | |
| 8225 | ST506 | 3.5" | 20 | 771 | 2 | 772 | 128 | RLL | 26 | |
| 8434F | ST506 | 3.5" | 32 | 615 | 4 | 616 | 128 | RLL | 26 | |
| 8438 | ST506 | 3.5" | 32 | 615 | 4 | 616 | 128 | RLL | 26 | |
| 8438F | ST506 | 3.5" | 32 | 615 | 4 | 616 | 128 | RLL | 26 | |
| 8450 | ST506 | 3.5" | 40 | 771 | 4 | 772 | 128 | RLL | 26 | |
| 3085E | ESDI | 5.25" HH | 72 | 1270 | 3 | 1271 | 512 | | | |
| 3130E | ESDI | 5.25" HH | 112 | 1250 | 5 | 1251 | 512 | | | |

| Model | Interface | Form Factor | Capacity | Cylinders | Heads | Read/Write Cylinders | Write/Precomp Cylinders | Encoding Cylinders | Sectors | Other |
|-------|-----------|-------------|----------|-----------|-------|----------------------|-------------------------|--------------------|---------|-------|
| **Miniscribe (*continued*)** | | | | | | | | | | |
| 3180E | ESDI | 5.25" HH | 157 | 1250 | 7 | 1251 | 512 | | | |
| 9380E | ESDI | 5.25" FH | 338 | 1224 | 14 | 1225 | 512 | | | |
| 9424E | ESDI | 5.25" FH | 360 | 1661 | 8 | 1662 | 512 | | | |
| 9780E | ESDI | 5.25" FH | 676 | 1661 | 15 | 1662 | 512 | | | |
| 9230 | ESDI/SCSI | 5.25" FH | 203 | 1224 | 9 | 1225 | 512 | RLL | | |
| 9380 | ESDI/SCSI | 5.25" FH | 338 | 1224 | 15 | 1225 | 512 | RLL | | |
| 3085S | SCSI | 5.25" HH | 72 | 1255 | 3 | 1256 | 512 | | | |
| 3130S | SCSI | 5.25" HH | 115 | 1255 | 5 | 1256 | 512 | | | |
| 3180S | SCSI | 5.25" HH | 160 | 1255 | 7 | 1256 | 512 | | | |
| 8051S | SCSI | 3.5" | 45 | 793 | 4 | 794 | 128 | | 28 | |
| 8425S | SCSI | 3.5" | 21 | 612 | 4 | 613 | 128 | RLL | | |
| 9380S | SCSI | 5.25" FH | 347 | 1224 | 15 | 1225 | 512 | | | |
| 9424S | SCSI | 5.25" FH | 355 | 1661 | 8 | 1662 | 512 | | | |
| 9780S | SCSI | 5.25" FH | 668 | 1661 | 15 | 1662 | 512 | | | |
| **Mitsubishi** | | | | | | | | | | |
| MR521 | ST506 | 5.25" HH | 10 | 612 | 2 | | | MFM | 17 | |
| MR522 | ST506 | 5.25" HH | 20 | 612 | 4 | | | MFM | 17 | |
| MR535 | ST506 | 5.25" HH | 42 | 977 | 5 | 300 | 300 | MFM | 17 | |
| MR535 | ST506 | 5.25" HH | 65 | 977 | 5 | 300 | 300 | RLL | 26 | |
| **MMI** | | | | | | | | | | |
| M 112 | ST506 | 3.5" | 10 | 306 | 4 | | 128 | MFM | 17 | |
| M 106 | ST506 | 3.5" | 5 | 306 | 2 | | 128 | MFM | 17 | |
| M 125 | ST506 | 3.5" | 20 | 306 | 8 | | 128 | MFM | 17 | |
| M 306 | ST506 | 3.5" | 5 | 306 | 2 | | 128 | MFM | 17 | |
| M 212 | ST506 | 5.25" HH | 10 | 306 | 4 | | 128 | MFM | 17 | |
| M 225 | ST506 | 5.25" HH | 20 | 306 | 8 | | 128 | MFM | 17 | |
| M 312 | ST506 | 5.25" HH | 10 | 306 | 4 | | 128 | MFM | 17 | |
| M 325 | ST506 | 5.25" HH | 20 | 306 | 8 | | 128 | MFM | 17 | |
| **NEC** | | | | | | | | | | |
| D3216 | ST506 | 3.5" | 20 | 615 | 4 | 616 | 256 | MFM | 17 | |
| D3146H | ST506 | 3.5" | 40 | 615 | 8 | | | MFM | 32 | |
| D3142 | ST506 | 3.5" | 42 | 642 | 8 | | | MFM | 32 | |
| D5124 | ST506 | 5.25" HH | 10 | 309 | 4 | 310 | 310 | MFM | 17 | |
| D5126 | ST506 | 5.25" HH | 20 | 612 | 4 | 613 | | MFM | 17 | |
| D5126H | ST506 | | 20 | 612 | 4 | 613 | | MFM | 17 | |
| D5146 | ST506 | 5.25" HH | 40 | 615 | 8 | 616 | | MFM | 17 | |
| D5146H | ST506 | 5.25" HH | 40 | 615 | 8 | 616 | | MFM | 17 | |
| D5652 | ESDI | 5.25" HH | 143 | 823 | 10 | | | R/N | | |
| D5655 | ESDI | 5.25" HH | 153 | 1224 | 7 | | | R/N | 35 | |

| Model | Interface | Form Factor | Capacity | Cylinders | Heads | Read/Write Cylinders | Write/Precomp Cylinders | Encoding Cylinders | Sectors | Other |
|-------|-----------|-------------|----------|-----------|-------|----------------------|-------------------------|--------------------|---------|-------|
| **Newbury Data** | | | | | | | | | | |
| NDR 340 | ST506 | 3.5" | 42 | 615 | 8 | | | MFM | | |
| NDR 1065 | ST506 | 5.25" FH | 55 | 918 | 7 | | | MFM | | |
| NDR 1085 | ST506 | 5.25" FH | 71 | 1025 | 8 | | | MFM | | |
| NDR 1105 | ST506 | 5.25" FH | 87 | 918 | 11 | | | MFM | | |
| NDR 1140 | ST506 | 5.25" FH | 105 | 918 | 15 | | | MFM | | |
| NDR 2190 | ST506 | 5.25" FH | 191 | 918 | 15 | | | MFM | | |
| NDR 4175 | ESDI | 5.25" FH | 179 | 1224 | 7 | | | R/N | 36 | |
| NDR 4380 | ESDI | 5.25" FH | 384 | 1224 | 15 | | | R/N | 36 | |
| NDR 3170S | SCSI | 5.25" FH | 146 | 1224 | 9 | | | RLL | 26 | |
| NDR 3280S | SCSI | 5.25" FH | 244 | 1224 | 15 | | | RLL | 26 | |
| NDR 4380S | SCSI | 5.25" FH | 319 | 1224 | 15 | | | RLL | 34 | |
| **Okidata** | | | | | | | | | | |
| OD526 | ST506 | | 31 | 640 | 4 | 651 | 651 | RLL | 26 | |
| OD540 | ST506 | | 47 | 640 | 6 | 651 | 651 | RLL | 26 | |
| **Olivetti** | | | | | | | | | | |
| HD662/11 | ST506 | | 10 | 612 | 2 | | | MFM | 17 | |
| HD662/12 | ST506 | | 20 | 612 | 4 | | | MFM | 17 | |
| **Otari** | | | | | | | | | | |
| C 214 | ST506 | | 10 | 306 | 4 | 128 | 128 | MFM | 17 | |
| C 519 | ST506 | | 15 | 306 | 6 | 128 | 128 | MFM | 17 | |
| C 526 | ST506 | | 21 | 306 | 8 | 128 | 128 | MFM | 17 | |
| **Panasonic** | | | | | | | | | | |
| JU-116 | ST506 | 3.5" | 20 | 615 | 4 | 616 | 616 | MFM | 17 | |
| JU-128 | ST506 | 3.5" | 42 | 733 | 7 | 734 | 734 | MFM | 17 | |
| **Priam** | | | | | | | | | | |
| V-150 | ST506 | | 42 | 987 | 5 | 988 | 988 | MFM | 17 | |
| V-160 | ST506 | | 50 | 1166 | 5 | 1167 | 1167 | MFM | 17 | |
| V-170 | ST506 | | 60 | 987 | 7 | 988 | 988 | MFM | 17 | |
| V-185 | ST506 | | 71 | 1166 | 7 | 1167 | 1167 | MFM | 17 | |
| 502 | ST506 | | 46 | 755 | 7 | 756 | 756 | MFM | 17 | |
| 504 | ST506 | | 46 | 755 | 7 | 756 | 756 | MFM | 17 | |
| 514 | ST506 | 5.25" FH | 117 | 1124 | 11 | 1225 | 1225 | MFM | 17 | |
| 519 | ST506 | 5.25" FH | 160 | 1224 | 15 | 1225 | 1225 | MFM | 17 | |
| ED40 | ST506 | 5.25" FH | 42 | 1018 | 5 | | | MFM | 17 | |
| ED60 | ST506 | 5.25" FH | 59 | 1018 | 7 | | | MFM | 17 | |
| ED130 | ST506 | 5.25" FH | 132 | 1218 | 15 | | | MFM | 17 | |
| ID45 | ST506 | 5.25" FH | 44 | 1018 | 5 | | | MFM | 17 | |
| ID45H | ST506 | 5.25" HH | 44 | 1018 | 5 | | | MFM | 17 | |

| Model | Interface | Form Factor | Capacity | Cylinders | Heads | Read/Write Cylinders | Write/Precomp Cylinders | Encoding | Cylinders | Sectors | Other |
|-------|-----------|-------------|----------|-----------|-------|---------------------|------------------------|----------|-----------|---------|-------|

## Priam (*continued*)

| Model | Interface | Form Factor | Capacity | Cylinders | Heads | Read/Write Cylinders | Write/Precomp Cylinders | Encoding Cylinders | Sectors | Other |
|-------|-----------|-------------|----------|-----------|-------|---------------------|------------------------|--------------------|---------|-------|
| ID62 | ST506 | 5.25" FH | 62 | 1018 | 7 | | | MFM | 17 | |
| ID130 | ST506 | 5.25" FH | 132 | 1218 | 15 | | | MFM | 17 | |
| V-130 | ST506 | | 39 | 987 | 3 | 988 | 988 | RLL | 26 | |
| V-170 | ST506 | | 91 | 987 | 7 | 988 | 988 | RLL | 26 | |
| 617 | ESDI | 5.25" FH | 153 | 1225 | 7 | 1226 | 1226 | R/N | | |
| 628 | ESDI | 5.25" FH | 241 | 1225 | 11 | 1226 | 1226 | R/N | | |
| 638 | ESDI | 5.25" FH | 329 | 1225 | 15 | 1226 | 1226 | R/N | | |
| ED120 | ESDI | 5.25" HH | 121 | 1017 | 7 | | | | 33/35 | |
| ED150 | ESDI | 5.25" HH | 159 | 1268 | 7 | | | | 33/35 | |
| ED250 | ESDI | 5.25" FH | 246 | 1218 | 11 | | | | 36 | |
| ED250 | ESDI | 5.25" FH | 241 | 1195 | 11 | | | | 36 | |
| ED330 | ESDI | 5.25" FH | 336 | 1218 | 15 | | | | 36 | |
| ED330 | ESDI | 5.25" FH | 330 | 1195 | 15 | | | | 36 | |
| ID120 | ESDI | 5.25" HH | 121 | 1017 | 7 | | | | 33/35 | |
| ID150 | ESDI | 5.25" HH | 159 | 1268 | 7 | | | | 33/35 | |
| ID160 | ESDI | 5.25" FH | 156 | 1218 | 7 | | | | 36 | |
| ID160 | ESDI | 5.25" FH | 152 | 1195 | 7 | | | | 36 | |
| ID250 | ESDI | 5.25" FH | 246 | 1218 | 11 | | | | 36 | |
| ID250 | ESDI | 5.25" FH | 241 | 1195 | 11 | | | | 36 | |
| ID330 | ESDI | 5.25" FH | 336 | 1218 | 15 | | | | 36 | |
| ID330 | ESDI | 5.25" FH | 330 | 1195 | 15 | | | | 36 | |
| 717 | SCSI | 5.25" FH | 153 | 1225 | 7 | 1226 | 1226 | RLL | | |
| 728 | SCSI | 5.25" FH | 241 | 1225 | 11 | 1226 | 1226 | RLL | | |
| 738 | SCSI | 5.25" FH | 329 | 1225 | 15 | 1226 | 1226 | RLL | | |
| ED160 | SCSI | 5.25" FH | 158 | 1218 | 7 | | | | 36 | |
| ED250 | SCSI | 5.25" FH | 248 | 1218 | 11 | | | | 36 | |
| ED330 | SCSI | 5.25" FH | 338 | 1218 | 15 | | | | 36 | |
| ID160 | SCSI | 5.25" FH | 158 | 1218 | 7 | | | | 36 | |
| ID250 | SCSI | 5.25" FH | 248 | 1218 | 11 | | | | 36 | |
| ID330 | SCSI | 5.25" FH | 338 | 1218 | 15 | | | | 36 | |

## PTI

| Model | Interface | Form Factor | Capacity | Cylinders | Heads | Read/Write Cylinders | Write/Precomp Cylinders | Encoding Cylinders | Sectors | Other |
|-------|-----------|-------------|----------|-----------|-------|---------------------|------------------------|--------------------|---------|-------|
| PT-238A | ST506 | 3.5" | 32 | 615 | 4 | | | RLL | | |
| PT-251A | ST506 | 3.5" | 43 | 820 | 4 | | | RLL | | |
| PT-357A | ST506 | 3.5" | 49 | 615 | 6 | | | RLL | | |
| PT-376A | ST506 | 3.5" | 65 | 820 | 6 | | | RLL | | |
| PT-225 | ST506 | 3.5" | 21 | 615 | 4 | | | MFM | 17 | |
| PT-234 | ST506 | 3.5" | 28 | 820 | 4 | | | MFM | 17 | |
| PT-338 | ST506 | 3.5" | 32 | 615 | 6 | | | MFM | 17 | |
| PT-351 | ST506 | 3.5" | 42 | 820 | 6 | | | MFM | 17 | |
| PT-238R | ST506 | 3.5" | 32 | 615 | 4 | | | RLL | 26 | |
| PT-251R | ST506 | 3.5" | 43 | 820 | 4 | | | RLL | 26 | |
| PT-357R | ST506 | 3.5" | 49 | 615 | 6 | | | RLL | 26 | |

| Model | Interface | Form Factor | Capacity | Cylinders | Heads | Read/Write Cylinders | Write/Precomp Cylinders | Encoding Cylinders | Sectors | Other |
|---|---|---|---|---|---|---|---|---|---|---|
| **PTI (*continued*)** | | | | | | | | | | |
| PT-376R | ST506 | 3.5" | 65 | 820 | 6 | | | RLL | 26 | |
| PT-4102R | ST506 | 3.5" | 87 | 820 | 8 | | | RLL | 26 | |
| PT-238S | SCSI | 3.5" | 32 | 615 | 4 | | | RLL | | |
| PT-251S | SCSI | 3.5" | 43 | 820 | 4 | | | RLL | | |
| PT-357S | SCSI | 3.5" | 49 | 615 | 6 | | | RLL | | |
| PT-376S | SCSI | 3.5" | 65 | 820 | 6 | | | RLL | | |
| **Quantum** | | | | | | | | | | |
| PRO40 | ST506 | 3.5" | 42 | | | | | | | |
| PRO80 | ST506 | 3.5" | 84 | | | | | | | |
| PRO120 | ST506 | 3.5" | 120 | | | | | | | |
| PRO170 | ST506 | 3.5" | 168 | | | | | | | |
| Q-510 | ST506 | | 8 | 512 | 2 | 256 | 256 | MFM | 17 | |
| Q-520 | ST506 | | 18 | 512 | 4 | 256 | 513 | MFM | 17 | |
| Q-530 | ST506 | | 27 | 512 | 6 | 256 | 513 | MFM | 17 | |
| Q-540 | ST506 | | 36 | 512 | 8 | 256 | 513 | MFM | 17 | |
| PRO100 | ESDI | 3.5" | 103 | | | | | | | |
| PRO145 | ESDI | 3.5" | 145 | | | | | | | |
| Q-250 | SCSI | 5.25" HH | 53 | 823 | 4 | | | RLL | | |
| Q-280 | SCSI | 5.25" HH | 80 | 823 | 6 | | | RLL | | |
| Q-160 | SCSI | 5.25" HH | | 200 | 12 | | | RLL | | |
| PRO40 | SCSI | 3.5" | 42 | | | | | | | |
| PRO80 | SCSI | 3.5" | 84 | | | | | | | |
| PRO120 | SCSI | 3.5" | 120 | | | | | | | |
| PRO170 | SCSI | 3.5" | 168 | | | | | | | |
| **Ricoh Systems** | | | | | | | | | | |
| RH-5130 | ST506 | | 10 | 612 | 2 | 613 | 400 | MFM | 17 | |
| RH-5260 | ST506 | | 10 | 615 | 2 | | | MFM | 17 | |
| RH-5261 | SCSI | | 10 | 612 | 2 | | | MFM | | |
| **Rodime** | | | | | | | | | | |
| RO 101 | ST506 | 5.25" FH | 6 | 192 | 2 | 96 | 192 | MFM | 17 | |
| RO 102 | ST506 | 5.25" FH | 12 | 192 | 4 | 96 | 192 | MFM | 17 | |
| RO 103 | ST506 | 5.25" FH | 18 | 192 | 6 | 96 | 192 | MFM | 17 | |
| RO 104 | ST506 | 5.25" FH | 24 | 192 | 8 | 96 | 192 | MFM | 17 | |
| RO 201 | ST506 | 5.25" FH | 5 | 321 | 2 | 132 | 300 | MFM | 17 | |
| RO 201E | ST506 | 5.25" FH | 11 | 640 | 2 | 264 | 300 | MFM | 17 | |
| RO 202 | ST506 | 5.25" FH | 10 | 321 | 4 | 132 | 300 | MFM | 17 | |
| RO 202E | ST506 | 5.25" FH | 21 | 640 | 4 | 264 | 300 | MFM | 17 | |
| RO 203 | ST506 | 5.25" FH | 15 | 321 | 6 | 132 | 300 | MFM | 17 | |
| RO 203E | ST506 | 5.25" FH | 32 | 640 | 6 | 264 | 300 | MFM | 17 | |
| RO 204 | ST506 | 5.25" FH | 21 | 320 | 8 | 132 | 300 | MFM | 17 | |

| Model | Interface | Form Factor | Capacity | Cylinders | Heads | Read/Write Cylinders | Write/Precomp Cylinders | Encoding Cylinders | Sectors | Other |
|---|---|---|---|---|---|---|---|---|---|---|
| **Rodime (*continued*)** | | | | | | | | | | |
| RO 204E | ST506 | 5.25" FH | 43 | 640 | 8 | 264 | 300 | MFM | 17 | |
| RO 251 | ST506 | 5.25" HH | 5 | 306 | 2 | 307 | 307 | MFM | 17 | |
| RO 252 | ST506 | 5.25" HH | 11 | 306 | 4 | 64 | 128 | MFM | 17 | |
| RO 351 | ST506 | | 5 | 306 | 2 | 307 | 307 | MFM | 17 | |
| RO 352 | ST506 | 3.5" | 11 | 306 | 4 | 64 | 128 | MFM | 17 | |
| RO 365 | ST506 | 3.5" | 21 | 612 | 4 | 613 | 613 | MFM | 17 | |
| RO 3045 | ST506 | | 37 | 872 | 5 | 873 | | MFM | 17 | |
| RO 3055 | ST506 | | 45 | 872 | 6 | 873 | | MFM | 17 | |
| RO 3065 | ST506 | | 53 | 872 | 7 | | | MFM | 17 | |
| RO 5065 | ST506 | 5.25" HH | 63 | | 5 | | | MFM | 17 | |
| RO 5090 | ST506 | 5.25" HH | 89 | 1224 | 7 | | | MFM | 17 | |
| RO 652A | SCSI | | 20 | | | | | | | |
| RO 652B | SCSI | | 20 | 306 | 4 | | | RLL | | |
| RO 752A | SCSI | 5.25" HH | 25 | | | | | | | |
| RO 3070S | SCSI | | 71 | | | | | | | |
| RO 3085S | SCSI | | 85 | 750 | 7 | | | RLL | | |
| RO 3057S | SCSI | | 45 | 680 | 5 | | | RLL | | |
| RO 5040 | SCSI | | 38 | | 3 | | | MFM | | |
| RO 5075S | SCSI | 5.25" HH | 76 | | | | | | | |
| RO 5125S | SCSI | 5.25" HH | 127 | 1219 | 5 | | | RLL | | |
| RO 5180S | SCSI | 5.25" HH | 178 | 1219 | 7 | | | | | |
| **SeaGate** | | | | | | | | | | |
| ST125A | AT IDE | 3.5" | 21 | 404 | 4 | | | | 26 | |
| ST138A | AT IDE | 3.5" | 32 | 604 | 4 | | | | 26 | |
| ST157A | AT IDE | 3.5" | 43 | 539 | 6 | | | | 26 | |
| ST325A | AT IDE | | 21 | 615 | 2 | | | | 34 | |
| ST1057A | AT IDE | | 49 | 1024 | 3 | | | | 31 | |
| ST1102A | AT IDE | | 84 | 1024 | 5 | | | | 32 | |
| ST1144A | AT IDE | | 125 | 1024 | 7 | | | | 34 | |
| ST3025A | AT IDE | | 21 | 1024 | 2 | | | | 41 | |
| ST3057A | AT IDE | | 49 | 1024 | 3 | | | | 31 | |
| ST3096A | AT IDE | | 84 | 1024 | 3 | | | | 51 | |
| ST124 | ST506 | | 21 | 615 | 4 | 616 | 616 | | 17 | |
| ST125 | ST506 | 3.5" | 21 | 615 | 4 | 616 | 616 | | 17 | |
| ST138 | ST506 | 3.5" | 32 | 615 | 6 | 616 | 616 | | 17 | |
| ST151 | ST506 | 3.5" | 42 | 977 | 5 | 978 | 978 | | 17 | |
| ST206 | ST506 | 5.25" FH | 5 | 306 | 2 | 307 | 128 | MFM | 17 | |
| ST212 | ST506 | 5.25" FH | 10 | 306 | 4 | 307 | 128 | MFM | 17 | |
| ST213 | ST506 | 5.25" FH | 10 | 615 | 2 | 613 | 307 | MFM | 17 | |
| ST225 | ST506 | 5.25" HH | 21 | 615 | 4 | 616 | 616 | MFM | 17 | |
| ST251 | ST506 | 5.25" HH | 42 | 820 | 6 | 821 | 821 | MFM | 17 | |
| ST252 | ST506 | | 42 | 820 | 6 | 820 | 820 | | 17 | |

| Model | Interface | Form Factor | Capacity | Cylinders | Heads | Read/Write Cylinders | Write/Precomp Cylinders | Encoding Cylinders | Sectors | Other |
|-------|-----------|-------------|----------|-----------|-------|----------------------|-------------------------|--------------------|---------|-------|
| **SeaGate** (*continued*) | | | | | | | | | | |
| ST253 | ST506 | | 43 | 989 | 5 | 989 | 989 | MFM | 17 | |
| ST406 | ST506 | 5.25" FH | 5 | 306 | 2 | 307 | 128 | MFM | 17 | |
| ST412 | ST506 | 5.25" FH | 10 | 306 | 4 | 307 | 128 | MFM | 17 | |
| ST419 | ST506 | 5.25" FH | 15 | 306 | 6 | 307 | 128 | MFM | 17 | |
| ST425 | ST506 | 5.25" FH | 21 | 306 | 8 | 307 | 128 | MFM | 17 | |
| ST506 | ST506 | 5.25" FH | 5 | 153 | 4 | 128 | 128 | MFM | 17 | |
| ST706 | ST506 | 5.25" FH | 5 | 306 | 2 | 307 | 128 | MFM | 17 | |
| ST1100 | ST506 | | 84 | 1072 | 9 | 1073 | 1073 | MFM | 17 | |
| ST4026 | ST506 | 5.25" FH | 21 | 615 | 4 | 616 | 307 | MFM | 17 | |
| ST4038 | ST506 | 5.25" HH | 31 | 733 | 5 | 734 | 367 | MFM | 17 | |
| ST4051 | ST506 | 5.25" FH | 42 | 977 | 5 | 978 | 498 | MFM | 17 | |
| ST4053 | ST506 | 5.25" FH | 44 | 1024 | 5 | 1025 | 1025 | MFM | 17 | |
| ST4085 | ST506 | | 71 | 1024 | 8 | 1023 | 1023 | MFM | 17 | |
| ST4086 | ST506 | | 72 | 925 | 9 | 926 | 926 | MFM | 17 | |
| ST4096 | ST506 | 5.25" FH | 80 | 1024 | 9 | 1025 | 1025 | MFM | 17 | |
| ST4097 | ST506 | | 80 | 1024 | 9 | 1025 | 1025 | MFM | 17 | |
| ST138R | ST506 | 3.5" | 32 | 615 | 4 | 616 | 616 | RLL | 26 | |
| ST157R | ST506 | 3.5" | 49 | 615 | 6 | 616 | 616 | RLL | 26 | |
| ST225R | ST506 | 5.25" HH | 21 | 667 | 2 | 668 | 668 | RLL | 31 | |
| ST238R | ST506 | 5.25" HH | 32 | 615 | 4 | 616 | 616 | RLL | 26 | |
| ST250R | ST506 | 5.25" HH | 43 | 667 | 4 | 668 | 668 | RLL | 31 | |
| ST251R | ST506 | 5.25" HH | 43 | 820 | 4 | 821 | 821 | RLL | 26 | |
| ST277R | ST506 | 5.25" HH | 65 | 820 | 6 | 821 | 821 | RLL | 26 | |
| ST278R | ST506 | | 65 | 820 | 6 | | | | 26 | |
| ST279R | ST506 | | 65 | 989 | 5 | | | | 26 | |
| ST1150R | ST506 | | 128 | 1072 | 9 | | | | 26 | |
| ST4077R | ST506 | 5.25" FH | 65 | 1024 | 5 | 1025 | 1025 | RLL | 26 | |
| ST4135R | ST506 | | 11 | 960 | 9 | | | | 26 | |
| ST4144R | ST506 | 5.25" FH | 122 | 1024 | 9 | 1025 | 1025 | RLL | 26 | |
| ST4192E | ESDI | 5.25" FH | 169 | 1147 | 8 | 1148 | 1148 | RLL | 36 | |
| ST125N | SCSI | 3.5" | 21 | 407 | 4 | 408 | 408 | | 26 | |
| ST138N | SCSI | 3.5" | 32 | 613 | 4 | 614 | 614 | RLL | 35 | |
| ST157N | SCSI | 3.5" | 48 | 613 | 6 | 614 | 614 | RLL | 26 | |
| ST177N | SCSI | 3.5" | 60 | 921 | 5 | 922 | 922 | | 26 | |
| ST224N | SCSI | | 21 | 615 | 2 | | | | 26 | |
| ST225N | SCSI | 5.25" HH | 21 | 615 | 4 | 616 | 616 | MFM | 17 | |
| ST250N | SCSI | | 43 | 820 | 4 | | | | 26 | |
| ST251N | SCSI | 5.25" HH | 43 | 820 | 4 | 821 | 821 | RLL | 26 | |
| ST251N-1 | SCSI | 5.25" HH | 43 | 630 | 4 | 631 | 631 | RLL | 34 | |
| ST277N | SCSI | 5.25" HH | 64 | 820 | 6 | 821 | 821 | RLL | 26 | |
| ST277N-1 | SCSI | 5.25" HH | 64 | 628 | 6 | 629 | 629 | RLL | 34 | |
| ST296N | SCSI | 5.25" HH | 85 | 820 | 6 | 821 | 821 | RLL | 34 | |
| ST325N | SCSI | | 21 | 615 | 2 | | | | 34 | |
| ST1057N | SCSI | | 49 | 1024 | 3 | | | | 31 | |

| Model | Interface | Form Factor | Capacity | Cylinders | Heads | Read/Write Cylinders | Write/Precomp Cylinders | Encoding Cylinders | Sectors | Other |
|---|---|---|---|---|---|---|---|---|---|---|
| **SeaGate (*continued*)** | | | | | | | | | | |
| ST1096N | SCSI | 3.5" | 83 | 906 | 7 | 907 | 907 | | 26 | |
| ST1102N | SCSI | | 84 | 1024 | 5 | | | | 32 | |
| ST1126N | SCSI | | 107 | N/A | 7 | | | | N/A | |
| ST1144N | SCSI | | 125 | 1024 | 7 | | | | 34 | |
| ST1162N | SCSI | | 137 | N/A | 9 | | | | 29 | |
| ST1201N | SCSI | | 171 | N/A | 9 | | | | N/A | |
| ST1239N | SCSI | | 204 | N/A | 9 | | | | N/A | |
| ST2106N | SCSI | | 91 | 1024 | 5 | | | | N/A | |
| ST2125N | SCSI | | 107 | 1544 | 3 | | | | N/A | |
| ST2209N | SCSI | | 179 | 1544 | 5 | | | | N/A | |
| ST2383N | SCSI | | 332 | 1261 | 7 | | | | N/A | |
| ST2502N | SCSI | | 435 | 1755 | 7 | | | | N/A | |
| ST3025N | SCSI | | 21 | 1024 | 1 | | | | 41 | |
| ST3057N | SCSI | | 49 | 1024 | 3 | | | | 31 | |
| ST3096N | SCSI | | 84 | 1024 | 3 | | | | 53 | |
| ST4077N | SCSI | 5.25" FH | 67 | 1024 | 5 | 1025 | 1025 | RLL | 26 | |
| ST4182N | SCSI | | 155 | 969 | 9 | | | | 34 | |
| ST4192N | SCSI | 5.25" FH | 168 | 1147 | 8 | 1148 | 1148 | RLL | 36 | |
| ST4350N | SCSI | | 300 | 1412 | 9 | | | | N/A | |
| ST4376N | SCSI | | 330 | 1549 | 9 | | | | N/A | |
| ST4385N | SCSI | | 330 | N/A | 16 | | | | N/A | |
| ST4702N | SCSI | | 601 | 1546 | 16 | | | | N/A | |
| ST4766N | SCSI | | 663 | 1632 | 15 | | | | 54 | |
| ST4767N | SCSI | | 665 | 1356 | 15 | | | | 64 | |
| ST41200N | SCSI | | 1037 | 1931 | 15 | | | | N/A | |
| **Shugart** | | | | | | | | | | |
| 604 | ST506 | | 5 | 160 | 4 | 128 | 128 | MFM | 17 | |
| 606 | ST506 | | 7 | 160 | 6 | 128 | 128 | MFM | 17 | |
| 612 | ST506 | | 11 | 306 | 4 | 307 | 128 | MFM | 17 | |
| 706 | ST506 | | 6 | 320 | 2 | 321 | 128 | MFM | 17 | |
| 712 | ST506 | | 11 | 320 | 4 | 321 | 128 | MFM | 17 | |
| **Siemens** | | | | | | | | | | |
| 1200 | ST506 | 5.25" FH | 174 | 1216 | 8 | | | R/N | | |
| 1300 | ST506 | 5.25" FH | 261 | 1216 | 12 | | | R/N | | |
| 2300 | ST506 | 5.25" FH | 261 | 1216 | 12 | | | R/N | | |
| 2200 | ST506 | 5.25" FH | 174 | 1216 | 8 | | | R/N | | |
| 5710 | ST506 | | 655 | | 15 | | | R/N | | |
| 5720 | SCSI | | 655 | | 15 | | | | | |

| Model | Interface | Form Factor | Capacity | Cylinders | Heads | Read/Write Cylinders | Write/Precomp Cylinders | Encoding Cylinders | Sectors | Other |
|---|---|---|---|---|---|---|---|---|---|---|
| **SyQuest** | | | | | | | | | | |
| SQ 306RD | ST506 | | 5 | 306 | 2 | 307 | 307 | MFM | 17 | |
| SQ 312RD | ST506 | | 10 | 615 | 2 | 616 | 616 | MFM | 17 | |
| SQ 325F | ST506 | | 20 | 615 | 4 | 616 | 616 | MFM | 17 | |
| SQ 338F | ST506 | | 30 | 615 | 6 | 616 | 616 | MFM | 17 | |
| SQ 340AF | ST506 | | 38 | 640 | 6 | 616 | 616 | MFM | 17 | |
| **Tandon WD** | | | | | | | | | | |
| TM 252 | ST506 | | 10 | 306 | 4 | 307 | 307 | MFM | 17 | |
| TM 261 | ST506 | | 10 | 615 | 2 | 616 | 616 | MFM | 17 | |
| TM 262 | ST506 | | 21 | 615 | 4 | 616 | 616 | MFM | 17 | |
| TM 361 | ST506 | | 10 | 615 | 2 | 616 | 616 | MFM | 17 | |
| TM 362 | ST506 | | 21 | 615 | 4 | 616 | 616 | MFM | 17 | |
| TM 501 | ST506 | | 5 | 306 | 2 | 128 | 153 | MFM | 17 | |
| TM 502 | ST506 | | 10 | 306 | 4 | 128 | 153 | MFM | 17 | |
| TM 503 | ST506 | | 15 | 306 | 6 | 128 | 153 | MFM | 17 | |
| TM 702 | ST506 | | 20 | 615 | 4 | 616 | 616 | RLL | 26 | |
| TM 703 | ST506 | | 31 | 733 | 5 | 734 | 734 | MFM | 17 | |
| TM 755 | ST506 | 5.25" FH | 43 | 981 | 5 | 982 | 982 | MFM | 17 | |
| TM 3085 | ST506 | | 71 | 024 | 8 | 1024 | 1024 | MFM | 17 | |
| TM 602S | ST506 | | 10 | 153 | 4 | 128 | 128 | MFM | 17 | |
| TM 603S | ST506 | | 10 | 153 | 6 | 128 | 128 | MFM | 17 | |
| TM 603SE | ST506 | | 21 | 230 | 6 | 128 | 128 | MFM | 17 | |
| TM 244 | ST506 | | 41 | 782 | 4 | 783 | 783 | RLL | 26 | |
| TM 246 | ST506 | | 62 | 782 | 6 | 783 | 783 | RLL | 26 | |
| TM 262R 186364-408 | ST506 | 3.5" | 20 | 782 | 2 | 783 | 783 | RLL | 26 | |
| TM 264 | ST506 | 3.5" | 41 | 782 | 4 | 783 | 783 | RLL | 26 | |
| TM 344 | ST506 | 3.5" | 41 | 782 | 4 | 783 | 783 | RLL | 26 | |
| TM 346 | ST506 | 3.5" | 62 | 782 | 6 | 783 | 783 | RLL | 26 | |
| TM 362R | ST506 | 3.5" | 20 | 782 | 2 | 783 | 783 | RLL | 26 | |
| TM 364 186364-102 | ST506 | 3.5" | 41 | 782 | 4 | 783 | 783 | RLL | 26 | |
| TM 3085 | ST506 | | 104 | 1024 | 8 | 1024 | 1024 | RLL | 26 | |
| TM 2085 | SCSI | | 74 | 1004 | 9 | 1005 | 1005 | MFM | | |
| TM 2128 | SCSI | | 115 | 1004 | 9 | 1005 | 1005 | RLL | | |
| TM 2170 | SCSI | | 154 | 1344 | 9 | 1345 | 1345 | RLL | | |
| **Teac** | | | | | | | | | | |
| SD 510 | ST506 | | 10 | 306 | 4 | 128 | 128 | MFM | 17 | |
| SD 520 | ST506 | | 20 | 615 | 4 | 128 | 128 | MFM | 17 | |
| **Toshiba** | | | | | | | | | | |
| MK 53FA | ST506 | 5.25" FH | 43 | 830 | 5 | 831 | 831 | MFM | 17 | |
| MK 53FB | ST506 | 5.25" FH | 64 | 830 | 5 | 831 | 831 | RLL | 17 | |

| Model | Interface | Form Factor | Capacity | Cylinders | Heads | Read/Write Cylinders | Write/Precomp Cylinders | Encoding Cylinders | Sectors | Other |
|---|---|---|---|---|---|---|---|---|---|---|
| **Toshiba (*continued*)** | | | | | | | | | | |
| MK 54FA | ST506 | 5.25" FH | 60 | 830 | 7 | 831 | 831 | MFM | 17 | |
| MK 54FB | ST506 | 5.25" FH | 90 | 830 | 7 | 831 | 831 | RLL | 17 | |
| MK 56FA | ST506 | 5.25" FH | 86 | 830 | 10 | 831 | 831 | MFM | 17 | |
| MK 56FB | ST506 | 5.25" FH | 130 | 830 | 10 | 831 | 831 | RLL | 17 | |
| MK134FA | ST506 | 3.5" | 44 | 733 | 7 | 734 | 734 | MFM | 17 | |
| MK 53FA | ST506 | 5.25" FH | 43 | 830 | 5 | 831 | 831 | MFM | 26 | |
| MK 53FB | ST506 | 5.25" FH | 64 | 830 | 5 | 831 | 831 | RLL | 26 | |
| MK 54FA | ST506 | 5.25" FH | 60 | 830 | 7 | 831 | 831 | MFM | 26 | |
| MK 54FB | ST506 | 5.25" FH | 90 | 830 | 7 | 831 | 831 | RLL | 26 | |
| MK 56FA | ST506 | 5.25" FH | 86 | 830 | 10 | 831 | 831 | MFM | 26 | |
| MK 56FB | ST506 | 5.25" FH | 130 | 830 | 10 | 831 | 831 | RLL | 26 | |
| MK153FA | ESDI | 5.25" FH | 74 | 830 | 5 | 831 | 831 | R/N | 35 | |
| MK154FA | ESDI | 5.25" FH | 104 | 830 | 7 | 831 | 831 | R/N | 35 | |
| MK156FA | ESDI | 5.25" FH | 148 | 830 | 10 | 831 | 831 | R/N | 35 | |
| MK250F | ESDI/SCSI | 5.25" FH | 382 | 1224 | 10 | | | R/N | 35 | |
| MK153FB | SCSI | 5.25" FH | 74 | 830 | 5 | 831 | 831 | RLL | 35 | |
| MK154FB | SCSI | 5.25" FH | 104 | 830 | 7 | 831 | 831 | RLL | 35 | |
| MK156FB | SCSI | 5.25" FH | 148 | 830 | 10 | 831 | 831 | RLL | 35 | |
| MK234FB | SCSI | 3.5" | 45 | 845 | 3 | | | | | |
| MK233FB | SCSI | 3.5" | 75 | 845 | 5 | | | | | |
| MK234FB | SCSI | 3.5" | 106 | 845 | 7 | | | | | |
| **Tulin** | | | | | | | | | | |
| 213 | ST506 | | 10 | 640 | 2 | 656 | 656 | MFM | 17 | |
| 226 | ST506 | | 22 | 640 | 4 | 656 | 656 | MFM | 17 | |
| 240 | ST506 | | 33 | 640 | 6 | 656 | 656 | MFM | 17 | |
| 326 | ST506 | | 22 | 640 | 4 | 641 | 641 | MFM | 17 | |
| 340 | ST506 | | 33 | 640 | 6 | 641 | 641 | MFM | 17 | |
| **Vertex (*see also* Priam)** | | | | | | | | | | |
| V130 | ST506 | | 26 | 987 | 3 | 988 | 988 | MFM | 17 | |
| V150 | ST506 | | 43 | 987 | 5 | 988 | 988 | MFM | 17 | |
| V170 | ST506 | | 60 | 987 | 7 | 988 | 988 | MFM | 17 | |
| **Western Digital** | | | | | | | | | | |

LBA = Logical Blocks Available

SCSI drives do not use CYL, H, RWC, or WPC

| Model | Interface | Form Factor | Capacity | Cylinders | Heads | Read/Write Cylinders | Write/Precomp Cylinders | Encoding Cylinders | Sectors | Other |
|---|---|---|---|---|---|---|---|---|---|---|
| WD93028-X | XT | 3.5" | 20 | 782 | 2 | 784 | 784 | RLL | 27 | |
| WD93038-X | XT | 3.5" | 30 | 782 | 3 | 784 | 784 | RLL | 27 | |
| WD93048-X | XT | 3.5" | 40 | 782 | 4 | 784 | 784 | RLL | 27 | |
| WD93024-X | XT | 3.5" | 20 | 782 | 2 | 783 | 783 | RLL | 27 | |
| WD93034-X | XT | 3.5" | 30 | 782 | 3 | 783 | 783 | RLL | 27 | |
| WD93044-X | XT | 3.5" | 40 | 782 | 4 | 783 | 783 | RLL | 27 | |

| Model | Interface | Form Factor | Capacity | Cylinders | Heads | Read/Write Cylinders | Write/Precomp Cylinders | Encoding Cylinders | Sectors |
|---|---|---|---|---|---|---|---|---|---|
| **Western Digital (*continued*)** | | | | | | | | | |
| WD95028-X | XT | 5.25" HH | 20 | 782 | 2 | 784 | 784 | RLL | 27 |
| WD95038-X | XT | 5.25" HH | 30 | 782 | 3 | 784 | 784 | RLL | 27 |
| WD95048-X | XT | 5.25" HH | 40 | 782 | 4 | 784 | 784 | RLL | 27 |
| WD95024-X | XT | 3.5" to 5.25" HH | 20 | 782 | 2 | 783 | 783 | RLL | 27 |
| WD95034-X | XT | 3.5" to 5.25" HH | 30 | 782 | 3 | 783 | 783 | RLL | 27 |
| WD95044-X | XT | 3.5" to 5.25" HH | 40 | 782 | 4 | 783 | 783 | RLL | 27 |
| WD93028-A | AT IDE | 3.5" | 20 | 782 | 2 | 783 | 783 | RLL | 27 |
| WD93048-A | AT IDE | 3.5" | 40 | 782 | 4 | 783 | 783 | RLL | 27 |
| WD95028-A | AT IDE | 5.25" HH | 20 | 782 | 2 | 783 | 783 | RLL | 27 |
| WD95048-A | AT IDE | 5.25" HH | 40 | 782 | 4 | 783 | 783 | RLL | 27 |
| WD93028-AD | AT IDE | 3.5" | 20 | 782 | 2 | 783 | 783 | RLL | 27 |
| WD93048-AD | AT IDE | 3.5" | 40 | 782 | 4 | 783 | 783 | RLL | 27 |
| WD95028-AD | AT IDE | 3.5" to 5.25" HH | 20 | 782 | 2 | 783 | 783 | RLL | 27 |
| WD95048-AD | AT IDE | 3.5" to 5.25" HH | 40 | 782 | 4 | 783 | 783 | RLL | 27 |
| WD93024-A | AT IDE | 3.5" | 20 | 782 | 2 | 783 | 783 | RLL | 27 |
| WD93044-A | AT IDE | 3.5" | 40 | 782 | 4 | 783 | 783 | RLL | 27 |
| WD95024-A | AT IDE | 3.5" to 5.25" HH | 20 | 782 | 2 | 783 | 783 | RLL | 27 |
| WD95044-A | AT IDE | 3.5" to 5.25" HH | 40 | 782 | 4 | 783 | 783 | RLL | 27 |
| WDAP4200 | AT IDE | 3.5" | 212 | 987 | 12 | 988 | 988 | RLL | 35 |
| WDAC140 | AT IDE | 3.5" | 42.7 | 980 | 5 | 981 | 981 | RLL | 17 |
| WDAC160 | AT IDE | 3.5" | 62.3 | 1024 | 7 | 1023 | 1023 | RLL | 17 |
| WDAC280 | AT IDE | 3.5" | 85.3 | 980 | 10 | 981 | 981 | RLL | 17 |
| WDAC2120 | AT IDE | 3.5" | 125 | 872 | 8 | 872 | 872 | RLL | 35 |
| WDAB130 | AT IDE | 3.5" | 31.9 | 733 | 5 | 734 | 734 | RLL | 17 |
| WDAH260 | AT IDE | 3.5" | 62.3 | 1024 | 7 | 1023 | 1023 | RLL | 17 |
| WDSPA200 | AT IDE | 3.5" | 209.7 | | 7 | 409,600 | LBA | RLL | N/A |
| WDSC8320 | AT IDE | 3.5" | 320.1 | | | 625,356 | LBA | RLL | N/A |
| WDSC8400 | AT IDE | 3.5" | 400.6 | | | 782,576 | LBA | RLL | N/A |
| WD 262 | ST506 | 3.5" to 5.25" HH | 20 | 615 | 4 | 616 | 616 | MFM | 17 |
| WD 362 | ST506 | 3.5" | 20 | 615 | 4 | 616 | 616 | MFM | 17 |
| WD 344R | ST506 | 3.5" | 40 | 782 | 4 | 783 | 783 | RLL | 26 |
| WD 382R TM262R | ST506 | 3.5" | 20 | 782 | 2 | 783 | 783 | RLL | 26 |
| WD 383R | ST506 | 3.5" | 30 | 615 | 4 | 616 | 616 | RLL | 26 |
| WD 384R TM364 | ST506 | 3.5" | 40 | 782 | 4 | 783 | 783 | RLL | 26 |
| WD 544R | ST506 | 3.5" to 5.25" HH | 40 | 782 | 4 | 783 | 783 | RLL | 26 |
| WD 582R | ST506 | 3.5" to 5.25" HH | 20 | 782 | 2 | 783 | 783 | RLL | 26 |
| WD 583R | ST506 | 3.5" | 30 | 615 | 4 | 616 | 616 | RLL | 26 |
| WD 584R | ST506 | 3.5" to 5.25" HH | 40 | 782 | 4 | 783 | 783 | RLL | 26 |
| **Xebec** | | | | | | | | | |
| OWL I | ST506 | 5.25" HH | 25 | | 4 | | | MFM | |
| OWL II | ST506 | 5.25" HH | 38 | | 4 | | | MFM | |
| OWL III | ST506 | 5.25" HH | 52 | | 4 | | | MFM | |

# APPENDIX F
# Data Sheets

This appendix shows the data sheets that Intel Corporation provides for many of the components discussed in this book. Figures F.1–F.33 are as follows:

# 8088
# 8-BIT HMOS MICROPROCESSOR
## 8088/8088-2

- 8-Bit Data Bus Interface
- 16-Bit Internal Architecture
- Direct Addressing Capability to 1 Mbyte of Memory
- Direct Software Compatibility with 8086 CPU
- 14-Word by 16-Bit Register Set with Symmetrical Operations
- 24 Operand Addressing Modes

- Byte, Word, and Block Operations
- 8-Bit and 16-Bit Signed and Unsigned Arithmetic in Binary or Decimal, Including Multiply and Divide
- Two Clock Rates:
  — 5 MHz for 8088
  — 8 MHz for 8088-2
- Available in EXPRESS
  — Standard Temperature Range
  — Extended Temperature Range

The Intel® 8088 is a high performance microprocessor implemented in N-channel, depletion load, silicon gate technology (HMOS), and packaged in a 40-pin CERDIP package. The processor has attributes of both 8- and 16-bit microprocessors. It is directly compatible with 8086 software and 8080/8085 hardware and peripherals.

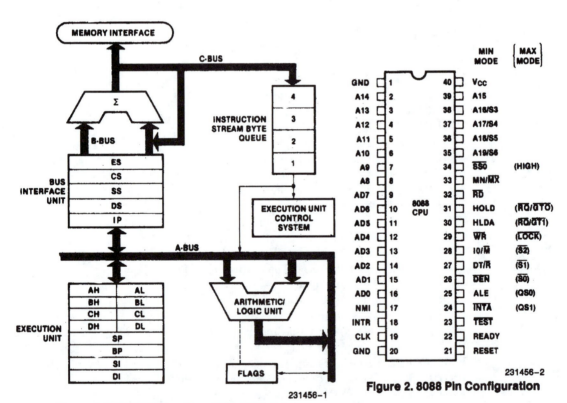

Figure 1. 8088 CPU Functional Block Diagram

Figure 2. 8088 Pin Configuration

---

**FIGURE F.1   8088 central processing unit (CPU),** reprinted by permission of Intel Corp. Copyright © 1994 by Intel Corp.

# 8087
# NUMERIC DATA COPROCESSOR
## 8087/8087-2/8087-1

- High Performance Numeric Data Coprocessor
- Adds Arithmetic, Trigonometric, Exponential, and Logarithmic Instructions to the Standard 8086/8088 and 80186/80188 Instruction Set for All Data Types
- CPU/8087 Supports 7 Data Types: 16-, 32-, 64-Bit Integers, 32-, 64-, 80-Bit Floating Point, and 18-Digit BCD Operands
- Compatible with IEEE Floating Point Standard 754

- Available in 5 MHz (8087), 8 MHz (8087-2) and 10 MHz (8087-1): 8 MHz 80186/80188 System Operation Supported with the 8087-1
- Adds 8 x 80-Bit Individually Addressable Register Stack to the 8086/8088 and 80186/80188 Architecture
- 7 Built-In Exception Handling Functions
- MULTIBUS® System Compatible Interface

The 8087 Numeric Data Coprocessor provides the instructions and data types needed for high performance numeric applications, providing up to 100 times the performance of a CPU alone. The 8087 is implemented in N-channel, depletion load, silicon gate technology (HMOS III), housed in a 40-pin package. Sixty-eight numeric processing instructions are added to the 8086/8088, 80186/80188 instruction sets and eight 80-bit registers are added to the register set. The 8087 is compatible with the IEEE Floating Point Standard 754.

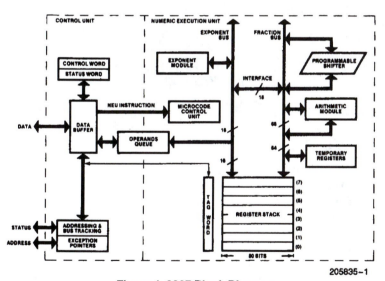

Figure 1. 8087 Block Diagram

205835–1

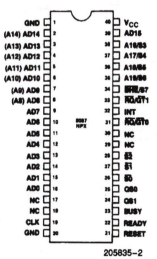

Figure 2. 8087 Pin Configuration

205835–2

---

FIGURE F.2    8087 numeric processing unit (NPU), reprinted by permission of Intel Corp. Copyright © 1994 by Intel Corp.

401

# 82C84A
## CHMOS CLOCK GENERATOR AND DRIVER
## FOR 80C86, 80C88 PROCESSORS

- **Generates the System Clock for the 80C86, 80C88 Processors:**
  - 82C84A-5 for 5 MHz
  - 82C84A for 8 MHz
- **Pin Compatible with Bipolar 8284A***
- **Uses a Crystal or an External Frequency Source**
- **Provides Local READY and MULTIBUS® READY Synchronization**

- **Generates System Reset Output from Schmitt Trigger Input**
- **Capable of Clock Synchronization with other 82C84As**
- **Low Power Consumption**
- **Single 5V Power Supply**
- **TTL Compatible Inputs/Outputs**
- **Available in 18-Lead Plastic DIP**
  (See Packaging Spec., Order #231369)

The Intel 82C84A is a high performance CHMOS clock generator-driver designed to service the requirements of the 80C86/88 and 8086/88. Power consumption is a fraction of that of equivalent bipolar circuits. The chip contains a crystal controlled oscillator, a divide-by-three counter and complete READY synchronization and reset logic. Crystal controlled operation up to 15, 25 MHz utilizes a parallel, fundamental mode crystal and two small load capacitors.

*The Bipolar 8284A requires two load resistors and a resonant crystal.

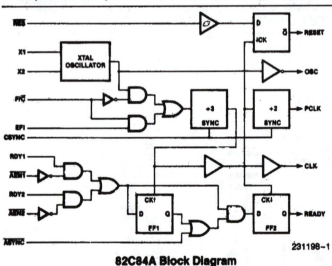

82C84A Block Diagram

231198-1

| Control Pin | Logical 1 | Logical 0 |
|---|---|---|
| F/C̄ | External Clock | Crystal Drive |
| RES̄ | Normal | Reset |
| RDY 1 RDY 2 | Bus Ready | Bus not ready |
| ĀEN 1 ĀEN 2 | Address Disabled | Address Enabled |
| ĀSYNC̄ | 1 Stage Ready Synchronization | 2 Stage Ready Synchronization |

**82C84A Pin Description**

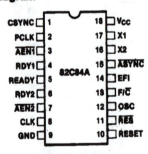

**82C84A 18-Lead DIP Configuration**

231198-2

---

**FIGURE F.3   8284 clock generator,** reprinted by permission of Intel Corp. Copyright © 1994 by Intel Corp.

# 82C88
# CHMOS BUS CONTROLLER

- Pin Compatible with Bipolar 8288
- Provides Support for 8086/88, 80C86/88, 80186, 80188
- Low Power Operation
  — $I_{CCS}$ = 100 $\mu$A
  — $I_{CC}$ = 10 mA
- Provides Advanced Commands for Multi-Master Busses

- 3-State Command Output Drivers
- High Drive Capability
- Configurable for Use with an I/O Bus
- Single 5V Power Supply
- 8 MHz Operation
  — 82C88-2

The Intel 82C88-2 is a high performance CHMOS version of the 8288 bipolar bus controller. The 82C88-2 provides command and control timing generation for 8086 architecture* systems. Static CHMOS circuit design ensures low operating power. The 82C88-2 high output drive capability eliminates the need for additional bus drivers.

*NOTE:
In this data sheet, all references to 8086 or 8086 architecture include: 8086/88, 80C86/88, 80186 and 80188.

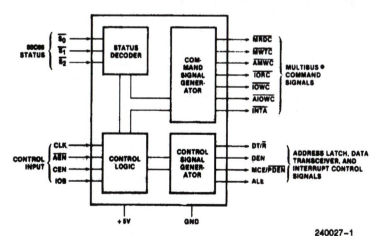

Figure 1. Block Diagram

240027-1

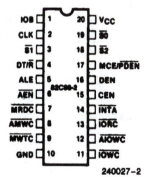

240027-2

Figure 2a. 82C88-2 20-Lead
DIP Configuration

---

FIGURE F.4  8288 bus controller, reprinted by permission of Intel Corp. Copyright © 1994 by Intel Corp.

# 8237A
# HIGH PERFORMANCE
# PROGRAMMABLE DMA CONTROLLER
## (8237A, 8237A-4, 8237A-5)

- Enable/Disable Control of Individual DMA Requests
- Four Independent DMA Channels
- Independent Autoinitialization of All Channels
- Memory-to-Memory Transfers
- Memory Block Initialization
- Address Increment or Decrement
- High Performance: Transfers up to 1.6M Bytes/Second with 5 MHz 8237A-5

- Directly Expandable to Any Number of Channels
- End of Process Input for Terminating Transfers
- Software DMA Requests
- Independent Polarity Control for DREQ and DACK Signals
- Available in EXPRESS — Standard Temperature Range
- Available in 40-Lead Cerdip and Plastic Packages

(See Packaging Spec, Order #231369)

The 8237A Multimode Direct Memory Access (DMA) Controller is a peripheral interface circuit for microprocessor systems. It is designed to improve system performance by allowing external devices to directly transfer information from the system memory. Memory-to-memory transfer capability is also provided. The 8237A offers a wide variety of programmable control features to enhance data throughput and system optimization and to allow dynamic reconfiguration under program control.

The 8237A is designed to be used in conjunction with an external 8-bit address latch. It contains four independent channels and may be expanded to any number of channels by cascading additional controller chips. The three basic transfer modes allow programmability of the types of DMA service by the user. Each channel can be individually programmed to Autoinitialize to its original condition following an End of Process (EOP). Each channel has a full 64K address and word count capability.

The 8273A-4 and 8237A-5 are 4 MHz and 5 MHz versions of the standard 3 MHz 8237A respectively.

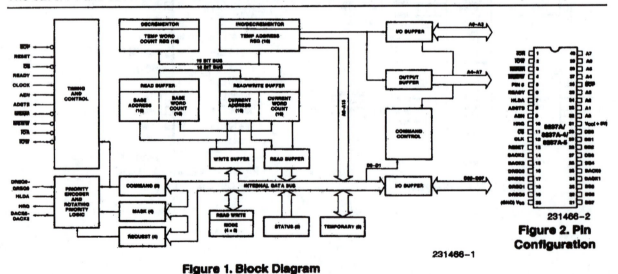

Figure 1. Block Diagram    231466-1

Figure 2. Pin Configuration    231466-2

---

FIGURE F.5    8237 DMA controller, reprinted by permission of Intel Corp. Copyright © 1994 by Intel Corp.

## 8259A
# PROGRAMMABLE INTERRUPT CONTROLLER
## (8259A/8259A-2/8259A-8)

- 8086, 8088 Compatible
- MCS-80®, MCS-85® Compatible
- Eight-Level Priority Controller
- Expandable to 64 Levels
- Programmable Interrupt Modes
- Individual Request Mask Capability

- Single +5V Supply (No Clocks)
- 28-Pin Dual-In-Line Package
- Available in EXPRESS
  — Standard Temperature Range
  — Extended Temperature Range

The Intel 8259A Programmable Interrupt Controller handles up to eight vectored priority interrupts for the CPU. It is cascadable for up to 64 vectored priority interrupts without additional circuitry. It is packaged in a 28-pin DIP, uses NMOS technology and requires a single +5V supply. Circuitry is static, requiring no clock input.

The 8259A is designed to minimize the software and real time overhead in handling multi-level priority interrupts. It has several modes, permitting optimization for a variety of system requirements.

The 8259A is fully upward compatible with the Intel 8259. Software originally written for the 8259 will operate the 8259A in all 8259 equivalent modes (MCS-80/85, Non-Buffered, Edge Triggered).

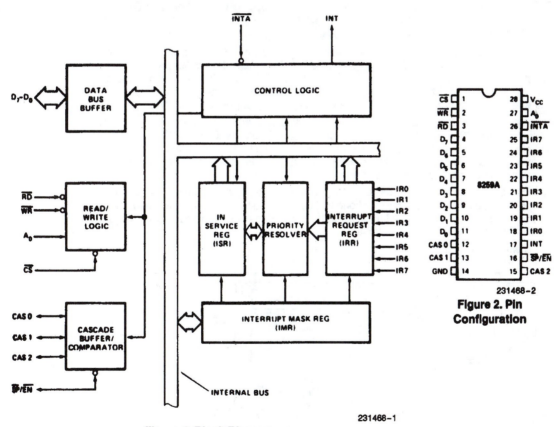

Figure 1. Block Diagram

Figure 2. Pin Configuration

---

**FIGURE F.6   8259 interrupt controller,** reprinted by permission of Intel Corp. Copyright © 1994 by Intel Corp.

# 80286
## High Performance Microprocessor
## with Memory Management and Protection
### (80286-12, 80286-10, 80286-8, 80286-6)

■ High Performance Processor (Up to six times 8086)

■ Large Address Space:
  — 16 Megabytes Physical
  — 1 Gigabyte Virtual per Task

■ Integrated Memory Management, Four-Level Memory Protection and Support for Virtual Memory and Operating Systems

■ High Bandwidth Bus Interface (12.5 Megabyte/Sec)

■ Industry Standard O.S. Support:
  — iRMX®
  — XENIX*
  — UNIX*
  — MS-DOS*

■ Optional Processor Extension:
  — 80287 High Performance 80-bit Numeric Data Processor

■ Two 8086 Upward Compatible Operating Modes:
  — 8086 Real Address Mode
  — Protected Virtual Address Mode

■ Range of Clock Rates
  — 12.5 MHz for 80286-12
  — 10 MHz for 80286-10
  — 8 MHz for 80826-8
  — 6 MHz for 80286-6

■ Complete System Development Support:
  — Development Software: Assembler, PL/M, Pascal, FORTRAN, and System Utilities
  — In-Circuit-Emulator (ICE™-286)

■ Available in 68 Pin Ceramic LCC (Leadless Chip Carrier), PGA (Pin Grid Array), and PLCC (Plastic Leaded Chip Carrier) Packages
  (See Packaging Spec., Order #231369)

The 80286 is an advanced, high-performance microprocessor with specially optimized capabilities for multiple user and multi-tasking systems. The 80286 has built-in memory protection that supports operating system and task isolation as well as program and data privacy within tasks. A 10 MHz 80286 provides five times or more throughput than the standard 5 MHz 8086. The 80286 includes memory management capabilities that map $2^{30}$ (one gigabyte) of virtual address space per task into $2^{24}$ bytes (16 megabytes) of physical memory.

The 80286 is upward compatible with 8086 and 88 software. Using 8086 real address mode, the 80286 is object code compatible with existing 8086, 88 software. In protected virtual address mode, the 80286 is source code compatible with 8086, 88 software and may require upgrading to use virtual addresses supported by the 80286's integrated memory management and protection mechanism. Both modes operate at full 80286 performance and execute a superset of the 8086 and 88 instructions.

The 80286 provides special operations to support the efficient implementation and execution of operating systems. For example, one instruction can end execution of one task, save its state, switch to a new task, load its state, and start execution of the new task. The 80286 also supports virtual memory systems by providing a segment-not-present exception and restartable instructions.

*XENIX and MS-DOS are trademarks of Microsoft Corp.
*UNIX is a trademark of Bell Labs or AT&T

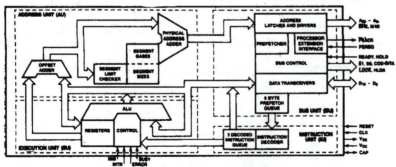

**Figure 1. 80286 Internal Block Diagram**

210253–1

---

**FIGURE F.7**   **80286 central processing unit (CPU),** reprinted by permission of Intel Corp. Copyright © 1994 by Intel Corp.

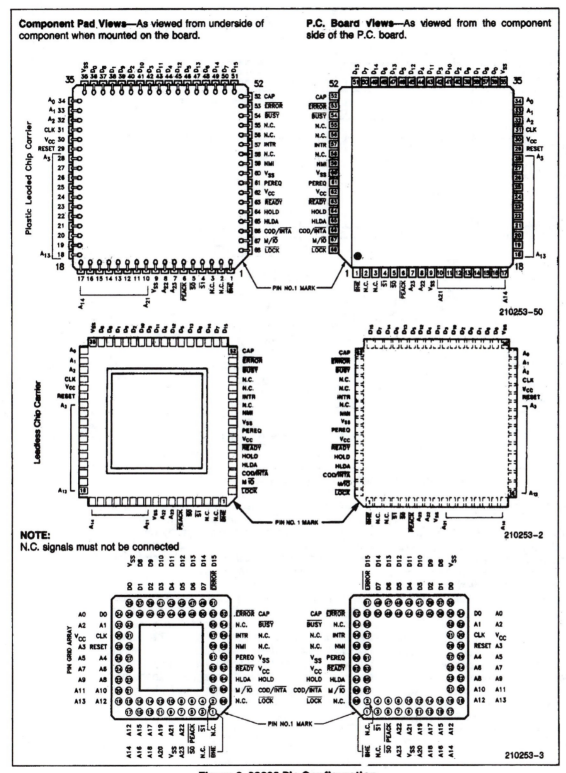

**Component Pad Views**—As viewed from underside of component when mounted on the board.

**P.C. Board Views**—As viewed from the component side of the P.C. board.

**NOTE:**
N.C. signals must not be connected

**Figure 2. 80286 Pin Configuration**

**FIGURE F.8    80286 pinout,** reprinted by permission of Intel Corp. Copyright © 1994 by Intel Corp.

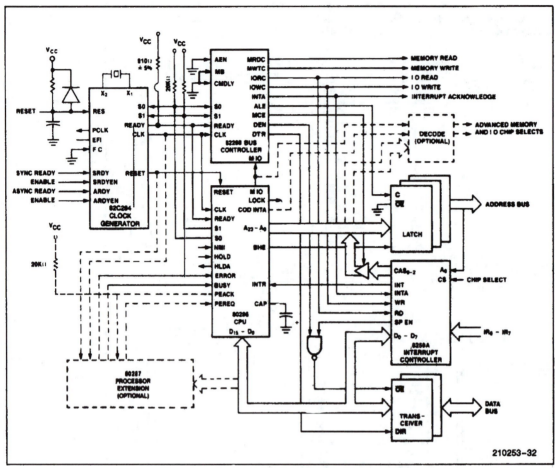

**Figure 31. Basic 80286 System Configuration**

## SYSTEM CONFIGURATIONS

The versatile bus structure of the 80286 microsystem, with a full complement of support chips, allows flexible configuration of a wide range of systems. The basic configuration, shown in Figure 31, is similar to an 8086 maximum mode system. It includes the CPU plus an 8259A interrupt controller, 82C284 clock generator, and the 82288 Bus Controller.

As indicated by the dashed lines in Figure 31, the ability to add processor extensions is an integral feature of 80286 microsystems. The processor extension interface allows external hardware to perform special functions and transfer data concurrent with CPU execution of other instructions. Full system integrity is maintained because the 80286 supervises all data transfers and instruction execution for the processor extension.

The 80287 has all the instructions and data types of an 8087. The 80287 NPX can perform numeric calculations and data transfers concurrently with CPU program execution. Numerics code and data have the same integrity as all other information protected by the 80286 protection mechanism.

The 80286 can overlap chip select decoding and address propagation during the data transfer for the previous bus operation. This information is latched by ALE during the middle of a $T_s$ cycle. The latched chip select and address information remains stable during the bus operation while the next cycle's address is being decoded and propagated into the system. Decode logic can be implemented with a high speed bipolar PROM.

---

**FIGURE F.9    80286 block diagram,** reprinted by permission of Intel Corp. Copyright © 1994 by Intel Corp.

# 80287
# 80-BIT HMOS
# NUMERIC PROCESSOR EXTENSION
## (80287-3, 80287-6, 80287-8, 80287-10)

- ■ High Performance 80-Bit Internal Architecture
- ■ Implements Proposed IEEE Floating Point Standard 754
- ■ Expands 80286 Data types to Include 32-, 64-, 80-Bit Floating Point, 32-, 64-Bit Integers and 18-Digit BCD Operands
- ■ Object Code Compatible with 8087
- ■ Built-In Exception Handling
- ■ Operates in Both Real and Protected Mode 80286 Systems
- ■ 8x80-Bit, Individually Addressable, Numeric Register Stack

- ■ Protected Mode Operation Completely Conforms to the 80286 Memory Management and Protection Mechanisms
- ■ Directly Extends 80286 Instruction Set to Trigonometric, Logarithmic, Exponential and Arithmetic Instructions for All Data types
- ■ Operates with 80386 CPU without Software Modification
- ■ Available in EXPRESS—Standard Temperature Range
- ■ Available in 40 pin-CERDIP package
  (see Packaging Spec: Order #231369)

The Intel 80287 is a high performance numerics processor extension that extends the 80286 architecture with floating point, extended integer and BCD data types. The 80286/80287 computing system fully conforms to the proposed IEEE Floating Point Standard. Using a numerics oriented architecture, the 80287 adds over fifty mnemonics to the 80286/80287 instruction set, making the 80286/80287 a complete solution for high performance numeric processing. The 80287 is implemented in N-channel, depletion load, silicon gate technology (HMOS) and packaged in a 40-pin cerdip package. The 80286/80287 is object code compatible with the 8086/8087 and 8088/8087.

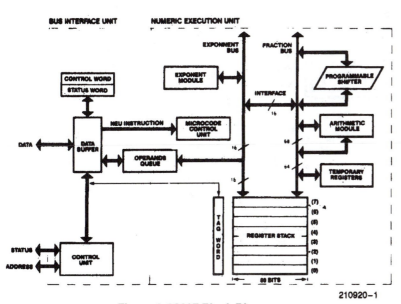

Figure 1. 80287 Block Diagram

210920-1

NOTE:
N/C Pins should not be connected

Figure 2.
80287 Pin Configuration

210920-2

---

FIGURE F.10    80287 numeric processing unit (NPU), reprinted by permission of Intel Corp. Copyright © 1994 by Intel Corp.

# 82C288
# BUS CONTROLLER FOR 80286 PROCESSORS
## (82C288-12, 82C288-10, 82C288-8)

- ■ Provides Commands and Controls for Local and System Bus
- ■ Wide Flexibility in System Configurations
- ■ Implemented in High Speed CHMOS III Technology
- ■ Fully Compatible with the HMOS 82288

- ■ Fully Static Device
- ■ Single +5V Supply
- ■ Available in 20 Pin PLCC (Plastic Leaded Chip Carrier) and 20 Pin Cerdip Packages
  (See Packaging Spec, Order #231369)

The Intel 82C288 Bus Controller is a 20-pin CHMOS III component for use in 80286 microsystems. The 82C288 is fully compatible with its predecessor the HMOS 82288. The bus controller is fully static and supports a low power mode. The bus controller provides command and control outputs with flexible timing options. Separate command outputs are used for memory and I/O devices. The data bus is controlled with separate data enable and direction control signals.

Two modes of operation are possible via a strapping option: MULTIBUS® I compatible bus cycles, and high speed bus cycles.

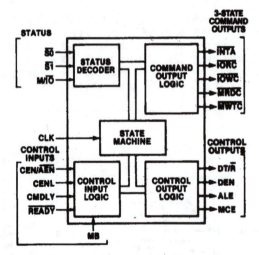

Figure 1. 82C288 Block Diagram

240042–1

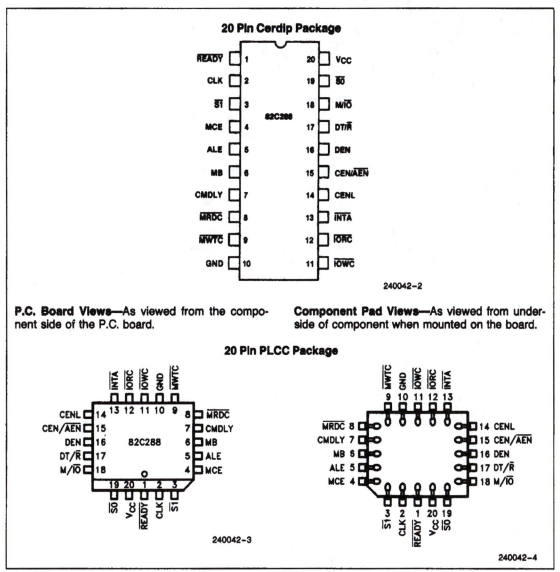

Figure 2. 82C288 Pin Configuration

**FIGURE F.12**   **82288 pinout,** reprinted by permission of Intel Corp. Copyright © 1994 by Intel Corp.

# 82C284
# CLOCK GENERATOR AND READY INTERFACE
# FOR 80286 PROCESSORS
# (82C284-12, 82C284-10, 82C284-8)

■ Generates System Clock for 80286 Processors

■ Uses Crystal or TTL Signal for Frequency Source

■ Provides Local READY and MULTIBUS®I READY Synchronization

■ Single +5V Power Supply

■ CHMOS III Technology

■ Generates System Reset Output from Schmitt Trigger Input

■ Available in 18-Lead Cerdip and 20-Pin PLCC (Plastic Leaded Chip Carrier) Packages

(See Packaging Spec, Order #231369)

The 82C284 is a clock generator/driver which provides clock signals for 80286 processors and support components. It also contains logic to supply READY to the CPU from either asynchronous or synchronous sources and synchronous RESET from an asynchronous input with hysteresis.

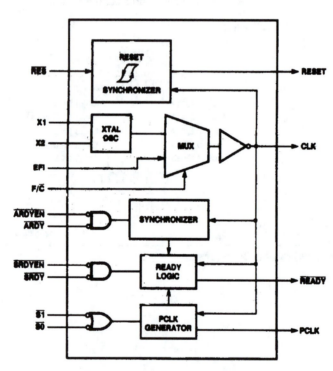

Figure 1. 82C284 Block Diagram

210453-1

---

FIGURE F.13   82284 clock generator, reprinted by permission of Intel Corp. Copyright © 1994 by Intel Corp.

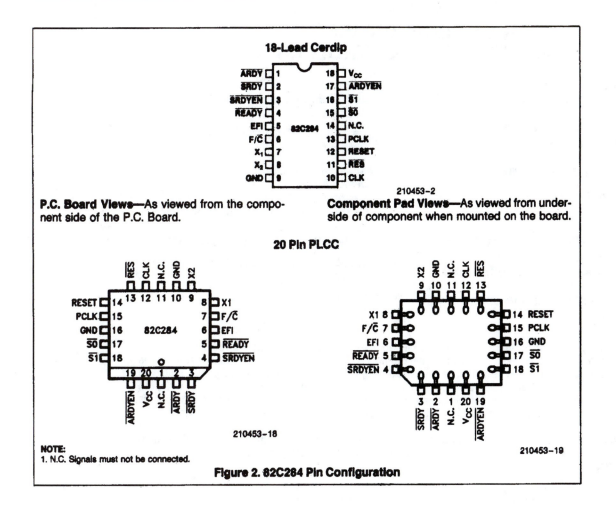

NOTE:
1. N.C. Signals must not be connected.

**Figure 2. 82C284 Pin Configuration**

# 82385
# HIGH PERFORMANCE
# 32-BIT CACHE CONTROLLER

- **Improves 80386 System Performance**
  - Reduces Average CPU Wait States to Nearly Zero
  - Zero Wait State Read Hit
  - Zero Wait State Posted Writes
  - Allows Other Masters to Access the System Bus More Readily
- **Hit Rates up to 99%**
- **Optimized as 80386 Companion**
  - Simple 80386 Interface
  - Part of 386-Based Compute Engine Including 80387 Numerics Coprocessor and 82380 Integrated System Peripheral
  - 16 MHz and 20 MHz Operation
- **Software Transparent**

- **Synchronous Dual Bus Architecture**
  - Bus Watching Maintains Cache Coherency
- **Maps Full 80386 Address Space (4 Gigabytes)**
- **Flexible Cache Mapping Policies**
  - Direct Mapped or 2-Way Set Associative Cache Organization
  - Supports Non-Cacheable Memory Space
  - Unified Cache for Code and Data
- **Integrates Cache Directory and Cache Management Logic**
- **High Speed CHMOS III Technology**
- **132-Pin PGA Package**

The 82385 Cache Controller is a high performance 32-bit peripheral for Intel's 80386 Microprocessor. It stores a copy of frequently accessed code and data from main memory in a zero wait state local cache memory. The 82385 enables the 80386 to run at its full potential by reducing the average number of CPU wait states to nearly zero. The dual bus architecture of the 82385 allows other masters to access system resources while the 80386 operates locally out of its cache. In this situation, the 82385's "bus watching" mechanism preserves cache coherency by monitoring the system bus address lines at no cost to system or local throughput.

The 82385 is completely software transparent, protecting the integrity of system software. High performance and board savings are achieved because the 82385 integrates a cache directory and all cache management logic on one chip.

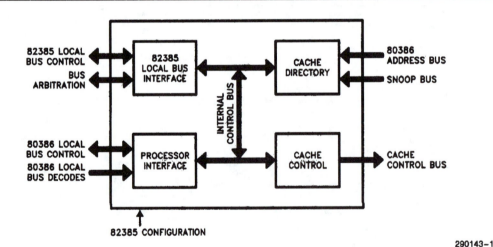

290143-1

**82385 Internal Block Diagram**

**FIGURE F.15** **82385 cache controller,** reprinted by permission of Intel Corp. Copyright © 1994 by Intel Corp.

# intel.

## Intel386™ DX MICROPROCESSOR
## HIGH PERFORMANCE 32-BIT CHMOS MICROPROCESSOR
## WITH INTEGRATED MEMORY MANAGEMENT

- **Flexible 32-Bit Microprocessor**
  - —8, 16, 32-Bit Data Types
  - —8 General Purpose 32-Bit Registers
- **Very Large Address Space**
  - —4 Gigabyte Physical
  - —64 Terabyte Virtual
  - —4 Gigabyte Maximum Segment Size
- **Integrated Memory Management Unit**
  - —Virtual Memory Support
  - —Optional On-Chip Paging
  - —4 Levels of Protection
  - —Fully Compatible with 80286
- **Object Code Compatible with All 8086 Family Microprocessors**
- **Virtual 8086 Mode Allows Running of 8086 Software in a Protected and Paged System**
- **Hardware Debugging Support**

- **Optimized for System Performance**
  - —Pipelined Instruction Execution
  - —On-Chip Address Translation Caches
  - —20, 25 and 33 MHz Clock
  - —40, 50 and 66 Megabytes/Sec Bus Bandwidth
- **Numerics Support via Intel387™ DX Math Coprocessor**
- **Complete System Development Support**
  - —Software: C, PL/M, Assembler System Generation Tools
  - —Debuggers: PSCOPE, ICE™-386
- **High Speed CHMOS IV Technology**
- **132 Pin Grid Array Package**
- **132 Pin Plastic Quad Flat Package**
  (See Packaging Specification, Order #231369)

The Intel386 DX Microprocessor is an entry-level 32-bit microprocessor designed for single-user applications and operating systems such as MS-DOS and Windows. The 32-bit registers and data paths support 32-bit addresses and data types. The processor addresses up to four gigabytes of physical memory and 64 terabytes (2**46) of virtual memory. The integrated memory management and protection architecture includes address translation registers, multitasking hardware and a protection mechanism to support operating systems. Instruction pipelining, on-chip address translation, ensure short average instruction execution times and maximum system throughput.

The Intel386 DX CPU offers new testability and debugging features. Testability features include a self-test and direct access to the page translation cache. Four new breakpoint registers provide breakpoit:t traps on code execution or data accesses, for powerful debugging of even ROM-based systems.

Object-code compatibility with all 8086 family members (8086, 8088, 80186, 80188, 80286) means the Intel386 DX offers immediate access to the world's largest microprocessor software base.

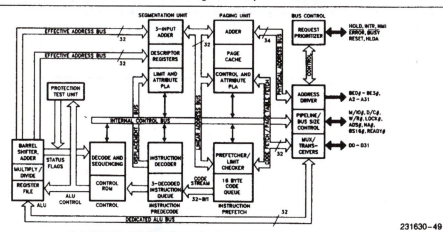

**Intel386™ DX Pipelined 32-Bit Microarchitecture**

231630–49

Intel386™ DX and Intel387™ DX are Trademarks of Intel Corporation.
MS-DOS and Windows are Trademarks of MICROSOFT Corporation.

---

**FIGURE F.16   80386DX central processing unit (CPU),** reprinted by permission of Intel Corp. Copyright © 1994 by Intel Corp.

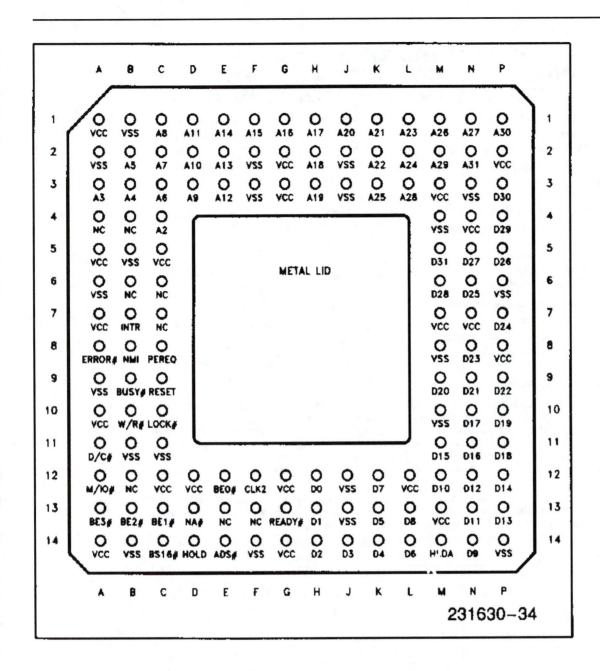

**FIGURE F.17** **80386DX pinout,** reprinted by permission of Intel Corp. Copyright © 1994 by Intel Corp.

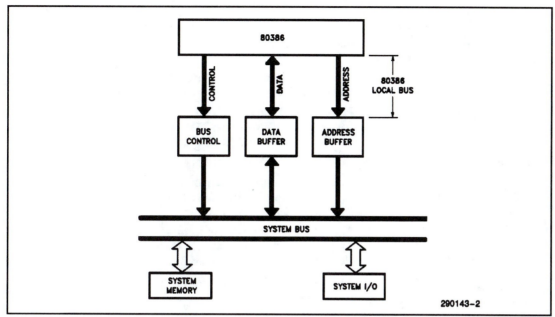

Figure 1-1. 80386 System Bus Structure

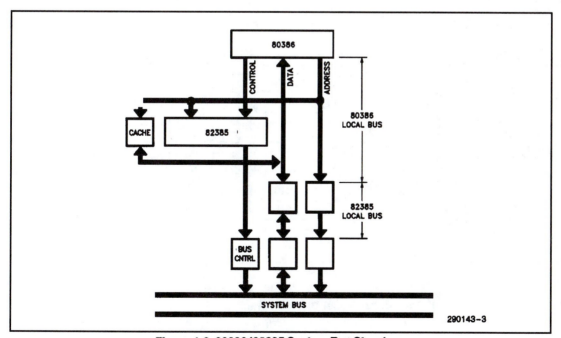

Figure 1-2. 80386/82385 System Bus Structure

**FIGURE F.18   80386DX system bus,** reprinted by permission of Intel Corp. Copyright © 1994 by Intel Corp.

# 387™ DX
# MATH COPROCESSOR

- **High Performance 80-Bit Internal Architecture**
- **Implements ANSI/IEEE Standard 754-1985 for Binary Floating-Point Arithmetic**
- **Six to Eleven Times 8087/80287 Performance**
- **Expands 386™ DX CPU Data Types to Include 32-, 64-, 80-Bit Floating Point, 32-, 64-Bit Integers and 18-Digit BCD Operands**
- **Directly Extends 386™ DX CPU Instruction Set to Include Trigonometric, Logarithmic, Exponential and Arithmetic Instructions for All Data Types**

- **Upward Object-Code Compatible from 8087 and 80287**
- **Full-Range Transcendental Operations for SINE, COSINE, TANGENT, ARCTANGENT and LOGARITHM**
- **Built-In Exception Handling**
- **Operates Independently of Real, Protected and Virtual-8086 Modes of the 386™ DX Microprocessor**
- **Eight 80-Bit Numeric Registers, Usable as Individually Addressable General Registers or as a Register Stack**
- **Available in 68-Pin PGA Package**
  (See Packaging Spec: Order #231369)

The Intel 387™ DX Math CoProcessor (MCP) is an extension to the Intel 386™ microprocessor architecture. The combination of the 387 DX with the 386™ DX Microprocessor dramatically increases the processing speed of computer application software which utilize mathmatical operations. This makes an ideal computer workstation platform for applications such as financial modeling and spreadsheets, CAD/CAM, or graphics.

The 387 DX Math CoProcessor adds over seventy mnemonics to the 386 DX Microprocessor instruction set. Specific 387 DX math operations include logarithmic, arithmetic, exponentional, and triginometric functions. The 387 DX supports integer, extended integer, floating point and BCD data formats, and fully conforms to the ANSI/IEEE floating point standard.

The 387 DX Math CoProcessor is object code compatible with the 80387SX, and upward object code compatible from the 80287 and 8087 math coprocessors. Object code for 386 DX/387 DX is also compatible with the Intel 486™ microprocessor. The 387 DX is manufactured on 1 micron, CHMOS IV technology and packaged in a 68-pin PGA package.

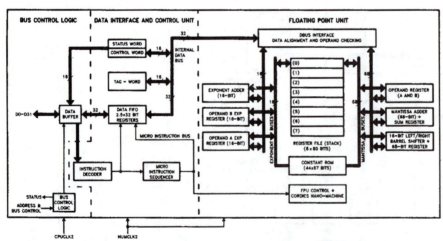

240448-1

**Figure 0.1. 387™ DX Math Coprocessor Block Diagram**

---

**FIGURE F.19    80387 numeric processing unit (NPU),** reprinted by permission of Intel Corp. Copyright © 1994 by Intel Corp.

# Intel386™ SX MICROPROCESSOR

- Full 32-Bit Internal Architecture
  - 8-, 16-, 32-Bit Data Types
  - 8 General Purpose 32-Bit Registers

- Runs Intel386™ Software in a Cost Effective 16-Bit Hardware Environment
  - Runs Same Applications and O.S.'s as the Intel386™ DX Processor
  - Object Code Compatible with 8086, 80186, 80286, and Intel386™ Processors

- High Performance 16-Bit Data Bus
  - 16, 20, 25 and 33 MHz Clock
  - Two-Clock Bus Cycles
  - Address Pipelining Allows Use of Slower/Cheaper Memories

- Integrated Memory Management Unit
  - Virtual Memory Support
  - Optional On-Chip Paging
  - 4 Levels of Hardware Enforced Protection
  - MMU Fully Compatible with Those of the 80286 and Intel386 DX CPUs

- Virtual 8086 Mode Allows Execution of 8086 Software in a Protected and Paged System

- Large Uniform Address Space
  - 16 Megabyte Physical
  - 64 Terabyte Virtual
  - 4 Gigabyte Maximum Segment Size

- Numerics Support with the Intel387™ SX Math CoProcessor

- On-Chip Debugging Support Including Breakpoint Registers

- Complete System Development Support
  - Software: C, PL/M, Assembler
  - Debuggers: PMON-386 DX, ICE™-386 SX

- High Speed CHMOS IV Technology

- Operating Frequency:
  - Standard
    (Intel386 SX -33, -25, -20, -16)
    Min/Max Frequency
    (4/33, 4/25, 4/20, 4/16) MHz
  - Low Power
    (Intel386 SX -33, -25, -20, -16, -12)
    Min/Max Frequency
    (2/33, 2/25, 2/20, 2/16, 2/12) MHz

- 100-Pin Plastic Quad Flatpack Package
  (See Packaging Outlines and Dimensions #231369)

The Intel386™ SX Microprocessor is an entry-level 32-bit CPU with a 16-bit external data bus and a 24-bit external address bus. The Intel386 SX CPU brings the vast software library of the Intel386™ Architecture to entry-level systems. It provides the performance benefits of a 32-bit programming architecture with the cost savings associated with 16-bit hardware systems.

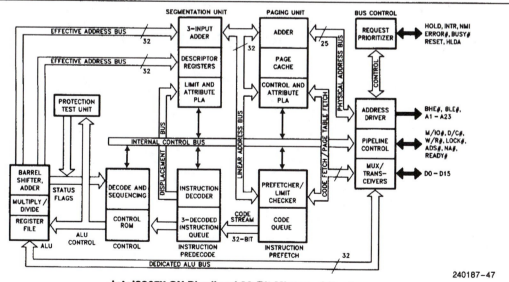

Intel386™ SX Pipelined 32-Bit Microarchitecture

240187-47

**FIGURE F.20    80386SX central processing unit (CPU),** reprinted by permission of Intel Corp. Copyright © 1994 by Intel Corp.

## 1.0  PIN DESCRIPTION

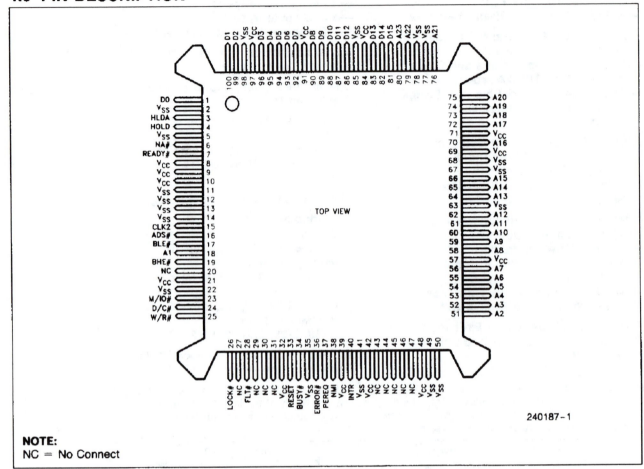

NOTE:
NC = No Connect

Figure 1.1. Intel386™ SX Microprocessor Pin out Top View

---

**FIGURE F.21**    **80386SX pinout,** reprinted by permission of Intel Corp. Copyright © 1994 by Intel Corp.

# Intel387™ SX
# MATH COPROCESSOR

- **New Automatic Power Management**
  - Low Power Consumption
  - Typically 100 mA in Dynamic Mode, and 4 mA in Idle Mode
- **Socket Compatible with Intel387 Family of Math CoProcessors**
  - Hardware and Software Compatible
  - Supported by Over 2100 Commercial Software Packages
  - 10% to 15% Performance Increase on Whetstone and Livermore Benchmarks

- **Compatible with the Intel386™ SX Microprocessor**
  - Extends CPU Instruction Set to Include Trigonometric, Logarithmic, and Exponential
- **High Performance 80-Bit Internal Architecture**
- **Implements ANSI/IEEE Standard 754-1985 for Binary Floating-Point Arithmetic**
- **Available in a 68-Pin PLCC Package**
  See Intel Packaging Specification, Order #231369

The Intel387™ SX Math CoProcessor is an extension to the Intel386™ SX microprocessor architecture. The combination of the Intel387™ SX with the Intel386™ SX microprocessor dramatically increases the processing speed of computer application software that utilizes high performance floating-point operations. An internal Power Management Unit enables the Intel387™ SX to perform these floating-point operations while maintaining very low power consumption for portable and desktop applications. The internal Power Management Unit effectively reduces power consumption by 95% when the device is idle.

The Intel387™ SX Math CoProcessor is available in a 68-pin PLCC package, and is manufactured on Intel's advanced 1.0 micron CHMOS IV technology.

240225-22

Intel386 and Intel387 are trademarks of Intel Corporation.

**FIGURE F.22   80387SX numeric processing unit (NPU),** reprinted by permission of Intel Corp. Copyright © 1994 by Intel Corp.

intel.

# Intel386™ SL MICROPROCESSOR
## Intel386™ Microprocessor Core, with
## Integrated Memory and Cache Controllers and
## System Power Management
## Fully-Static CHMOS IV Technology

- **Static Intel386™ SL CPU**
  - Optimized and Compatible with Standard Operating System Software such as: MS-DOS, WINDOWS, OS/2 and UNIX
  - Object Code Compatible with Intel 8086, 80286 and Intel386™ Microprocessors
  - Runs All Desk-Top Applications, 16- or 32-Bit
  - D.C. to 25 MHz Operation
  - 20 Megabytes Physical Memory/ 64 Terabytes Virtual Memory
  - 4 Gigabyte Maximum Segment Size
  - High Integration, Low Power Intel CHMOS IV Process Technology

- **Transparent Power-Management System Architecture**
  - System Management Mode Architecture Extension for Truly Compatible Systems
  - Power Management Transparent to Operating Systems and Application Programs
  - Programmable Hardware Supports Custom Power-Control Methods

- **Direct Drive Bus Interfaces**
  - Full ISA Bus Interface
  - High Speed Peripheral Interface Bus

- **Integrated Cache Controller and Tag RAM**
  - No-Glue Cache SRAM Interface
  - 16k, 32k, or 64 kByte Cache Size
  - Direct, 2-Way or 4-Way Set Associative Organization
  - Write Posting—Posted Memory Writes
  - 16-Bit Line Size—Reduces Bus Utilization for Cache Line Fills
  - Write-Thru, with SmartHit Algorithm for Reduced Main Memory Power Consumption

- **Programmable Memory Control**
  - No-Glue, Page-Mode DRAM Interface
  - SRAM Support for Lowest Power (Standard 5V Mode Only)
  - 1, 2, or 4 Banks Interleaved, with Programmable Wait States
  - 512k to 20 MBytes
  - Advanced, Flexible Address-Map Configuration
  - Full Hardware LIM EMS 4.0 Address Translation to 32 Megabytes without Waitstate Penalty

- **3.3 CPU core with FlexibleVoltage Interfaces (Optional)**
  - Pin-Compatible with Standard 5V CPU
  - Flexible 5V/3.3V ISA/PI Bus Interface
  - 50% Power Savings

---

**FIGURE F.23    80386SL central processing unit (CPU),** reprinted by permission of Intel Corp. Copyright © 1994 by Intel Corp.

# Intel486™ SX MICROPROCESSOR

**IMPORTANT**—Read This Section Before Reading The Rest Of The Data Sheet

This data sheet describes the Intel486 SX microprocessor, the Intel OverDrive™ Processor, and the Intel487™ SX Math CoProcessor. All normal text describes the functionality for the Intel486 SX microprocessor, the Intel OverDrive Processor, and the Intel487 SX Math CoProcessor unless explicitly stated otherwise. All sections of the data sheet which describe the Intel487 SX Math CoProcessor and Intel OverDrive Processor functionality only are highlighted as shown in the following example:

> This is an example of what the highlighted sections look like. The highlighted sections describe functionality which apply to the Intel487 SX Math CoProcessor and Intel OverDrive Processor only.

All references to the Intel OverDrive Processor are also applicable to the Intel487 SX Math CoProcessor unless otherwise stated.

Section 13.0, OverDrive Processor, contains additional information specific to the OverDrive Processor.

- **Binary Compatible with Large Software Base**
  - **MS-DOS\*, OS/2\*\*, Windows**
  - **UNIX\*\*\* System V/386**
  - **iRMX®, iRMK™ Kernels**
- **High Integration Enables On-Chip**
  - **8 Kbyte Code and Data Cache**
  - **Paged, Virtual Memory Management**
- **Easy To Use**
  - **Built-In Self Test**
  - **Hardware Debugging Support**
  - **Intel Software Support**
  - **Extensive Third Party Software Support**
- **196-Lead PQFP and 168-Pin Grid Array Package for Intel486™ SX Microprocessor**
- **169-Pin Grid Array Package for Intel OverDrive™ Processor and Intel487™ SX Math CoProcessor**

- **High Performance Design**
  - **Intel486 One Clock Instruction Core**
  - **33, 25, 20 and 16 MHz Clock Frequencies at 5V**
  - **25, 20 and 16 MHz Clock Frequencies at 3.3V**
  - **106 Mbyte/sec Burst Bus at 33 MHz**
  - **CHMOS IV and CHMOS V Process Technology**
  - **Dynamic Bus Sizing for 8-, 16- and 32-Bit Busses**
- **Complete 32-Bit Architecture**
  - **Address and Data Busses**
  - **Registers**
  - **8-, 16- and 32-Bit Data Types**
- **Multiprocessor Support**
  - **Multiprocessor Instructions**
  - **Cache Consistency Protocols**
  - **Support for Second Level Cache**
- **Optional Intel OverDrive™ Processor/ Intel487™ SX Math CoProcessor (16–25 MHz only)**
- **IEEE 1149.1 Boundary Scan Compatibility**
  - **Available on PQFP Intel486 SX CPU Only**

iRMX® is a registered trademark of Intel Corporation.
iRMK™, Intel386™, Intel387™, i386™, i387™, Intel486™, Intel487™, i486™, OverDrive™ and i487™ are trademarks of Intel Corporation.
\*MS-DOS®, WINDOWS are registered trademarks of Microsoft Corporation.
\*\*OS/2™ is a trademark of Microsoft Corporation.
\*\*\*UNIX™ is a trademark of Unix Systems Labs.

**FIGURE F.24   80486SX central processing unit (CPU),** reprinted by permission of Intel Corp. Copyright © 1994 by Intel Corp.

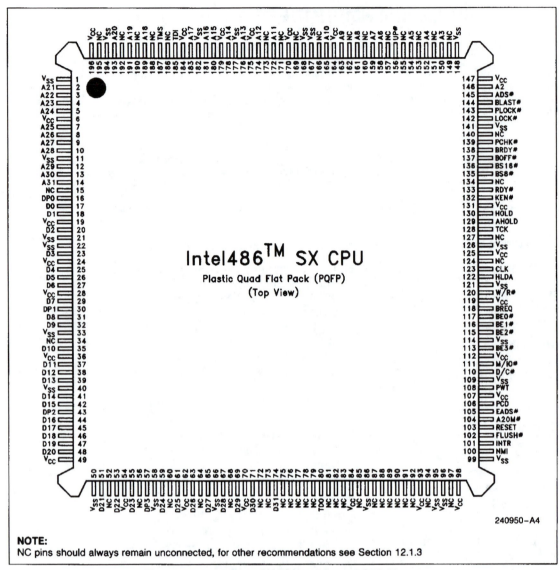

Intel486™ SX CPU

Plastic Quad Flat Pack (PQFP)

(Top View)

240950–A4

NOTE:
NC pins should always remain unconnected, for other recommendations see Section 12.1.3

**Figure 1.3. Intel486™ SX CPU 196 Lead PQFP Pinout**

**FIGURE F.25    80486SX pinout,** reprinted by permission of Intel Corp. Copyright © 1994 by Intel Corp.

# Intel486™ DX
# MICROPROCESSOR

- **Binary Compatible with Large Software Base**
  - MS-DOS*, OS/2**, Windows*
  - UNIX*** System V/386
  - iRMX®, IRMK™ Kernels
- **High Integration Enables On-Chip**
  - 8 Kbyte Code and Data Cache
  - Floating Point Unit
  - Paged, Virtual Memory Management
- **Easy To Use**
  - Built-In Self Test
  - Hardware Debugging Support
  - Intel Software Support
  - Extensive Third Party Software Support
- **IEEE 1149.1 Boundary Scan Compatibility**
  - Available on 50 MHz Version Only
- **Upgradable to Intel OverDrive™ Processor**

- **168-Pin Grid Array Package**
- **High Performance Design**
  - RISC Integer Core with Frequent Instructions Executing In One Clock
  - 25 MHz, 33 MHz, and 50 MHz Clock
  - 80, 106, 160 Mbyte/sec Burst Bus
  - CHMOS IV and CHMOS V Process Technology
  - Dynamic Bus Sizing for 8-, 16-, and 32-Bit Busses
- **Complete 32-Bit Architecture**
  - Address and Data Busses
  - Registers
  - 8-, 16- and 32-Bit Data Types
- **Multiprocessor Support**
  - Multiprocessor Instructions
  - Cache Consistency Protocols
  - Support for Second Level Cache

The Intel486 CPU offers the highest performance for DOS, OS/2, Windows, and UNIX System V/386 applications. It is 100% binary compatible with the Intel386™ CPU. Over one million transistors integrate the RISC integer core, 8 Kbyte cache memory, floating point hardware, and memory management on-chip while retaining binary compatibility with previous members of the Intel386/Intel486 architectural family. The RISC integer core executes frequently-used instructions in one cycle, providing leadership performance levels. An 8 Kbyte unified code and data cache allow the high performance levels to be sustained. A 160 MByte/sec burst bus at 50 MHz ensures high system throughput even with inexpensive DRAMs.

## Intel486™ Microprocessor Pipelined 32-Bit Microarchitecture

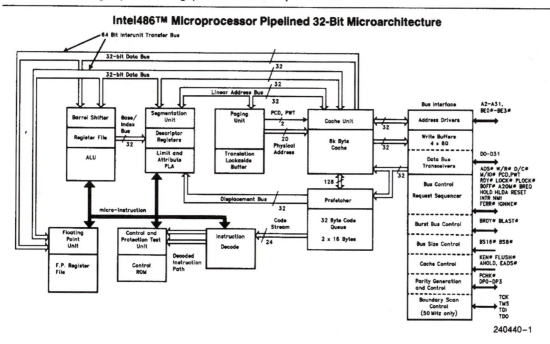

240440-1

---

**FIGURE F.26   80486DX central processing unit (CPU),** reprinted by permission of Intel Corp. Copyright © 1994 by Intel Corp.

**Intel486™ MICROPROCESSOR 25 MHz AND 33 MHz VERSIONS — PIN SIDE VIEW**

| | 1 | 2 | 3 | 4 | 5 | 6 | 7 | 8 | 9 | 10 | 11 | 12 | 13 | 14 | 15 | 16 | 17 | |
|---|---|---|---|---|---|---|---|---|---|---|---|---|---|---|---|---|---|---|
| S | A27 | A26 | A23 | NC | A14 | VSS | A12 | VSS | VSS | VSS | VSS | VSS | A10 | VSS | A6 | A4 | ADS# | S |
| R | A28 | A25 | VCC | VSS | A18 | VCC | A15 | VCC | VCC | VCC | VCC | A11 | A8 | VCC | A3 | BLAST# | NC | R |
| Q | A31 | VSS | A17 | A19 | A21 | A24 | A22 | A20 | A16 | A13 | A9 | A5 | A7 | A2 | BREQ | PLOCK# | PCHK# | Q |
| P | D0 | A29 | A30 | | | | | | | | | | | | HLDA | VCC | VSS | P |
| N | D2 | D1 | DP0 | | | | | | | | | | | | LOCK# | M/IO# | W/R# | N |
| M | VSS | VCC | D4 | | | | | | | | | | | | D/C# | VCC | VSS | M |
| L | VSS | D6 | D7 | | | | | | | | | | | | PWT | VCC | VSS | L |
| K | VSS | VCC | D14 | | | | | | | | | | | | BE0# | VCC | VSS | K |
| J | VCC | D5 | D16 | | | | | | | | | | | | BE2# | BE1# | PCD | J |
| H | VSS | D3 | DP2 | | | | | | | | | | | | BRDY# | VCC | VSS | H |
| G | VSS | VCC | D12 | | | | | | | | | | | | NC | VCC | VSS | G |
| F | DP1 | D8 | D15 | | | | | | | | | | | | KEN# | RDY# | BE3# | F |
| E | VSS | VCC | DI0 | | | | | | | | | | | | HOLD | VCC | VSS | E |
| D | D9 | D13 | D17 | | | | | | | | | | | | A20M# | BS8# | BOFF# | D |
| C | D11 | D18 | CLK | VCC | VCC | D27 | D26 | D28 | D30 | NC | NC | NC | NC | FERR# | FLUSH# | RESET | BS16# | C |
| B | D19 | D21 | VSS | VSS | VSS | D25 | VCC | D31 | VCC | NC | VCC | NC | NC | NC | NMI | NC | EADS# | B |
| A | D20 | D22 | NC | D23 | DP3 | D24 | VSS | D29 | VSS | NC | VSS | NC | NC | NC | IGNNE# | INTR | AHOLD | A |
| | 1 | 2 | 3 | 4 | 5 | 6 | 7 | 8 | 9 | 10 | 11 | 12 | 13 | 14 | 15 | 16 | 17 | |

240440-2

**FIGURE F.27   80486DX pinout,** reprinted by permission of Intel Corp. Copyright © 1994 by Intel Corp.

# Intel486™ DX2 MICROPROCESSOR

- **Binary Compatible with Large Software Base**
  - MS-DOS*, OS/2**, Windows
  - UNIX*** System V/Intel386
  - iRMX®, iRMK™ Kernels

- **High Integration Enables On-Chip**
  - 8 Kbyte Code and Data Cache
  - Floating Point Unit
  - Paged, Virtual Memory Management

- **Easy To Use**
  - Built-In Self Test
  - Hardware Debugging Support

- **168-Pin Grid Array Package**
  - Pin Compatible with Intel486™ DX Microprocessor

- **IEEE 1149.1 Boundary Scan Compatibility**

- **High Performance Design**
  - 50 MHz/66 MHz Core Speed Using 25 MHz/33 MHz Bus Clocks
  - RISC Integer Core with Frequent Instructions Executing In One Core Clock
  - 80, 106 Mbyte/sec Burst Bus
  - Dynamic Bus Sizing for 8-, 16-, and 32-Bit Busses
  - Complete 32-Bit Architecture

- **Multiprocessor Support**
  - Cache Consistency Protocols
  - Support for Second Level Cache

The Intel486 DX2 CPU offers the highest performance for DOS, OS/2, Windows, and UNIX System V/Intel386 applications. It is 100% binary compatible with the Intel386™ CPU. Over one million transistors integrate the RISC integer core, 8 Kbyte cache memory, floating point hardware, and memory management on-chip while retaining binary compatibility with previous members of the Intel386/Intel486 architectural family. The RISC integer core executes frequently-used instructions in one core clock cycle, providing leadership performance levels. An 8 Kbyte unified code and data cache allow the high performance levels to be sustained. A 106 MByte/sec burst bus at 33 MHz bus clock ensures high system throughput even with inexpensive DRAMs.

New features enhance multiprocessing systems; new instructions speed manipulation of memory-based semaphores; and on-chip hardware ensures cache consistency and provides hooks for multilevel caches.

The built-in self-test extensively tests on-chip logic, cache memory, and the on-chip paging translation cache. Debug features include breakpoint traps on code execution and data accesses.

## Intel486™ DX2 Microprocessor Pipelined 32-Bit Microarchitecture

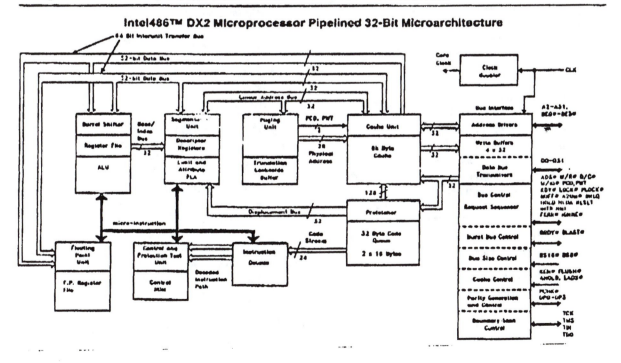

**FIGURE F.28   80486DX2 central processing unit (CPU),** reprinted by permission of Intel Corp. Copyright © 1994 by Intel Corp.

# IntelDX4™ Processor Product Brief

**Product Highlights**

- Fastest Intel486™ processor with an iCOMP™ Index rating of 435 (100 MHz)

- Available versions:
  - 100 MHz with a 50 or a 33 MHz external bus
  - 75 MHz with a 25 MHz external bus

- 3.3V operation results in low power consumption making the IntelDX4™ processor ideal for mobile systems

- 5V tolerant input buffers with TTL compatible outputs allow for direct interface with either 3.3V or 5V system logic and memory components

- Speed-multiplying technology enables the internal CPU core to operate two to three times faster than the external memory bus, delivering high performance with low-cost system design

- Intel standard SL Enhanced features provide intelligent power management

- On-chip 16 Kbyte unified code and data cache

- Thermally enhanced, 208-lead, SQFP package and 168-lead PGA package

- Compatible with an entire library of software based on operating systems such as MS-DOS*, Windows*, OS/2*, UNIX SVR4*, iRMX* operating system, iRMX* Real-Time Kernal, Windows NT*, NeXTSTEP 486*, Solaris 2.0*

- Common core instructions execute in a single internal clock cycle

- On-chip floating-point unit compatible with Intel486™ DX2 processor

- Burst data bus capable of a 160 Mbyte/sec transfer rate at 50 MHz bus clock

- Supported by Intel Verification Program for upgradability to a future Pentium™ OverDrive™ Processor

---

**FIGURE F.29    80486DX4 central processing unit (CPU),** reprinted by permission of Intel Corp. Copyright © 1994 by Intel Corp.

## 4.0   INTEL Pentium® OverDrive® PROCESSOR

- Powerful CPU Upgrade for most Intel486™ CPU-Based Systems
  - Makes Intel Procesor-Based Systems Run Faster
  - Significantly Accelerates All Software Applications
- Designed for Systems Based on:
  - Intel486 SX Processors
  - Intel486 DX Processors
  - IntelSX2™ Processors
  - IntelDX2™ Processors
- Compatible with Installed Base of Thousands of Applications
- Based on Intel Pentium® Processor Technology
  - Superscalar Architecture

- Branch Prediction
- Faster Floating Point Unit
- Enhancements to Core Pentium Processor Silicon
  - Separate Code and Data Caches
  - 16 KB Code Cache
  - 16 KB Write-Back Data Cache
  - 32-Bit Bus Interface
- Package Innovations
  - On-Package Voltage Regulation
  - Integrated Fan Heat Sink
- Incorporates SMM Power Saving Features

---

## 3.0  IntelDX4™ OverDrive® PROCESSOR FOR Intel486™ SX AND DX MICROPROCESSOR-BASED SYSTEMS

- Processor Upgrade for most Intel486™ SX and DX Processor-Based Systems
  - Single-Chip Upgrade
  - Increases Both Integer and Floating Point Performance
- Two Package Variations to Support Systems with and without an OverDrive® Processor Socket
- 169-Lead Pin Grid Array Package
  - Pin Compatible with Intel487™ SX Math CoProcessor
  - 169th Alignment Pin Ensures Proper Chip Orientation
- 168-Lead Pin Grid Array Package
  - Pin Compatible with Intel486™ DX Processor
- High Integration Enables On-Chip
  - 16 KByte Code and Data Cache
  - Paged, Virtual Memory Management
- Floating Point Math Unit Included On-Chip

- Utilizes IntelDX4 Speed-Tripling Technology
  - Processor Core Runs at Three Times the Frequency of the System Bus
  - Compatible with 33, 25, 20 and 16 MHz Systems
- Binary Compatible with Large Installed Software and Operating System Base
  - MS-DOS, OS/2™, Windows
  - UNIX System V/386
  - IRMX, IRMK™ Kernals
- High Performance Design
  - Core Clock Speed up to 100 MHz
  - CHMOS V Process Technology
- Complete 32-Bit Architecture
  - Address and Data Busses
  - Registers
  - 8-, 16-, 32-Bit Data Types
- SL Enhanced Intel486™ Microprocessor Features Included On-Chip

---

**FIGURE F.30   Overdrive,** reprinted by permission of Intel Corp. Copyright © 1994 by Intel Corp.

# PENTIUM™ PROCESSOR at iCOMP™ INDEX 735\90 MHz
# PENTIUM™ PROCESSOR at iCOMP™ INDEX 815\100 MHz

- Compatible with Large Software Base
  - MS-DOS‡, Windows‡, OS/2‡, UNIX‡
- 32-Bit CPU with 64-Bit Data Bus
- Superscalar Architecture
  - Two Pipelined Integer Units Are Capable of 2 Instructions/Clock
  - Pipelined Floating Point Unit
- Separate Code and Data Caches
  - 8K Code, 8K Write Back Data
  - MESI Cache Protocol
- Advanced Design Features
  - Branch Prediction
  - Virtual Mode Extensions
- 3.3V BiCMOS Silicon Technology
- 4M Pages for Increased TLB Hit Rate
- IEEE 1149.1 Boundary Scan

- Dual Processing Configuration
- Multi-Processor Support
  - Multiprocessor Instructions
  - Support for Second Level Cache
- On-Chip Local APIC Controller
  - MP Interrupt Management
  - 8259 Compatible
- Internal Error Detection Features
- Upgradable with a Future Pentium™ OverDrive™ Processor
- Power Management Features
  - System Management Mode
  - Clock Control
- Fractional Bus Operation
  - 100-MHz Core / 66-MHz Bus
  - 100-MHz Core / 50-MHz Bus
  - 90-MHz Core / 60-MHz Bus

The Pentium processor (735\90, 815\100) extends the Pentium processor family, providing performance needed for mainstream desktop applications as well as for workstations and servers. The Pentium processor is compatible with the entire installed base of applications for DOS, Windows, OS/2, and UNIX. The Pentium processor (735\90, 815\100) superscalar architecture can execute two instructions per clock cycle. Branch prediction and separate caches also increase performance. The pipelined floating point unit delivers workstation level performance. Separate code and data caches reduce cache conflicts while remaining software transparent. The Pentium processor (735\90, 815\100) has 3.3 million transistors and is built on Intel's advanced 3.3V BiCMOS silicon technology. The Pentium processor (735\90, 815\100) has on-chip dual processing support, a local multiprocessor interrupt controller, and SL power management features.

‡ Other brands and trademarks are the property of their respective owners.

**FIGURE F.31** **Pentium™ central processing unit (CPU),** reprinted by permission of Intel Corp. Copyright © 1994 by Intel Corp.

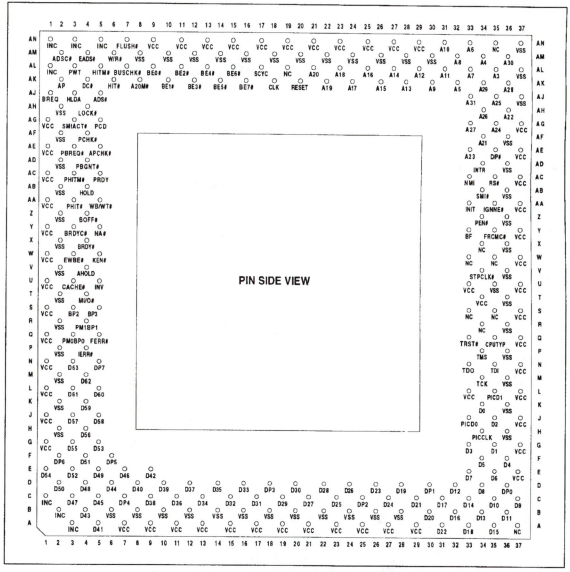

**Pentium™ Processor (735\90, 815\100) (Pin Side View)**

**FIGURE F.32   Pentium™ pinout,** reprinted by permission of Intel Corp. Copyright © 1994 by Intel Corp.

®

# PENTIUM® PRO PROCESSOR AT
# 150 MHz, 166 MHz, 180 MHz and
# 200 MHz

- Available at 150 MHz, 166 MHz, 180 MHz and 200MHz core speeds

- Binary compatible with applications running on previous members of the Intel microprocessor family

- Optimized for 32-bit applications running on advanced 32-bit operating systems

- Dynamic Execution microarchitecture

- Single package includes Pentium® Pro processor CPU, cache and system bus interface

- Scalable up to four processors and 4 GB memory

- Separate dedicated external system bus, and dedicated internal full-speed cache bus

- 8 KB / 8 KB separate data and instruction, non-blocking, level one cache

- Available with integrated 256 KB or 512 KB, non-blocking, level two cache on package

- Data integrity and reliability features include ECC, Fault Analysis/Recovery, and Functional Redundancy Checking

- Upgradable to a Future OverDrive® processor

The Pentium® Pro processor family is Intel's next generation of performance for high-end desktops, workstations and servers. The family consists of processors at 150 MHz and higher and is easily scalable to up to four microprocessors in a multiprocessor system. The Pentium Pro processor delivers more performance than previous generation processors through an innovation called Dynamic Execution. This is the next step beyond the superscalar architecture implemented in the Pentium processor. This makes possible the advanced 3D visualization and interactive capabilities required by today's high-end commercial and technical applications and tomorrow's emerging applications. The Pentium Pro processor also includes advanced data integrity, reliability, and serviceability features for mission critical applications.

The Pentium Pro processor may contain design defects or errors known as errata. Current characterized errata are available upon request.

**FIGURE F.33  Pentium Pro Processor,** reprinted by permission of Intel Corp. Copyright © 1996 by Intel Corp.

432

# APPENDIX G
# Chapter Review/Quiz Answers

**CHAPTER 1**

1. d
2. b
3. c
4. b
5. b
6. d
7. a
8. c
9. b
10. d
11. d
12. b
13. d
14. d
15. b
16. c
17. a
18. d
19. c
20. d
21. b
22. c
23. c
24. c
25. b
26. b

## CHAPTER 2

1. c
2. d
3. b
4. c
5. b
6. a
7. c
8. d
9. c
10. b
11. b
12. b
13. a
14. b
15. c
16. c

## CHAPTER 3

1. b
2. c
3. d
4. b
5. c
6. c
7. c
8. b
9. b
10. b
11. b
12. c
13. b
14. b
15. b
16. c
17. c
18. d
19. d
20. d
21. d
22. d
23. d
24. d
25. d
26. b
27. a
28. c
29. b
30. b
31. c
32. c
33. c
34. d
35. d

36. c
37. b
38. a
39. b
40. c
41. b
42. b
43. a
44. b
45. c

## CHAPTER 4

1. c
2. a
3. b
4. a
5. a
6. b
7. c
8. a
9. d
10. b
11. a
12. d
13. a
14. b
15. c
16. d
17. c
18. b
19. b
20. c
21. c
22. d
23. a
24. a
25. b
26. a
27. a
28. c
29. c
30. d

## CHAPTER 5

1. b
2. c
3. b
4. d
5. b
6. b
7. c
8. a
9. a
10. d

11. b
12. d
13. b
14. a
15. d
16. c
17. d
18. b
19. a

## CHAPTER 6

1. a
2. b
3. d
4. b
5. b
6. d
7. c
8. d
9. b
10. d
11. d
12. b
13. a
14. b
15. d
16. b
17. d
18. a
19. b
20. d
21. d
22. b
23. a
24. a
25. c
26. c
27. c
28. b
29. b
30. c
31. b

## CHAPTER 7

1. b
2. c
3. d
4. b
5. a
6. d
7. b
8. a
9. c
10. a

11. b
12. a
13. c
14. d
15. b
16. c
17. b
18. c

## CHAPTER 8

1. b
2. a
3. c
4. b
5. d
6. b
7. c
8. a
9. b
10. c
11. d
12. a
13. b
14. a

## CHAPTER 9

1. b
2. a
3. e
4. a
5. c
6. d
7. c
8. c
9. b
10. e
11. d
12. e
13. c
14. a
15. b
16. e
17. d
18. a
19. c
20. b
21. d
22. b
23. b
24. a
25. a
26. a
27. c

28. c
29. d
30. c
31. c
32. c
33. b
34. b

## CHAPTER 10

1. d
2. c
3. b
4. d
5. c
6. c
7. c
8. b
9. c
10. d
11. b
12. c
13. a
14. a
15. b
16. b
17. c
18. b
19. d
20. b
21. a
22. c
23. d
24. d
25. a

# GLOSSARY

**Absolute address**   The actual binary number address of the memory location being addressed. It is the addition of the segment shifted four digits left, and the offset.

**Address bus**   The conductors, stripes, or lines that transfer the memory address in the form of a binary number.

**ANSI**   The ANSI.SYS file has drivers for video conforming to the American National Standards Institute standard.

**ASCII (American Standard Code for Information Interchange)**   A standard code of numbers assigned to characters for computer systems.

**Autosensing**   The ability of some keyboards to sense whether they are attached to an XT or an AT system.

**Azimuth**   Alignment for disk heads is around a vertical axis through the head. Actually, it is almost never adjustable.

**Bandwidth**   The range of frequencies to which the circuitry is sensitive. Higher resolution requires higher bandwidth.

**Basic card swapping**   Repair by replacing a printed circuit board as opposed to the components on that board. It is easier and faster than the alternatives.

**Binary**   The number system using 2 as the base, the way decimal uses 10.

**BIOS (Basic Input–Output System)**   Stored in ROM of the system. It is the program that allows the computer to read the keyboard, print on the display, and boot a drive.

**Bit**   One digit of a binary number. The decimal number 5 is 0000 0101 in binary. This is given in 8 bits.

**Boot**   To activate the computer by running a segment of a program that loads a much larger program into memory and runs it. It is like "pulling yourself up by the bootstraps," and it is sometimes called bootstrapping the system. The program in ROM that accomplishes this is called the bootstrap loader.

**Borders**   In Windows, the thin double lines around all sides of a window. They are analogous to the frame of a picture.

**Buffer**   A chip between two others to protect one or to control communication between the two.

**Bus**   A group of conductors that transfer related information from one point to another.

**Byte**   Eight bits. The binary number of 8 bits is 1 byte.

**Cache**   *RAM cache* uses SRAM (which is faster than DRAM) as a small holding area for instructions (code) and data used by the microprocessor, thus increasing the speed of the system. *Hard disk cache* uses RAM to store information that is often referenced by the hard drive, thus minimizing the number of times the hard drive has to be accessed and thereby increasing system speed.

**Card**   A printed circuit board. Adapter cards are used to expand the basic computer to operate drives, video displays, printers, and other external devices.

**CCD (Charge Coupled Device)**   Usually a chip using capacitance to capture light falling on it. It is used in optical cameras and scanners.

**Chips**   Tiny, thin wafers of silicon on which electronic circuitry is etched and lithographed. These die are attached to a larger plastic or ceramic package and sealed. This package has the leads that finally are attached to the printed circuit board.

**CISC (Complex Instruction Set Computing)**   Typical of the 8088 family of microprocessors. CISC needs many clock cycles to perform most instructions. This architecture is what most of the microprocessors you hear about are based on.

**CMOS memory (Complementary Metal Oxide Semiconductor)**   RAM that requires very little power to retain data. It is powered by a lithium battery and holds the system configuration (number and type of drives, etc.).

**Cold boot**   Performed by turning on the power of a system that is turned off. This is done in order to start all over when a problem has stopped system functioning. It eliminates TSR program conflicts. Often AUTOEXEC.BAT and CONFIG.SYS are renamed so they do not interfere with a "clean" system start. See *warm boot.*

**Control bus**   The conductors, stripes, or lines that transfer the signals that control chips and circuits to perform certain operations.

**Control menu**   In Windows, the list of choices of actions to perform on the current window. It includes restore, move, size, minimize, maximize, close, and switch to. It is controlled by clicking the mouse in the box in the top left corner of a window.

**Conventional memory**   A block of RAM from 0 to 640K.

**Crystal**   A small, thin wafer of material, which oscillates at a very precise frequency when electronically stimulated.

**Decimal**   The number system based on the number 10.

**Default**   The assumed value or device. The default drive is the drive given in the DOS prompt, such as A>.

**Dhrystone**   A program to "benchmark" a system's speed. It runs a mixture of instructions and outputs the number of times the program can be run in a second.

**DIP (Dual Inline Package)**   The typical "chip" or integrated circuit you see, with two rows of leads. Resistors, transistors, and other components come in DIP packages also.

**Directories**   Listings of the files and subdirectories the user has organized together. The system boots into the root directory. From there, all others are subdirectories.

**DOS (Disk Operating System)**   The computer BIOS boots a drive and loads in the DOS. The DOS then allows future disk access.

**Dot pitch**   The distance between pixels. The closer together they are, the clearer the tiny details in the picture will be.

**Double word**   Two words, or 4 bytes, or 32 bits, or a *longword*.

**Dual inline header (DIH)**   A connector with two rows of pins usually 1/10 inch apart from each other and the other row.

**Dynamic RAM (DRAM)**   Consists basically of a capacitor and a resistor, having an RC time constant. It holds a charge for a short while, then slowly discharges. This is why dynamic RAM must be "refreshed," or read and then written on a regular basis. See also *static RAM*.

**Edit**   In DOS, a word processing program that is included in the DOS package. In Windows, a choice in many menu bars allowing the user to change the current work. Choices under edit may include undo, cut, copy, and paste.

**Encoding**   Converting data from one form to another, usually requiring less space or fewer conductor paths. The technique used to organize data on a disk.

**Expanded memory**   Memory on a RAM expansion card conforming to the Lotus/Intel/Microsoft (LIM) standard.

**Extended memory**   A block of RAM starting at 1 MB.

**Extended memory specification (XMS)**   Defines *HMA*.

**FAT (File Allocation Table)**   Holds the locations of the sectors of files stored on the disk, and bad track locations. Two copies reside on track 0, which is the reason the disk is considered unusable if track 0 is bad.

**File**   In DOS, a unit of data or commands. A business letter would be saved as a file. The word processing program used to create the letter is also a file. In Windows, the first choice on most menu bars. Choices under File may include New, Open, Close, Save, Save as, Print, and Exit.

**Firmware**   Programs in ROM on the system. Because they require more work to change (unplugging a chip and installing a new one) than a program on a floppy, they are "firmer" than "software" programs.

**Fixed disk**   Same as a hard, or Winchester disk.

**Floptical drive**   The largest capacity floppy drive available. Using optical recording, a 21-MB drive is said to be three times faster than a standard floppy drive, and it can also read and write standard disks.

**Flyback transformer**   An electronic device used to transform one voltage or current to another. In a computer display or television, as the horizontal sweep *flies back* from the right to the left side, the rapid change of voltage is used as an input to the flyback transformer to create the high voltage used on the anode of the CRT (picture tube).

**Form factor**   The shape, or physical, form of a device. Particularly, the power supply form factor must match the case for a proper fit. Modifying brackets may be required to adapt drives to the case.

**Fragmented**   The adding and deleting of files from the disk leaves gaps at certain cylinders and sectors. A new, larger file being added may be broken up into small pieces to fit the existing gaps. Since the heads must now jump here and there, reading small pieces of this fragmented file, this slows down file access.

**Freeze spray**   A chemical compressed in a spray can used to cool parts thought to be failing from heat. If the cooling causes the part to function again, it is heat sensitive, and must be replaced.

**GUI (Graphical User Interface)**   With a GUI the user sees graphical windows (different sections of the screen used for different functions or programs), icons (little pictures that look like the action you want to do), menus, and pointing devices (mice, trackballs, touchscreens). These help make it easier for the user to get real work done.

**Hardcard**   A very thin drive mounted right on the controller card. Hardcards fit in standard expansion slots and are very easy to install.

**Hard disk**   A disk with nonremovable media sealed in an airtight chamber to protect against dust. It is at least ten times faster than a floppy disk and holds more data.

**Hard sectored**   A disk that has an index hole for each sector, giving an absolute physical mark of each sector.

**Hardware**   The actual machinery of the system: the case, power supply, motherboard, drives, adapter cards, keyboard, monitor, and so on.

**Heater**   This filament of a tube used to heat the cathode to increase the energy level of the electrons. This facilitates breaking the valence bond and forming free electrons to be used for the electron beam in the tube.

**Help**   In DOS, a program that gives information on DOS commands. In Windows, a common menu bar choice that gives information on operation of the current program or window.

**Hexadecimal (Hex)**   The number system based on the number 16.

**High memory area (HMA)**   A 64K block of RAM starting at 1 MB.

**Integrated circuits**   Electronic circuits that have been standardized, interconnected, and miniaturized. They are constructed as chips.

**Interleave**   On a hard disk, the number of sectors skipped before reading another. An interleave of three means the drive reads sector 0, 3, 6, . . . . The purpose is to give slow electronics a chance to keep up with a rapidly spinning hard drive. An interleave of one means the controller board is quick enough to read consecutive sectors.

**Interrupt**   How does the microprocessor know when a device needs attention? The device signals on a line called an interrupt. The different interrupt lines have priority levels, so the processor can attend to more critical situations first.

**Isolation test**   A test with a meter to confirm the windings of a transformer or the lines of a bus are not electrically connected (shorted).

**Kilobyte (K)**   1024 bytes or 8192 bits.

**Logic pulser**   A testing instrument that sends a signal of pulses in a conductor. The presence of pulses or the lack of them as shown by a logic probe is data for troubleshooting digital circuits.

**Longword**   Two binary words, or a doubleword.

**Megabyte (MB)**   1,048,576 bytes or 1024K.

**Monitor**   The video display. It is also called the screen, the console, or the CRT.

**Multimedia**   Using many different types of media in a presentation or program. Music, speech, and full-motion video can all be used to create interactive programs that respond in different ways to varying inputs.

**Network**   A system to connect computers together to transfer or share information. A card is installed in each system, and they are connected together with cable.

**Nibble**   Four bits, or half a byte.

**Object Linking and Embedding (OLE)**   Some possible objects include a chart or graph, a spreadsheet, music files, a graphics picture, or sound or voice. The "object" can be "linked" with another program and "embedded" so that when one program runs, it shows pictures and plays sounds in one smooth presentation. These can be quite impressive.

**Offset**   The location of data in memory relative to the segment.

**Options**   In Windows, a choice on some menu bars. In the Program Manager, choices under options are Auto arrange (icons), Minimize, and Save settings. In Paint, choices include image attributes, edit/get/save colors, and brush shapes.

**Optoisolator**   A component, usually a chip, that isolates the input from the output by having the input drive an LED, and the light from the LED drives an LDR (Light Dependent Resistor) or optotransistor. Isolation can be to tens of thousands of volts. The main use in the computer is the power supply, to send feedback from the system DC back to the AC line circuits.

**Password**   Protection built into some BIOS ROMs. The password is saved in the CMOS RAM. If it is forgotten, disconnecting the CMOS battery will allow the system to come up. It is wise to save all other CMOS settings first.

**Peripherals**   Circuits added to the motherboard. This may include the keyboard, display, printer, fax/modem, network, scanner, mouse, and the like.

**Pick**   In Windows, a tool in the Paint program to select an area of a painting.

**Pixels**   The individual "dots" that make up the picture on the display screen. More pixels allow more definition and, therefore, create a better picture.

**Port**   Connectors on the computer system through which the outside world can be sensed or controlled. Ports are used for printers, modems, mice, video scanners, plotters, voice synthesizers, and much more.

**POST (Power-On Self-Test)**   This automatic internal diagnostic is part of the BIOS program. It gives an audio error code in the form of beeps or a video error code number on the display.

**Pulse width modulator**   An electronic circuit that gives a digital output of pulses of varying periods relative to the input. If filtered the output is an analog average voltage.

**Radial disk alignment**   Adjusting the position of the head as it travels from track 0 to the final track, so that it is exactly over the proper track.

**RAID (Redundant Array of Inexpensive Disks)**   A set of parallel disks to keep data safely in more than one place for absolute integrity of data.

**Raster**   The illuminated lines covering the screen of a display. A totally white screen.

**Register**   An electronic circuit that stores information.

**Reserved memory**   A block of RAM from 640K to 1 MB. Also called the *upper memory block*.

**RISC (Reduced Instruction Set Computing)**   Architecture that runs many instructions in one clock cycle.

**ROM (Read-Only Memory)**   Memory that is programmed or written (blown or burned) at the factory or by technicians with a ROM burner. EPROMS can be erased by ultraviolet light and rewritten. EEPROMS can be electrically erased and rewritten.

**Scroll**   To move the contents of the screen of the display. In Windows, if there is more information than can be shown in the current window, horizontal and vertical scroll bars will be shown. At both ends will be scroll arrows to click on, and in the bar, a horizontal scroll box. Clicking and dragging this scrolls the screen up or down, or back or forth.

**Segment**   A 64K section of memory. 16 bits, or 4 hex digits, are needed to address 64K memory locations, so the fifth hex digit changes the segment.

**Shadow RAM**   RAM used to hold the information already stored in ROM, either in the system BIOS or in the video adapter ROM. Reading the faster RAM now stops system slowdown on reading ROMs and speeds up system throughput.

**SIMM (Single Inline Memory Module)**   A number of surface mount chips in a single row mounted on a little PC board that clips into a socket. It is one package used for RAM.

**Single inline header**   Pins .1 on center in a row as a connector plug

**Simple signal checking**   Using a logic probe, meter, or oscilloscope to check the voltages, currents, or waveforms of the circuit under test.

**SIP (Single Inline Package)**   A number of surface mount chips in a single row mounted on a little PC board with leads or legs that plug into a socket that is a row of round holes. It is one package used for RAM.

**Sizing buttons**   In Windows, arrow buttons at the top right corner to maximize, minimize, or restore the window.

**SMT (Surface Mount Technology)**   Chips that do not have leads that go through holes in the board, but leads that sit on the surface of the Printed Circuit Board (PCB). Often leads are .05 inches on center.

**Soft sectored**   Disks with one index hole to identify position. The 9, 15, 18, or 36 sectors are timed.

**Software**   The programs that are run on the computer system.

**States**   In digital electronics, only two logic levels are allowed, high and low. These are the states of each binary bit. More bits allow more states. One bit allows two states. Two bits allow four states. Three bits allow eight states. The equation for possible states is $2^N$ where $N$ is the number of bits. The states always start at zero and go to $2^{N-1}$.

**Static RAM (SRAM)**   Composed basically of a flip-flop. Once set, it stays set until it receives a reset signal. It does not lose a charge like dynamic RAM, and it does not need refreshing. It does, however, require more power to hold its data, and therefore is only used in specific circuits. All RAM loses its data when power is removed.

**Subdirectories**   All directories other than the root directory. The root directory may have an entry as the subdirectory "c:\games." Under the "c:\games" subdirectory may be another subdirectory "c:\games\board." That subdirectory may have an even lower subdirectories such as "c:\games\board\chess" and "c:\games\board\checkers."

**Surface mount chips**   Standard chips or integrated circuits have leads that pass through holes in the printed circuit board and are soldered on the other side. Surface mount chips have leads that are soldered on the same side of the PCB the chip sits on.

**Terminated**   Having a resistance to stop reflection of signals from the end of the bus and the spurious signals that result.

**Text**   A file of only ASCII characters that almost any program can read. No special characters are allowed.

**Title bar**   In Windows, the top line of the window with the name of the program displayed in that window.

**Transceiver**   A chip that transmits and receives a signal. It is used to control the flow of information in either direction. Control lines are DIR and ENABLE.

**Transistor**   An electronic device that uses a tiny input signal to control a much larger output signal. Transistors are used to build many circuits including amplifiers and oscillators. It is the basis for transistor–transistor logic (TTL).

**TSR (Terminate and Stay Resident)**   Programs that attach themselves to DOS when run the first time, then terminate or end. But as DOS detects a certain sequence of keystrokes (hot keys), it activates or runs the program. This way the program seems hidden away, but it is resident, always prepared to spring into action when called.

**Upper memory block (UMB)**   A block of RAM from 640K to 1 MB. Also called *reserved memory*.

**View**   In Windows Paint program, the menu choice giving selections of zoom in/out, view tools and linesize, view (color) palette, view cursor position.

**Virtual memory**   The appearance of large RAM memory created by a technique of swapping information in physical RAM to the hard drive and loading new information in the physical RAM. The information is swapped back and forth automatically as needed.

**VLSI (Very Large Scale Integration)**   More than 10,000 transistors in one integrated circuit.

**Warm boot**   Usually accomplished by holding down ALT+CTRL+DEL. This tells the computer to start all over. It usually skips the POST and starts at rebooting the drives. It does not start at the absolute beginning as a *cold boot* does.

**Whetstone**   The same type of benchmark program as a Dhrystone, except that it tests only floating-point operations.

**WIMP interface**   Old DOS diehards call the GUI a WIMP (Windows, Icons, Menus, Pointers) interface.

**Winchester**   A hard or fixed disk drive, sealed, with nonremovable media.

**Word**   Two bytes.

**Work area**   In Windows, the main area of the window in which your work will be done.

**Zero**   In a meter, to adjust the reading for exactly 0.00 when there is no input.

# BIBLIOGRAPHY

Brenner, Robert C., *IBM PC Advanced Troubleshooting and Repair*. Howard W. Sams & Co., Indianapolis, Ind., 1988.

Brooks, Charles, *IBM PC Peripheral Troubleshooting and Repair*. Howard W. Sams & Co., Indianapolis, Ind., 1987.

Dowden, Tony, *Inside the EISA Computers*. Addison-Wesley, Reading, Mass., 1990.

Margolis, Art, *Troubleshooting and Repairing the New Personal Computers*. Tab Books, Blue Ridge Summit, Pa., 1987.

Minasi, Mark, *Maintaining, Upgrading, and Troubleshooting IBM PCs, Compatibles, and PS/2 Personal Computers*. Compute Books, Greensboro, N.C., 1990.

Mueller, Scott, *Upgrading and Repairing PCs*. Que Corp., Carmel, Ind., 1988.

Perozzo, James, *Assembling and Troubleshooting Microcomputers*, 2d ed. Delmar Publishers, Albany, N.Y., 1991.

Pilgrim, Aubrey, *Build Your Own 386/386SX Compatible and Save a Bundle*. McGraw-Hill/Windcrest/Tab, Blue Ridge Summit, Pa., 1992.

Skerrett, P. J., "Future Computers: Next Generation PCs, Part Two." *Popular Science,* April 1992.

Skerrett, P. J., "Future Computers: The Teraflops Race, Part One." *Popular Science,* March 1992.

Stankiewicz, Steve, "Super Density Floppies," *Popular Science,* August 1992.

Stephenson, John, and Bob Cahill, *Microcomputer Troubleshooting and Repair*. Howard W. Sams & Co., Indianapolis, Ind., 1989.

Woram, John, *The PC Configuration Handbook,* 2d ed. Bantam Books, N.Y., 1990.

# INDEX